16.95

SOLUTIONS MANUAL

for the

CIVIL ENGINEERING REFERENCE MANUAL

Sixth Edition

Michael R. Lindeburg, P.E.

PROFESSIONAL PUBLICATIONS, INC.
Belmont, CA 94002

In the ENGINEERING REFERENCE MANUAL SERIES

Engineer-In-Training Reference Manual
  Engineering Fundamentals Quick Reference Cards
  Engineer-In-Training Sample Examinations
  Mini-Exams for the E-I-T Exam
  1001 Solved Engineering Fundamentals Problems
  E-I-T Review: A Study Guide
Civil Engineering Reference Manual
  Civil Engineering Quick Reference Cards
  Civil Engineering Sample Examination
  Civil Engineering Review Course on Cassettes
  Seismic Design of Building Structures
  Seismic Design Fast
  Timber Design for the Civil P.E. Examination
  Fundamentals of Reinforced Masonry Design
  246 Solved Structural Engineering Problems
Mechanical Engineering Reference Manual
  Mechanical Engineering Quick Reference Cards
  Mechanical Engineering Sample Examination
  101 Solved Mechanical Engineering Problems
  Mechanical Engineering Review Course on Cassettes
  Consolidated Gas Dynamics Tables
Electrical Engineering Reference Manual
  Electrical Engineering Quick Reference Cards
  Electrical Engineering Sample Examination
Chemical Engineering Reference Manual
  Chemical Engineering Quick Reference Cards
  Chemical Engineering Practice Exam Set
Land Surveyor Reference Manual
Petroleum Engineering Practice Problem Manual
Expanded Interest Tables
Engineering Law, Design Liability, and Professional Ethics
Engineering Unit Conversions

In the ENGINEERING CAREER ADVANCEMENT SERIES

How to Become a Professional Engineer
The Expert Witness Handbook—A Guide for Engineers
Getting Started as a Consulting Engineer
Intellectual Property Protection—A Guide for Engineers
E-I-T/P.E. Course Coordinator's Handbook
Becoming a Professional Engineer
Engineering Your Start-Up

SOLUTIONS MANUAL FOR THE
CIVIL ENGINEERING REFERENCE MANUAL
Sixth Edition

Printed in the United States of America

ISBN: 0-912045-43-4

Professional Publications, Inc.
1250 Fifth Avenue, Belmont, CA 94002
(415) 593-9119

Current printing of this edition (last number): 6 5 4 3 2 1

PREFACE TO THE SIXTH EDITION

As did previous editions, this solutions manual contains solutions to all of the end-of-chapter problems in the *Civil Engineering Reference Manual*, now in its sixth edition. To simplify your review, solutions are keyed to pages, equations, and data in the reference manual. Because of changes I have made in some of the end-of-chapter problems, this solutions manual should not be used with earlier editions of the reference manual.

The sixth edition differs from previous solutions manuals in several ways. The obvious differences concern the outward appearance: there are more pages, the solutions have been typeset, and all of the illustrations have been professionally rendered. My handwritten solutions and hand-drawn figures have finally bitten the dust, and no one is happier to be rid of them than I.

Other new features include updated solutions and additional references to current concrete, steel, and traffic codes and standards, more frequent use of units in calculations and answers, and more explanatory text. I have also included alternate solutions in several problems that can be solved in more than one way.

The solutions have been carefully prepared, and most have withstood the test of time. However, time invariably moves on, leaving an old solution to a newly defined problem behind. Codes and laws, methods, and traditional practices change, and when conceptual errors and omissions are added to routine technological changes in the real world, it is inevitable that you will find something that can be improved upon. My promise to you is that I will listen if you talk. You can use the postage-paid card in the back for that purpose, and I will send you a reply.

Michael R. Lindeburg, P.E.
Belmont, CA
July 1992

TABLE OF CONTENTS

Notice To Examinees

Do not copy, memorize, or distribute problems from the Principles and Practice of Engineering (P&P) Examination. These acts are considered to be exam subversion.

The P&P examination is copyrighted by the National Council of Examiners for Engineering and Surveying. Copying and reproducing P&P exam problems for commercial purposes is a violation of federal copyright law. Reporting examination problems to other examinees invalidates the examination process and threatens the health and welfare of the public.

MATHEMATICS AND RELATED SUBJECTS

Untimed

1. The range of speeds is 48 mph − 20 mph = 28 mph. Since there are not a lot of observations, 10 cells would be best. Choose the cell width as 28 mph/10 ≈ 3 mph.

(a) and (d)

interval	midpoint	freq.	cum. freq.	cum. fraction
20–22	21	1	1	0.03
23–25	24	3	4	0.10
26–28	27	5	9	0.23
29–31	30	8	17	0.43
32–34	33	3	20	0.50
35–37	36	4	24	0.60
38–40	39	3	27	0.68
41–43	42	8	35	0.88
44–46	45	3	38	0.95
47–49	48	2	40	1.00

(b) and (c)

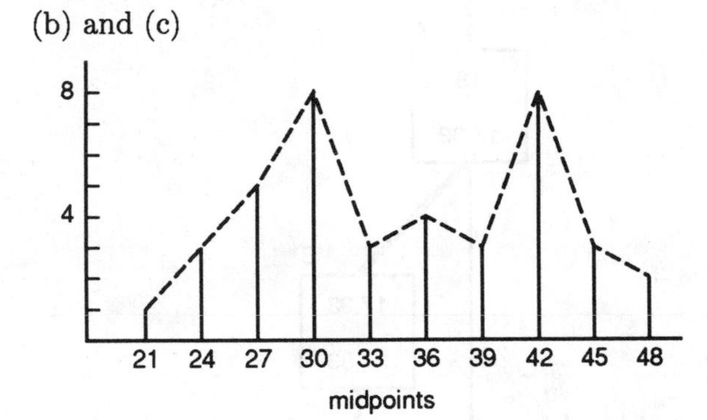

(e)

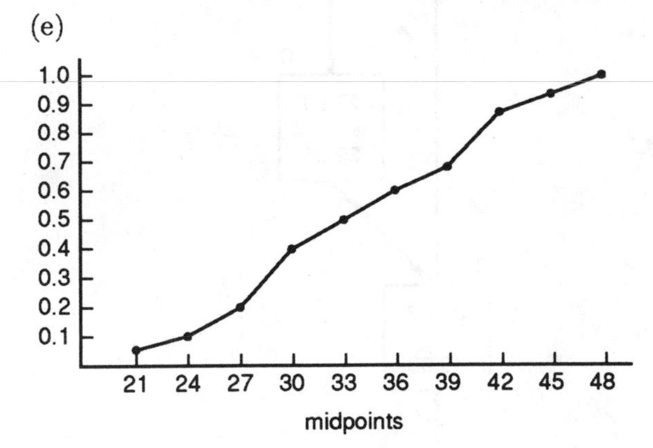

(f) Use the cumulative distribution graph. For 75% (0.75), the cell midpoint is approximately 40 mph.

(g) Use the cumulative distribution graph to find the midpoint for 50%. This occurs at approximately 33 mph.

The distribution is bimodal (i.e., has two peaks). The two modes are 30 mph and 42 mph.

$$\sum x_i = 1390$$

$$\overline{x} = \frac{\sum x_i}{n} = \frac{1390 \text{ mph}}{40} = \boxed{34.75 \text{ mph}}$$

(h)

$$\sum x^2 = 50{,}496$$

$$\sigma = \sqrt{\frac{50{,}496}{40} - \left(\frac{1390}{40}\right)^2}$$

$$= \boxed{7.405 \text{ mph}}$$

(i) $s = \sigma\sqrt{\frac{n}{n-1}} = 7.405\sqrt{\frac{40}{39}} = \boxed{7.500 \text{ mph}}$

(j) $s^2 = (7.500)^2 = \boxed{56.25 \text{ mph}^2}$

2. No contract deadline was given, so assume 36 as a scheduled time. Look for a path where $LS - ES = 0$ everywhere.

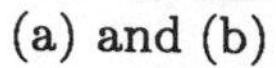
(a) and (b)

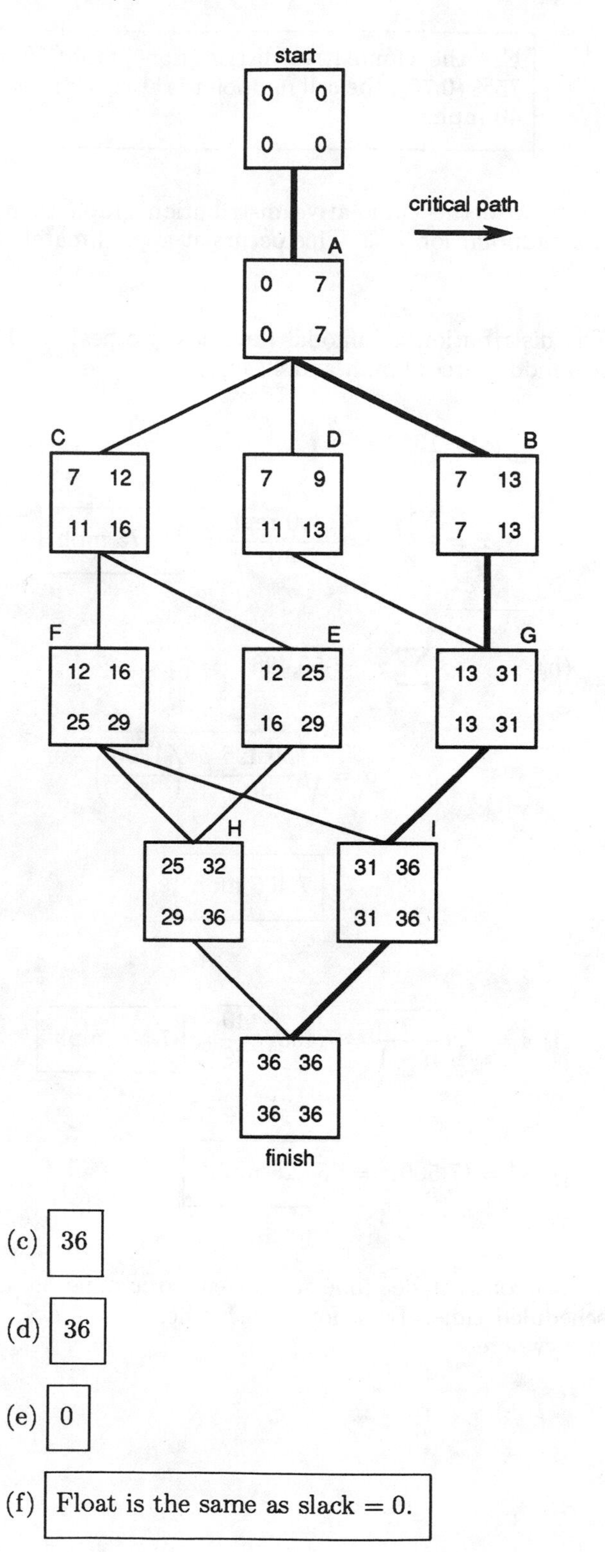

(c) 36

(d) 36

(e) 0

(f) Float is the same as slack = 0.

3. To solve this as a regular CPM problem, it is necessary to calculate t_{mean} and σ for each activity. For activity A,

$$t_{\text{mean}} = \left(\tfrac{1}{6}\right)\left[1 + (4)(2) + 5\right] = 2.33$$

$$\sigma_A = \left(\tfrac{1}{6}\right)(5 - 1) = 0.67$$

The following table is generated in the same manner.

activity	t_{mean}	σ
start	0	0
A	2.33	0.67
B	10.5	2.17
C	11.83	2.17
D	4.17	0.83
finish	0	0
	28.83	

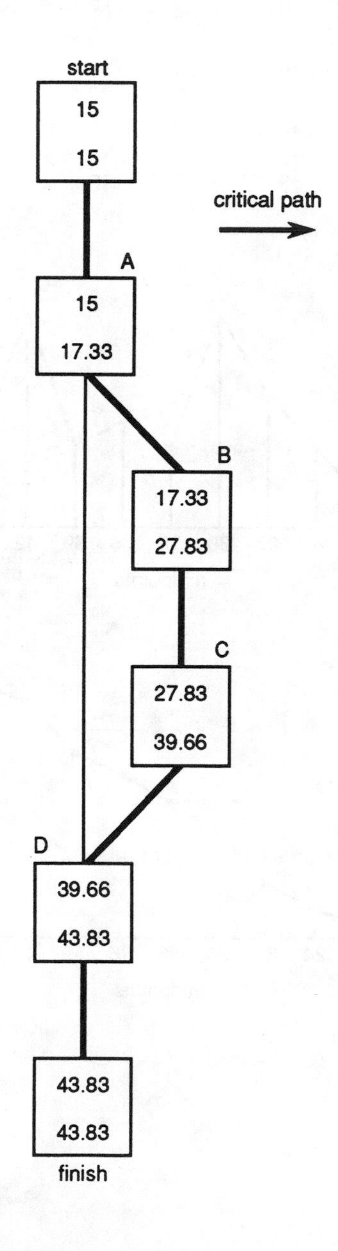

By observation, the critical path is start-A-B-C-D-finish. The project variance is

$$\sigma^2 = (0.67)^2 + (2.17)^2 + (2.17)^2 + (0.83)^2 = 10.56$$

The project standard deviation is

$$\sigma = \sqrt{10.56} = 3.25$$

Assume the completion times are normally distributed with a mean of 28.83 and a standard deviation of 3.25. The standard normal variable is

$$z = \frac{28.83 + 15 - 42}{3.25} = 0.56$$

The area under the tail of the normal curve for $z = 0.56$ is 0.2123, so

$$0.5 - 0.2123 = \boxed{0.2877 \ \ (28.77\%)}$$

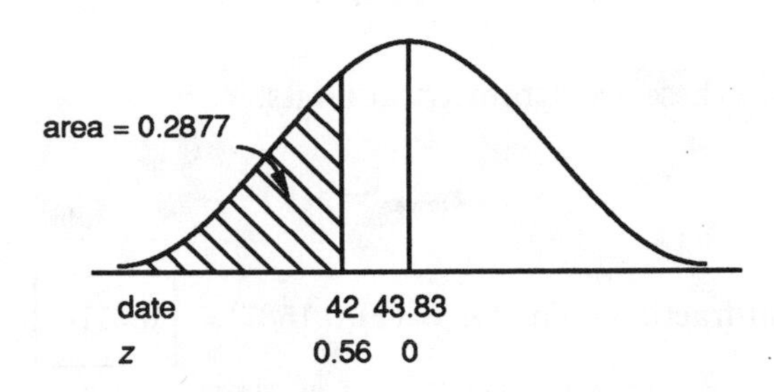

4. *method 1:* From Eqs. 1.3 and 1.4,

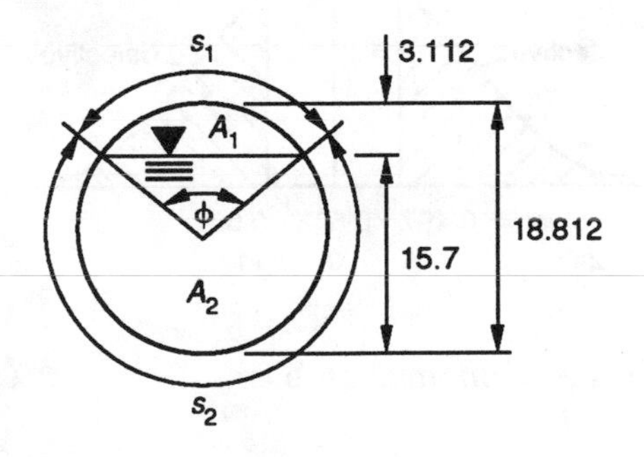

$$r = \frac{18.812 \text{ in}}{2} = 9.406 \text{ in}$$

$$\phi = (2)\left(\arccos \frac{9.406 - 3.112}{9.406}\right) = 1.675 \text{ rad}$$

$$\sin\phi = 0.9946$$

$$A_1 = \left(\tfrac{1}{2}\right)(9.406)^2(1.675 - 0.9946) = 30.1 \text{ in}^2$$

$$A_{\text{total}} = A_1 + A_2 = \left(\frac{\pi}{4}\right)(18.812)^2 = 277.95 \text{ in}^2$$

$$A_2 = 277.95 - 30.1 = 247.85 \text{ in}^2$$

$$S_1 = r\phi = (9.406)(1.675) = 15.76 \text{ in}$$

$$S_{\text{total}} = S_1 + S_2 = \pi(18.812) = 59.1 \text{ in}$$

$$S_2 = 59.1 - 15.76 = 43.34 \text{ in}$$

$$r_H = \frac{247.85 \text{ in}^2}{43.34 \text{ in}} = \boxed{5.719 \text{ in}}$$

method 2: Use Table 3.7.

$$\frac{d}{D} = \frac{15.7 \text{ in}}{18.812 \text{ in}} = 0.83$$

From the table, $r_H/D = 0.3041$.

$$r_H = (0.3041)(18.812 \text{ in}) = \boxed{5.72 \text{ in}}$$

5. Expand by cofactors of the first row.

$$D = 8\begin{vmatrix} 8 & 2 & 0 \\ 2 & 8 & 2 \\ 0 & 2 & 4 \end{vmatrix} - 2\begin{vmatrix} 2 & 0 & 0 \\ 2 & 8 & 2 \\ 0 & 2 & 4 \end{vmatrix} + 0 - 0$$

$$\begin{vmatrix} 8 & 2 & 0 \\ 2 & 8 & 2 \\ 0 & 2 & 4 \end{vmatrix} \begin{array}{l} \text{by first row:} \\ = (8)[(8)(4)-(2)(2)]-(2)[(2)(4)-(2)(0)] \\ = (8)(28)-(2)(8) = 208 \end{array}$$

$$\begin{vmatrix} 2 & 0 & 0 \\ 2 & 8 & 2 \\ 0 & 2 & 4 \end{vmatrix} \begin{array}{l} \text{by first row:} \\ = (2)[(8)(4)-(2)(2)] \\ = 56 \end{array}$$

$$D = (8)(208) - (2)(56) = \boxed{1552}$$

6. $\lambda = 20$ cars.

From Eq. 1.205,

(a) $p\{x = 17\} = f(17) = \dfrac{e^{-20}(20)^{17}}{17!} = 0.076$

(b) $p\{x \leq 3\} = f(0) + f(1) + f(2) + f(3)$

$$\begin{aligned} &= \frac{e^{-20}(20)^0}{0!} + \frac{e^{-20}(20)^1}{1!} \\ &\quad + \frac{e^{-20}(20)^2}{2!} + \frac{e^{-20}(20)^3}{3!} \\ &= (2 \times 10^{-9}) + (4.12 \times 10^{-8}) \\ &\quad + (4.12 \times 10^{-7}) + (2.75 \times 10^{-6}) \\ &= \boxed{3.2 \times 10^{-6}} \end{aligned}$$

7. From Eq. 1.208,

$$\mu = \frac{1}{23 \text{ sec}}$$

$$p\{x \geq 25\} = 1 - F(25) = e^{-(\frac{1}{23})(25)}$$

$$= \boxed{0.337}$$

8. Plot the data to see if it is linear.

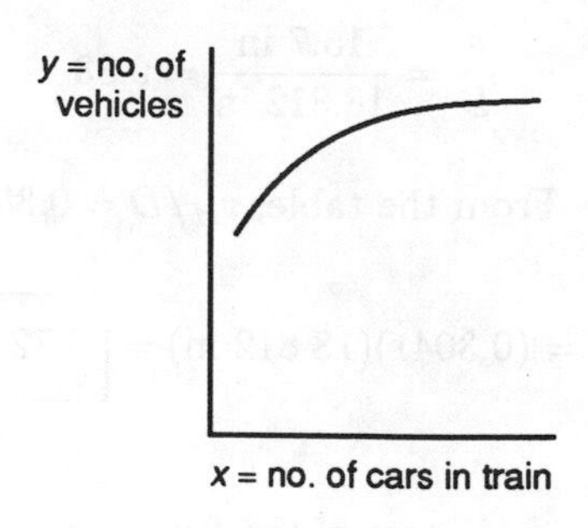

This is not linear. From Fig. 1.5, try the form $y = a + bz$ where $z = \sqrt{x}$.

y	$z = \sqrt{x}$
14.8	1.414
18.0	2.236
20.4	2.828
23.0	3.464
29.9	5.196

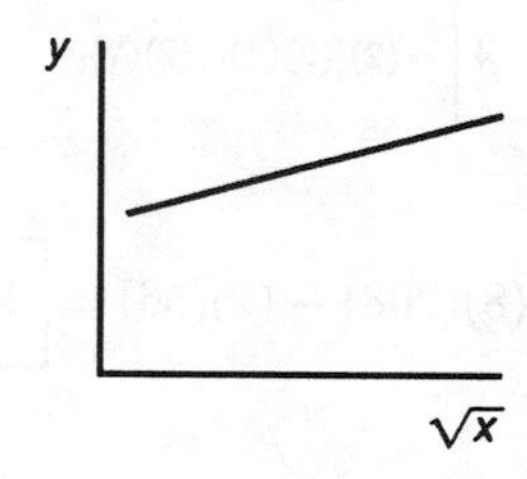

This looks linear.

From p. 1-13,

$$\Sigma z_i = 15.14 \qquad \Sigma y_i = 106.1$$
$$(\Sigma z_i)^2 = 229.2 \qquad (\Sigma y_i)^2 = 11{,}257$$
$$\Sigma z_i^2 = 54 \qquad \Sigma y_i^2 = 2382.2$$
$$\bar{z} = 3.028 \qquad \bar{y} = 21.22$$
$$\Sigma z_i y_i = 353.9$$

From Eq. 1.129,

$$m = \frac{(5)(353.9) - (15.14)(106.1)}{(5)(54) - 229.2} = 4.00$$

From Eq. 1.130,

$$b = 21.22 - (4.00)(3.028) = 9.11$$

The equation of the line is

$$\boxed{y = 4\sqrt{x} + 9.11}$$

The correlation coefficient is given by Eq. 1.131.

$$r = \frac{(5)(353.9) - (15.14)(106.1)}{\sqrt{[(5)(54) - 229.2]\,[(5)(2382.2) - 11{,}257]}} = 0.999$$

9. (a) From Eq. 1.209,

$$z_{\text{upper}} = \frac{0.507 \text{ in} - 0.502 \text{ in}}{0.005 \text{ in}} = +1$$

From Table 1.5, the area outside $z = +1$ is

$$0.5 - 0.3413 = 0.1587$$

Since these are symmetrical limits,

$$z_{\text{lower}} = -1$$

$$\text{total fraction defective} = (2)(0.1587) = \boxed{0.3174}$$

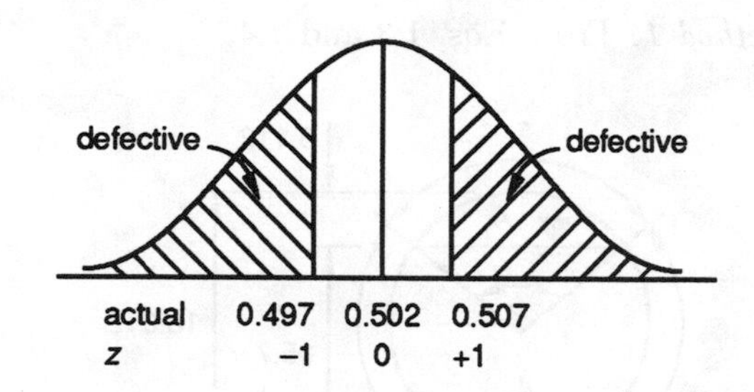

(b) This is a binomial problem.

$$p = p\{\text{defective}\} = 0.3174$$
$$q = 1 - p = 0.6826$$

From Eq. 1.204,

$$f(2) = \binom{15}{2}(0.3174)^2(0.6826)^{13}$$
$$= \left(\frac{15!}{13!\,2!}\right)(0.3174)^2(0.6826)^{13} = \boxed{0.0739}$$

10. (a) Use the characteristic equation method to solve the homogeneous case. (It is much quicker to use Laplace transforms, however.)

$$x'' + 2x' + 2x = 0 \quad \text{[differential equation]}$$
$$R^2 + 2R + 2 = 0 \quad \text{[characteristic equation]}$$

Complete the square to find R.

$$R^2 + 2R = -2$$
$$(R+1)^2 = -2+1$$
$$R+1 = \pm\sqrt{-1}$$
$$R = -1 \pm i$$

$$x(t) = A_1 e^{-t}\cos t + A_2 e^{-t}\sin t$$

Use the initial conditions to find A_1 and A_2.

$$x(0) = 0$$
$$0 = A_1(1)(1) + A_2(1)(0)$$
$$A_1 = 0$$

Differentiating the solution,

$$x'(t) = A_2(e^{-x}\cos x - \sin x\, e^{-x})$$

Using $x'(0) = 1$,

$$1 = A_2[(1)(1) - (0)(1)]$$
$$A_2 = 1$$

The solution is

$$x(t) = \boxed{e^{-2}\sin t}$$

(b) With no damping, the differential equation would be

$$x'' + 2x = 0$$

This has a solution $x = A\sin\sqrt{2}\,t$, so

$$\omega_{\text{natural}} = \boxed{\sqrt{2}}$$

(c)

$$x(t) = e^{-t}\sin t$$
$$x'(t) = e^{-t}\cos t - \sin t\, e^{-t}$$
$$= e^{-t}(\cos t - \sin t)$$

For x to be maximum, $x'(t) = 0$. Since e^{-t} is not 0 unless t is very large, $\cos t - \sin t$ must be 0.

$$\cos t = \sin t$$
$$\cos t = \sqrt{1-\cos^2 t} \quad [\text{since } \sin^2 t + \cos^2 t = 1]$$
$$t = \arccos\sqrt{\frac{1}{2}} = 45° \quad [0.785 \text{ radians}]$$
$$x(0.785) = e^{-0.785}\sin(0.785)$$
$$= \boxed{0.322}$$

(d) Use the Laplace transform method.

$$x'' + 2x' + 2x = \sin t$$
$$\mathcal{L}(x'') + 2\mathcal{L}(x') + 2\mathcal{L}(x) = \mathcal{L}(\sin t)$$
$$s^2\mathcal{L}(x) - 1 + 2s\mathcal{L}(x) + 2\mathcal{L}(x) = \frac{1}{s^2+1}$$
$$\mathcal{L}(x)(s^2+2s+2) - 1 = \frac{1}{s^2+1}$$

$$\mathcal{L}(x) = \frac{1}{s^2+2s+2} + \frac{1}{(s^2+1)(s^2+2s+2)}$$
$$= \frac{1}{(s+1)^2+1} + \frac{1}{(s^2+1)(s^2+2s+2)}$$

Use partial fractions to expand the second term.

$$\frac{1}{(s^2+1)(s^2+2s+2)} = \frac{A_1+B_1 s}{s^2+1} + \frac{A_2+B_2 s}{s^2+2s+2}$$

Cross multiplying,

$$= \frac{A_1 s^2 + 2A_1 s + 2A_1 + B_1 s^3 + 2B_1 s^2 + 2B_1 s + A_2 s^2 + A_2 + B_2 s^3 + B_2 s}{(s^2+1)(s^2+2s+2)}$$

$$= \frac{s^3(B_1+B_2) + s^2(A_1+A_2+2B_1) + s(2A_1+2B_1+B_2) + 2A_1 + A_2}{(s^2+1)(s^2+2s+2)}$$

Comparing numerators, the following four simultaneous equations are obtained.

$$\begin{aligned} B_1 + B_2 &= 0 \\ A_1 + A_2 + 2B_1 &= 0 \\ 2A_1 + 2B_1 + B_2 &= 0 \\ 2A_1 + A_2 &= 1 \end{aligned}$$

Use Cramer's rule to find A_1.

$$A_1 = \frac{\begin{vmatrix} 0 & 0 & 1 & 1 \\ 0 & 1 & 2 & 0 \\ 0 & 0 & 2 & 1 \\ 1 & 1 & 0 & 0 \end{vmatrix}}{\begin{vmatrix} 0 & 0 & 1 & 1 \\ 1 & 1 & 2 & 0 \\ 2 & 0 & 2 & 1 \\ 2 & 1 & 0 & 0 \end{vmatrix}} = \frac{-1}{-5} = \frac{1}{5}$$

The rest of the coefficients are found similarly.

$$A_1 = \tfrac{1}{5}$$
$$A_2 = \tfrac{3}{5}$$
$$B_1 = -\tfrac{2}{5}$$
$$B_2 = \tfrac{2}{5}$$

Then,

$$\mathcal{L}(x) = \frac{1}{(s+1)^2+1} + \frac{\frac{1}{5}}{s^2+1} + \frac{-\frac{2}{5}s}{s^2+1} + \frac{\frac{3}{5}}{s^2+2s+2} + \frac{\frac{2}{5}s}{s^2+2s+2}$$

Taking the inverse transform,

$$\begin{aligned} x(t) &= \mathcal{L}^{-1}\{\mathcal{L}(x)\} \\ &= e^{-t}\sin t + \tfrac{1}{5}\sin t - \tfrac{2}{5}\cos t + \tfrac{3}{5}e^{-t}\sin t \\ &\quad + \tfrac{2}{5}(e^{-t}\cos t - e^{-t}\sin t) \\ &= \boxed{\tfrac{6}{5}e^{-t}\sin t + \tfrac{2}{5}e^{-t}\cos t + \tfrac{1}{5}\sin t - \tfrac{2}{5}\cos t} \end{aligned}$$

11. Refer to p. 1-5.

$$f(x) = x^3 + 2x^2 + 8x - 2$$

Try to find an interval in which there is a root.

x	$f(x)$
0	−2
1	9

A root exists in the interval [0,1].

Try $x = \left(\frac{1}{2}\right)(0+1) = 0.5$.

$$f(0.5) = (0.5)^3 + (2)(0.5)^2 + (8)(0.5) - 2 = 2.625$$

A root exists in [0,0.5].

Try $x = 0.25$.

$$f(0.25) = (0.25)^3 + (2)(0.25)^2 + (8)(0.25) - 2 = 0.1406$$

A root exists in [0,0.25].

Try $x = 0.125$.

$$\begin{aligned} f(0.125) &= (0.125)^3 + (2)(0.125)^2 + (8)(0.125) - 2 \\ &= -0.967 \end{aligned}$$

A root exists in [0.125,0.25].

Try $x = \left(\frac{1}{2}\right)(0.125 + 0.25) = 0.1875$.

Continuing,

$$\begin{aligned} f(0.1875) &= -0.42 \quad [0.1875,0.25] \\ f(0.21875) &= -0.144 \quad [0.21875,0.25] \\ f(0.234375) &= -0.002 \quad [\text{This is close enough.}] \end{aligned}$$

One root is

$$x_1 \approx \boxed{0.234375}$$

Try to find the other two roots. Use long division to factor the polynomial.

$$\begin{array}{r} x^2 + 2.234375x + 8.52368 \\ x - 0.234375 \;\big)\; x^3 + 2x^2 + 8x - 2 \\ -(x^3 - 0.23475x^2) \\ \hline 2.23475x^2 + 8x \\ -(2.23475x^2 - 0.52368x) \\ \hline 8.52368x - 2 \\ -(8.52368x - 1.9977) \\ \hline \approx 0 \end{array}$$

Use Eq. 1.43 to find the roots of $x^2 + 2.234375x + 8.52368$.

$$\begin{aligned} x_2, x_3 &= \frac{-2.234375 \pm \sqrt{(2.234375)^2 - (4)(1)(8.52368)}}{(2)(1)} \\ &= \boxed{-1.1172 \pm i2.697} \quad [\text{both imaginary}] \end{aligned}$$

12. Plot the data to verify that it is linear.

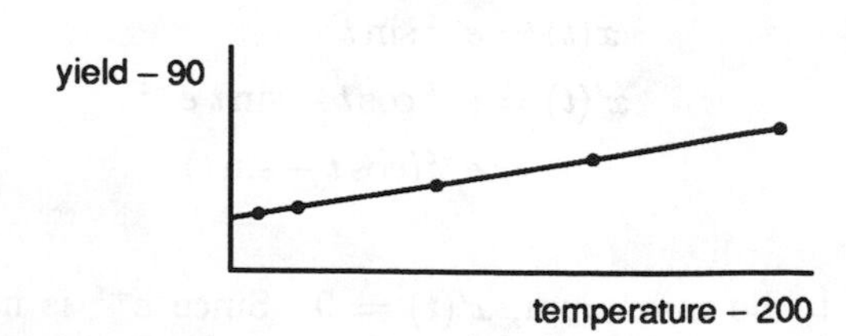

x $T - 200$	y $y - 90$
7.1	2.30
10.3	2.58
0.4	1.56
1.1	1.63
3.4	1.83

step 1:

$$\Sigma x_i = 22.3 \qquad \Sigma y_i = 9.9$$

$$\Sigma x_i^2 = 169.43 \qquad \Sigma y_i^2 = 20.39$$

$$(\Sigma x_i)^2 = 497.29 \qquad (\Sigma y_i)^2 = 98.01$$

$$\overline{x} = \frac{22.3}{5} = 4.46 \qquad \overline{y} = 1.98$$

$$\Sigma x_i y_i = 51.54$$

step 2: From Eq. 1.129, the slope is

$$m = \frac{(5)(51.54) - (22.3)(9.9)}{(5)(169.43) - 497.29} = 0.1055$$

step 3: From Eq. 1.130, the y-intercept is

$$b = 1.98 - (0.1055)(4.46) = 1.509$$

The equation of the line is

$$y = 0.1055x + 1.509$$

$$\text{yield} - 90 = 0.1055(\text{temperature} - 200) + 1.509$$

$$\boxed{\text{yield} = 0.1055 \text{ temperature} + 70.409}$$

step 4: Use Eq. 1.131 to get the correlation coefficient.

$$r = \frac{(5)(51.54) - (22.3)(9.9)}{\sqrt{[(5)(169.43) - 497.29][(5)(20.39) - 98.01]}}$$
$$= 0.995$$

13. Plot the data to see if it is linear.

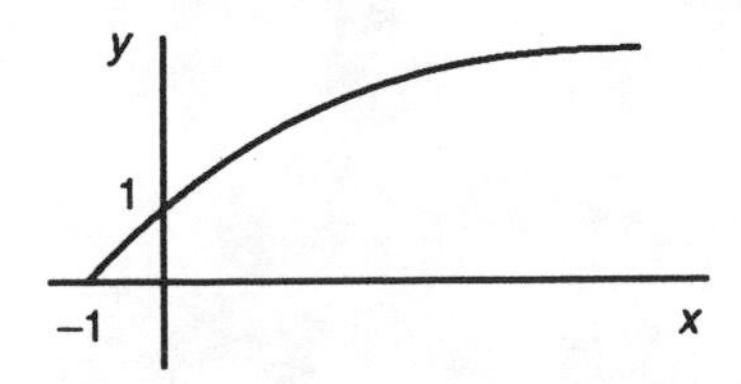

This looks like it could be of the form

$$y = a + b\sqrt{x}$$

However, when x is negative (as in the first point), the function is imaginary. Try shifting the curve to the right, replacing x with $x + 1$.

$$y = a + bz$$

$$z = \sqrt{x + 1}$$

z	y
0	0
1	1
1.414	1.4
1.732	1.7
2	2
2.236	2.2
2.45	2.4
2.65	2.6
2.83	2.8
3	3

Since $y \approx z$, the relationship is

$$\boxed{y = \sqrt{x + 1}}$$

In this problem, the answer was found accidentally. Usually, regression would be necessary.

14. (a) There are 4! = 24 different possible outcomes. By enumeration, there are nine completely wrong combinations.

$$p\{\text{all wrong}\} = \frac{9}{24} = \boxed{0.375}$$

(b) If three recruits get the correct size, the fourth recruit will also since there will be only one pair remaining.

$$p\{\text{exactly 3}\} = \boxed{0}$$

15. (a)

R	f	fR	fR^2
0.200	1	0.200	0.0400
0.210	3	0.630	0.1323
0.220	5	1.100	0.2420
0.230	10	2.300	0.5290
0.240	17	4.080	0.9792
0.250	40	10.000	2.5000
0.260	13	3.380	0.8788
0.270	6	1.620	0.4374
0.280	3	0.840	0.2352
0.290	2	0.580	0.1682
	100	24.730	6.1421

(b)

$$\overline{R} = \frac{\Sigma fR}{\Sigma f} = \frac{24.730}{100} = \boxed{0.2473\ \Omega}$$

(c) The sample standard deviation is given by Eq. 1.215.

$$s = \sqrt{\frac{\Sigma f R^2 - \frac{(\Sigma f R)^2}{n}}{n-1}}$$

$$= \sqrt{\frac{6.1421 - \frac{(24.73)^2}{100}}{99}}$$

$$= \boxed{0.0163\ \Omega}$$

(d)

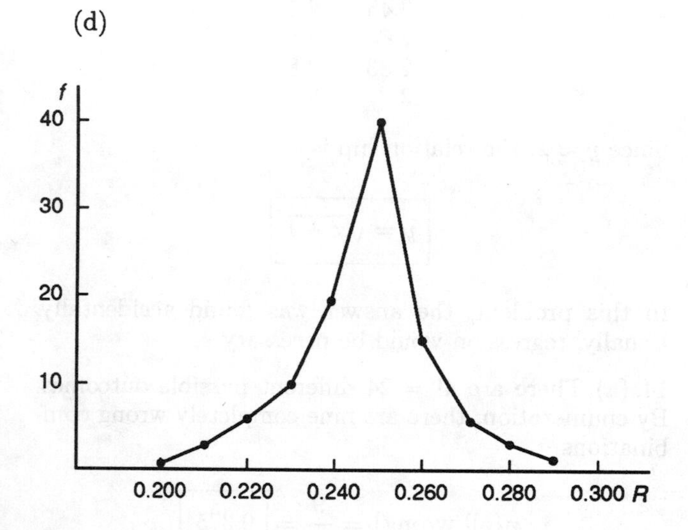

(e) 50% of the observations are below the median. 36 observations are below 0.240. 14 more are needed.

$$0.240 + \left(\frac{14}{40}\right)(0.250 - 0.240) = \boxed{0.2435\ \Omega}$$

(f) $s^2 = (0.0163)^2 = \boxed{0.0002656\ \Omega^2}$

16.

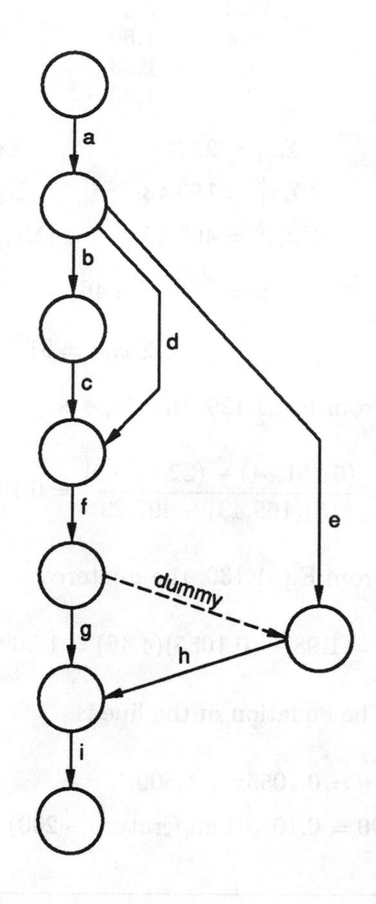

(Notice that a dummy node cannot be replaced with any other activity letter without duplicating that activity.)

17.

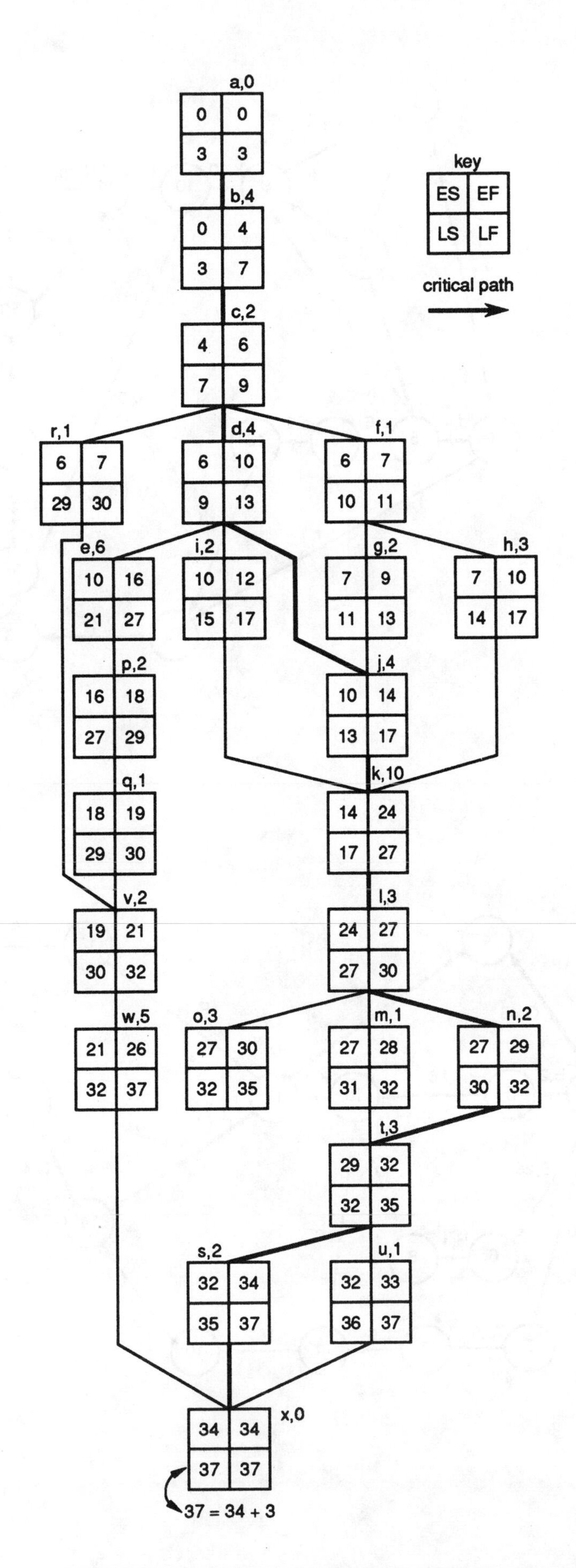
a,0
0 0
3 3
key
ES EF
LS LF
critical path
b,4
0 4
3 7
c,2
4 6
7 9
r,1
6 7
29 30
d,4
6 10
9 13
f,1
6 7
10 11
e,6
10 16
21 27
i,2
10 12
15 17
g,2
7 9
11 13
h,3
7 10
14 17
p,2
16 18
27 29
j,4
10 14
13 17
q,1
18 19
29 30
k,10
14 24
17 27
v,2
19 21
30 32
l,3
24 27
27 30
w,5
21 26
32 37
o,3
27 30
32 35
m,1
27 28
31 32
n,2
27 29
30 32
t,3
29 32
32 35
s,2
32 34
35 37
u,1
32 33
36 37
x,0
34 34
37 37
37 = 34 + 3

18.

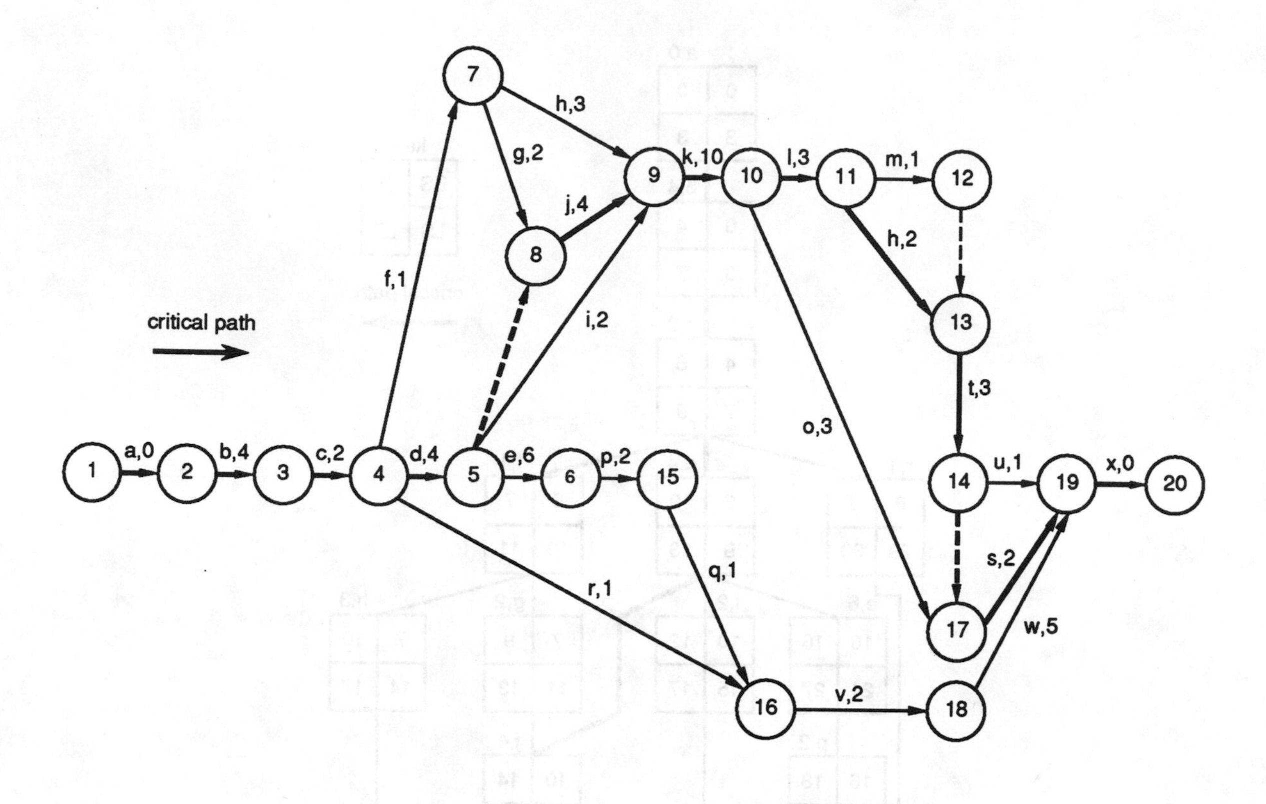

critical path
a,0
b,4
c,2
d,4
e,6
p,2
f,1
g,2
h,3
j,4
i,2
k,10
l,3
m,1
h,2
t,3
o,3
u,1
x,0
s,2
w,5
q,1
r,1
v,2

19.

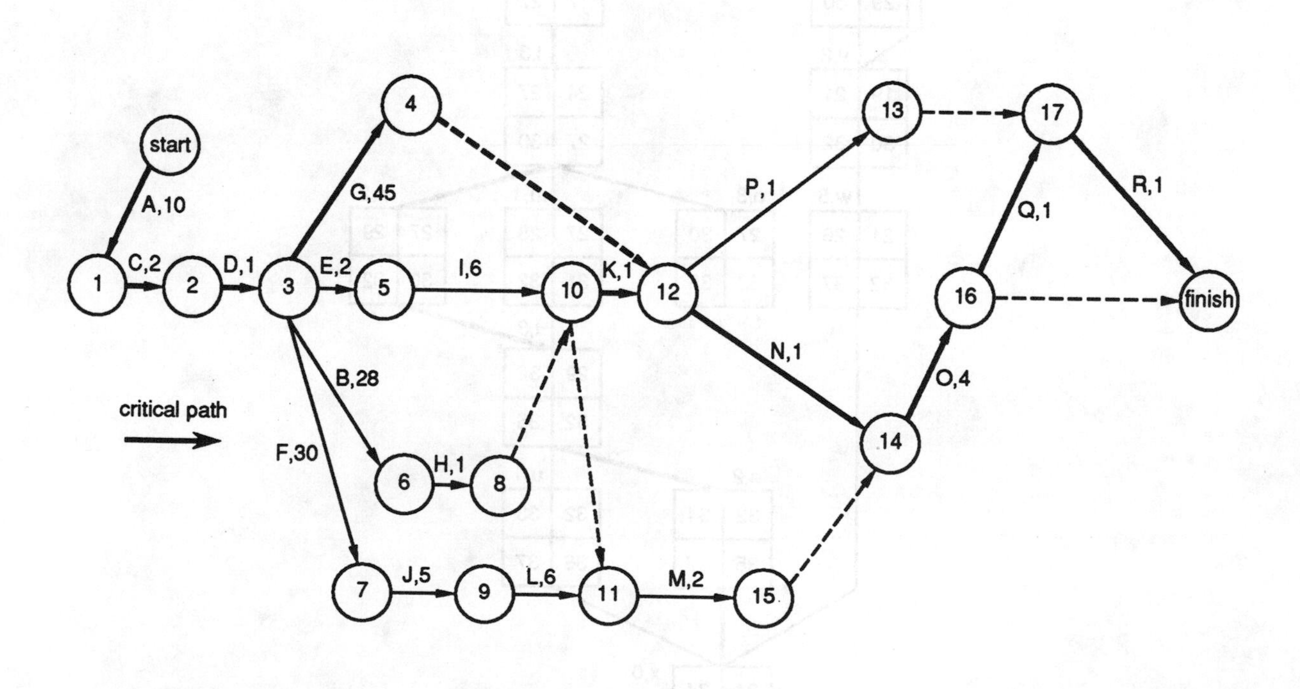

start
A,10
C,2
D,1
E,2
G,45
I,6
K,1
P,1
N,1
B,28
F,30
H,1
J,5
L,6
M,2
O,4
Q,1
R,1
finish
critical path

(b)

activity	variance (square of Eq. 1.315)	mean time (Eq. 1.314)
A	0.44	10
B	2.77	28
C	0.11	2
D	0.027	1
E	0.17	2
F	4.0	30
G	11.11	45
H	0	1
I	0.44	6
J	0.44	5
K	0.62	1
L	0.69	6
M	0.11	2
N	0.027	1
O	0.25	4
P	0	1
Q	0	1
R	0	1

Activities B, F, and G have the three largest variances, so these activities have the greatest uncertainties.

(c)

- Activity B is not on the critical path, so shortening it will not shorten the project.
- Activity G is on the critical path, so it should be shortened. Shortening it 5 days may change the critical path, however.

  path D-G-K: $1 + 45 = 46$ days

  path D-E-I-K: $1 + 2 + 6 = 9$ days

  path D-B-H-I-K: $1 + 28 + 1 + 6 = 36$ days

  Since the shortened path D-G-K has a length of 46 days − 5 days = 41 days and is still the longest path from D to K, it should be shortened.
- Activity O is on the critical path, but the cost of the crash schedule exceeds the benefit.
- The current path length is

  N-O-Q-R: $1 + 4 + 1 = 6$ days

  The new paths would be

  N-P-Q-R: $3 + 1 + 1 = 5$ days

  N-O-Q-R: $3 + 2 + 1 = 6$ days

  N-O-R: $(3 + 2) + (4 - 2) = 7$ days

  N-O-R would become part of the critical path and would be longer than the original critical path. Therefore, it is not acceptable.

20. Let

x = volume of 8% solution

y = volume of 10% solution

z = volume of 20% solution

The three conditions that must be satisfied are

$$x + y + z = 100 \text{ ml}$$

$$0.08x + 0.10y + 0.20z = (0.12)(100 \text{ ml}) = 12 \text{ ml}$$

$$0.08x = \left(\tfrac{1}{2}\right)(0.10y + 0.20z)$$

Simplifying these equations,

$$\begin{aligned} x + y + z &= 100 \\ 4x + 5y + 10z &= 600 \\ 8x - 5y - 10z &= 0 \end{aligned}$$

Adding the second and third equations gives

$$12x = 600$$

$$x = \boxed{50 \text{ ml}}$$

Work with the first two equations to get

$$\begin{aligned} y + z &= 100 - 50 = 50 \\ 5y + 10z &= 600 - (4)(50) = 400 \end{aligned}$$

Multiplying the top equation by −5 and adding to the bottom equation,

$$5z = 150$$

$$z = \boxed{30 \text{ ml}}$$

From the first equation,

$$y = \boxed{20 \text{ ml}}$$

21. Let

$m(t)$ = mass of salt in tank at time t

m_0 = 60 lbm

$m'(t)$ = rate at which salt content is changing

2 lbm of salt enter each minute, and 3 gal leave each minute. The amount of salt leaving each minute is

$$\left(3\ \frac{\text{gal}}{\text{min}}\right)\left(\text{concentration in } \frac{\text{lbm}}{\text{gal}}\right) = \left(3\ \frac{\text{gal}}{\text{min}}\right) \times \left(\frac{\text{salt content}}{\text{volume}}\right) = \left(3\ \frac{\text{gal}}{\text{min}}\right)\left(\frac{m(t)}{100 - t}\right)$$

$$m'(t) = 2 - (3)\left(\frac{m(t)}{100 - t}\right) \quad \text{or} \quad m'(t) + \frac{3m(t)}{100 - t} = 2$$

This is a first-order linear differential equation. From p. 1-32, the integrating factor is

$$\begin{aligned} m &= \exp\left[3\int \frac{dt}{100-t}\right] \\ &= \exp\left[(3)(-\ln(100-t))\right] \\ &= (100-t)^{-3} \\ m(t) &= (100-t)^3\left[2\int \frac{dt}{(100-t)^3} + k\right] \\ &= 100 - t + (k)(100-t)^3 \end{aligned}$$

But $m = 60$ at $t = 0$, so $k = -0.00004$.

$$m(t) = 100 - t - (0.00004)(100-t)^3$$

At $t = 60$,

$$\begin{aligned} m &= 100 - 60 - (0.00004)(100-60)^3 \\ &= \boxed{37.44 \text{ lbm}} \end{aligned}$$

Timed

1. μ is estimated by $\bar{x}$.

$$\begin{aligned} \bar{x} = \frac{\Sigma x_i}{n} &= 1249 + \frac{0.529 + 0.494 + 0.384 + 0.348}{4} \\ &= 1249.4388 \end{aligned}$$

Since there are only four samples, σ' is estimated by s (Eq. 1.215). Work with the fractional parts only.

$$\begin{aligned} \Sigma x_i^2 &= 0.792437 \\ (\Sigma x_i)^2 &= 3.080025 \end{aligned}$$

$$\begin{aligned} s &= \sqrt{\frac{0.792437 - \frac{3.080025}{4}}{3}} \\ &= 0.0865 \end{aligned}$$

This is a two-tail selection. 5% of the readings on either side of the limits should be rejected.

$$z = 1.645 \quad \text{[Table 1.5]}$$

The 90% confidence limits are

$$1249.4388 \pm (1.645)(0.0865) = 1249.5811,\ 1249.2965$$

(a) All readings are within the limits, and all are acceptable.

(b) None

(c) To be unacceptable, readings must be outside of $1.645s$ for 90% confidence.

(d) The unbiased estimate is $\bar{x} = 1249.4388$.

(e) There is 90% certainty that the actual distance is within the interval from 1249.2965 to 1249.5811. Since $\bar{x}$ is at the midpoint of the interval, the maximum error (at 90%) would be

$$|1249.5811 - 1249.4388| = \boxed{0.1423}$$

(f) See Eq. 17.4.

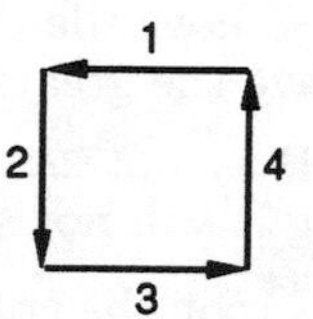

The maximum error in both legs 1 and 3 is 0.1423.

$$E_x = \sqrt{(0.1423)^2 + (0.1423)^2} = 0.2012$$

Similarly, E_y is 0.2012.

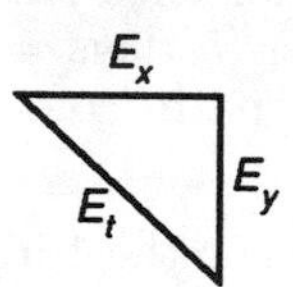

$$E_t = \sqrt{(0.2012)^2 + (0.2012)^2} = \boxed{0.2845}$$

(g) The ratio of error or precision is

$$\frac{\text{error of closure}}{\text{traverse perimeter}} = \frac{0.2845}{(4)(1249.4388)}$$

$$0.00005693 = \boxed{\frac{1}{17{,}564}}$$

(h) From Table 17.1, second-order accuracy must have an error of less than 1/10,000. This condition is satisfied.

(i) An average value (mean distance) that is close to the true value is accurate, regardless of the variability among readings. A precise value will have low variability, but it could have great bias.

(j) A systematic error is always present and unchanged in magnitude and direction. For example, a steel tape that is 0.01 ft too short will introduce a systematic error.

2. This is a fluid mixture problem. Refer to p. 1-34. The water mass in the lagoon is

$$\left(\frac{\pi}{4}\right)(120)^2(10)(62.4) = 7.06 \times 10^6 \text{ lbm}$$

The flow rate is

$$\phi(t) = \left(30\ \frac{\text{gal}}{\text{min}}\right)\left(8.345\ \frac{\text{lbm}}{\text{gal}}\right) = 250.4 \text{ lbm/min}$$

The mass of chemicals in the lagoon when $C = 1$ parts/billion is

$$m = (1 \times 10^{-9})(7.06 \times 10^6) = 7.06 \times 10^{-3} \text{ lbm}$$

$$S_{\text{in}}(t) = 0$$

$$S_{\text{out}}(t) = C(t)\phi(t)$$

$$= \frac{m(t)(250.4)}{7.06 \times 10^6} = 3.55 \times 10^{-5}\, m(t)$$

The differential equation is

$$m'(t) = S_{\text{in}}(t) - S_{\text{out}}(t)$$

$$= -3.55 \times 10^{-5}\, m(t)$$

$$m'(t) + 3.55 \times 10^{-5}\, m(t) = 0$$

The characteristic equation is

$$R + 3.55 \times 10^{-5} = 0$$

The solution to the differential equation is

$$m(t) = Ae^{-3.55\times 10^{-5} t}$$

Since $m = 90$ at $t = 0$, $A = 90$.

$$m(t) = 90e^{-3.55\times 10^{-5} t}$$

Solve for t when $m(t) = 7.06 \times 10^{-3}$.

$$7.06 \times 10^{-3} = 90e^{-3.55\times 10^{-5} t}$$

$$-9.45 = -3.55 \times 10^{-5} t \quad \text{[take the natural log]}$$

$$t = 2.66 \times 10^5 \text{ min} = \boxed{185 \text{ days}}$$

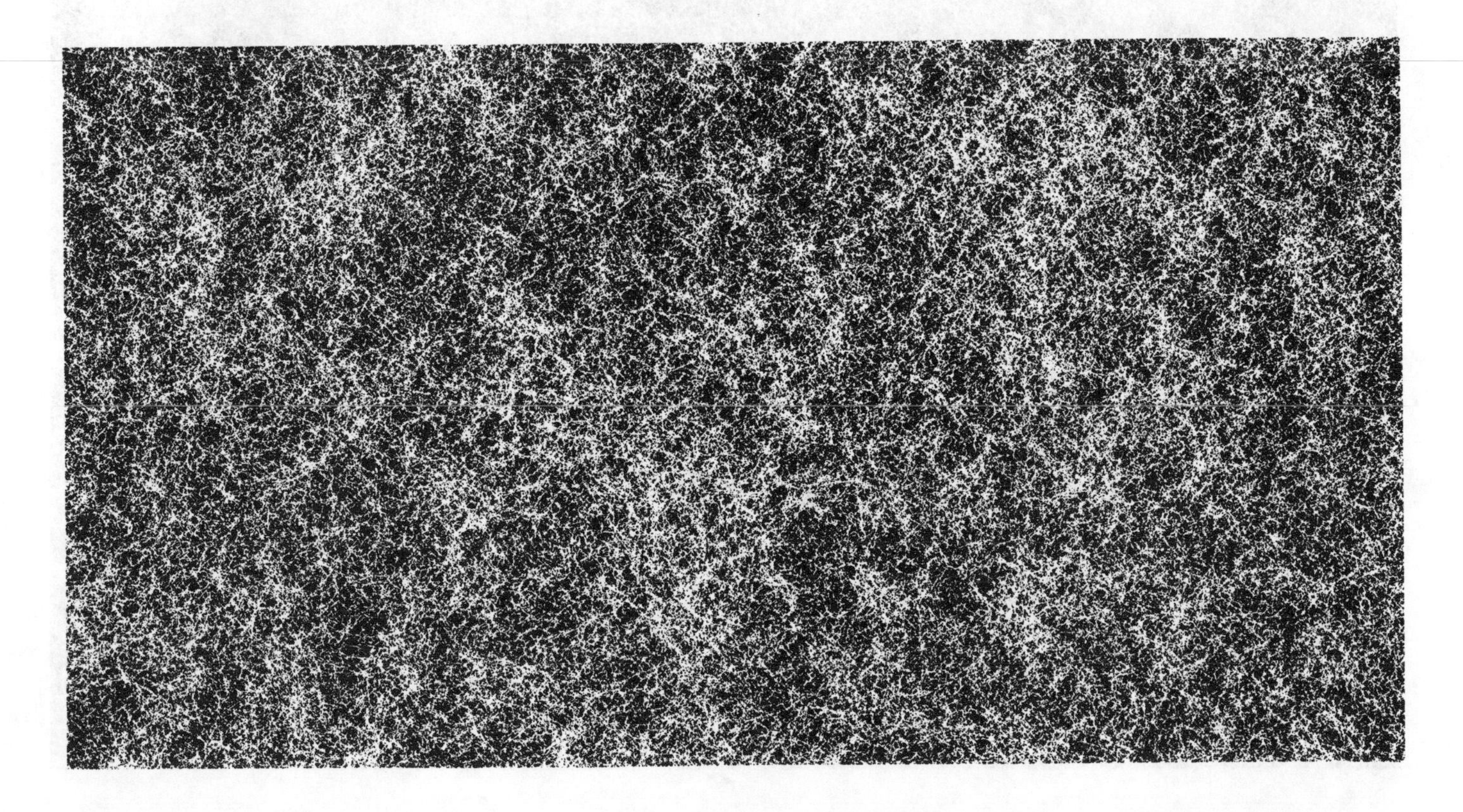

ENGINEERING ECONOMIC ANALYSIS

Untimed

1. (a) If the structure is kept for 1 year, the annual cost will be:

$$\begin{aligned}\text{EUAC}(1) &= (10{,}000)(A/P, 20\%, 1) \\ &\quad + (2000)(A/F, 20\%, 1) \\ &\quad - (8000)(A/F, 20\%, 1) \\ &= (10{,}000)(1.200) + (2000)(1.000) \\ &\quad - (8000)(1.000) \\ &= \$6000\end{aligned}$$

Similarly, if the structure is kept for 2 years or longer,

$$\begin{aligned}\text{EUAC}(2) &= (10{,}000)(A/P, 20\%, 2) \\ &\quad + (2000)(P/F, 20\%, 1)(A/P, 20\%, 2) \\ &\quad + (3000)(A/F, 20\%, 2) \\ &\quad - (7000)(A/F, 20\%, 2) \\ &= (10{,}000)(0.6545) + (2000)(0.8333)(0.6545) \\ &\quad + (3000)(0.4545) - (7000)(0.4545) \\ &= \$5818\end{aligned}$$

$$\begin{aligned}\text{EUAC}(3) &= [10{,}000 + (2000)(P/F, 20\%, 1) \\ &\quad + (3000)(P/F, 20\%, 2)](A/P, 20\%, 3) \\ &\quad + (4000 - 6000)(A/F, 20\%, 3) \\ &= [10{,}000 + (2000)(0.8333) + (3000)(0.6944)] \\ &\quad \times (0.4747) - (2000)(0.2747) \\ &= \$5978\end{aligned}$$

Since EUAC(2) is lowest, sell the structure after keeping it for 2 years.

(b) The cost is comprised of maintenance (6000), decrease in salvage value (5000 − 4000), and interest on 5000 at 20% ($1000).

$$6000 + 1000 + 1000 = \boxed{\$8000}$$

2. (a) The man should only charge his company for the additional costs incurred by the business travel. The annual increase in expenses and drop in salvage are

$$\begin{aligned}&= \left(\frac{5000}{15}\right)(0.60) + 100 + 50 + (500)(A/F, 10\%, 5) \\ &= 200 + 100 + 50 + (500)(0.1638) \\ &= \$431.90 \text{ per year}\end{aligned}$$

The cost per mile is

$$\frac{431.90}{5000} = 0.0864\ \$/\text{mi} \quad [8.64\ ¢/\text{mi}]$$

10¢ per mile is acceptable.

(b) Let x = no. of business miles

Option 1: $\text{EUAC} = 0.10x$

$$\begin{aligned}\text{Option 2: EUAC} &= (5000)(A/P, 10\%, 5) + 250 + 200 \\ &\quad + \left(\frac{x}{15}\right)(0.60) - (800)(A/F, 10\%, 5) \\ &= (5000)(0.2638) + 250 + 200 \\ &\quad + \left(\frac{x}{15}\right)(0.60) - (800)(0.1638) \\ &= 1637.96 + 0.04x\end{aligned}$$

If these costs are equal,

$$0.10x = 1637.96 + 0.04x$$

$$x = \boxed{27{,}299 \text{ miles per year}}$$

3. EEUAC = expected equivalent uniform annual cost

$$\begin{aligned}\text{EEUAC}(7) &= (0.15)(25{,}000) \\ &= \$3750\end{aligned}$$

$$\begin{aligned}\text{EEUAC}(8) &= (15{,}000)(A/P, 10\%, 20) \\ &\quad + (0.10)(25{,}000) \\ &= (15{,}000)(0.1175) + (0.10)(25{,}000) \\ &= \$4262.50\end{aligned}$$

$$\begin{aligned}\text{EEUAC}(9) &= (20{,}000)(A/P, 10\%, 20) \\ &\quad + (0.07)(25{,}000) \\ &= (20{,}000)(0.1175) + (0.07)(25{,}000) \\ &= \$4100\end{aligned}$$

$$\begin{aligned}\text{EEUAC}(10) &= (30{,}000)(A/P, 10\%, 20) \\ &\quad + (0.03)(25{,}000) \\ &= (30{,}000)(0.1175) + (0.03)(25{,}000) \\ &= \$4275\end{aligned}$$

It is cheapest to do nothing.

4.
$$\begin{aligned}\text{EUAC} &= (17{,}000 + 5000)(A/P, 6\%, 5) \\ &\quad - (14{,}000 + 2500)(A/F, 6\%, 5) + 200 \\ &= (22{,}000)(0.2374) - (16{,}500)(0.1774) + 200 \\ &= \boxed{\$2495.70}\end{aligned}$$

5. Assume the old bridge will be there forever.

• Keep the old bridge:

The generally accepted method is to consider the salvage value as a lost benefit (or cost). (See p. 2-6.)

$$\begin{aligned}\text{EUAC} &= (9000 + 13{,}000)(A/P, 8\%, 20) \\ &\quad - (10{,}000)(A/F, 8\%, 20) + 500 \\ &= (22{,}000)(0.1019) - (10{,}000)(0.0219) + 500 \\ &= \$2522.80\end{aligned}$$

• Replace:

$$\begin{aligned}\text{EUAC} &= (40{,}000)(A/P, 8\%, 25) \\ &\quad - (15{,}000)(A/F, 8\%, 25) + 100 \\ &= (40{,}000)(0.0937) - (15{,}000)(0.0137) + 100 \\ &= \$3642.50\end{aligned}$$

Since \$3642 > \$2522, **keep the old bridge.**

6.

$$\begin{aligned}D &= \frac{150{,}000}{15} = \$10{,}000 \\ 0 &= -150{,}000 + (32{,}000)(1 - 0.48) \\ &\quad (P/A, i\%, 15) - (7530)(1 - 0.48) \\ &\quad \times (P/A, i\%, 15) \\ &\quad + (10{,}000)(0.48)(P/A, i\%, 15) \\ &= -150{,}000 + (16{,}640 - 3915.60 + 4800) \\ &\quad \times (P/A, i\%, 15)\end{aligned}$$

$$(P/A, i\%, 15) = \frac{150{,}000}{17{,}524.40} = 8.5595$$

Searching the tables,

$$i = \boxed{8\%}$$

7. (a) $\dfrac{1{,}500{,}000 - 300{,}000}{1{,}000{,}000} = \boxed{1.2}$

(b) $1{,}500{,}000 - 300{,}000 - 1{,}000{,}000 = \boxed{\$200{,}000}$

8. (a) SL:

$$D = \frac{500{,}000 - 100{,}000}{25} = \boxed{\$16{,}000}$$

(b) SOYD:

$$T = \left(\frac{1}{2}\right)(25)(26) = 325$$

$$\begin{aligned}D_1 &= \left(\frac{25}{325}\right)(500{,}000 - 100{,}000) \\ &= \boxed{\$30{,}769}\end{aligned}$$

$$D_2 = \left(\frac{24}{325}\right)(400{,}000) = \boxed{\$29{,}538}$$

$$D_3 = \left(\frac{23}{325}\right)(400{,}000) = \boxed{\$28{,}308}$$

(c) DDB:

$$D_1 = \left(\frac{2}{25}\right)(500{,}000) = \boxed{\$40{,}000}$$

$$D_2 = \left(\frac{2}{25}\right)(500{,}000 - 40{,}000) = \boxed{\$36{,}800}$$

$$\begin{aligned}D_3 &= \left(\frac{2}{25}\right)(500{,}000 - 40{,}000 - 36{,}800) \\ &= \boxed{\$33{,}856}\end{aligned}$$

9.

$$\begin{aligned}P &= -12{,}000 + (2000)(P/F, 10\%, 10) \\ &\quad - (1000)(P/A, 10\%, 10) \\ &\quad - (200)(P/G, 10\%, 10) \\ &= -12{,}000 + (2000)(0.3855) - (1000)(6.1446) \\ &\quad - (200)(22.8913) \\ &= \boxed{-\$21{,}951.86}\end{aligned}$$

$$\begin{aligned}\text{EUAC} &= (21{,}951.86)(A/P, 10\%, 10) \\ &= (21{,}951.86)(0.1627) \\ &= \boxed{\$3571.57}\end{aligned}$$

10. $\text{EUAC}(9) = (1500)(A/P, 6\%, 20) + \left(\frac{1}{9}\right)(0.35)(1500) + (0.04)(1500)$
$= (1500)\left[0.0872 + \left(\frac{1}{9}\right)(0.35) + 0.04\right]$
$= \$249.13$

$\text{EUAC}(14) = (1600)\left[0.1272 + (0.35)\left(\frac{1}{14}\right)\right] = \$243.52$

$\text{EUAC}(30) = (1750)\left[0.1272 + (0.35)\left(\frac{1}{30}\right)\right] = \$243.02$

$\text{EUAC}(52) = (1900)\left[0.1272 + (0.35)\left(\frac{1}{52}\right)\right] = \$254.47$

$\text{EUAC}(86) = (2100)\left[0.1272 + (0.35)\left(\frac{1}{86}\right)\right] = \$275.67$

Choose the 30-year pipe since its EUAC is the smallest.

11. $P(A) = (4800)(P/A, 12\%, 25) = (4800)(7.8431) = \$37{,}646.88$

4 quarters × 25 years = 100 compounding periods

$P(B) = (1200)(P/A, 3\%, 100) = (1200)(31.5989) = \$37{,}918.68$

Choose B.

12. $\text{EUAC}(A) = (1500)(A/P, 15\%, 5) + 800 = (1500)(0.2983) + 800 = \$1247.45$

$\text{EUAC}(B) = (2500)(A/P, 15\%, 8) + 650 = (2500)(0.2229) + 650 = \$1207.25$

Choose B.

13. The data given implies that both investments return 4% or more. However, the increased investment of $30,000 may not be cost effective. Do an incremental analysis.

incremental cost = 70,000 − 40,000 = $30,000

incremental income = 5620 − 4075 = $1545

$0 = -30{,}000 + (1545)(P/A, i\%, 20)$

$(P/A, i\%, 20) = 19.417$

$i \approx 0.25\% < 4\%$

Choose B.

(The same conclusion could be reached by taking present worths of both alternatives.)

14. (a) $P(A) = (-3000)(P/A, 25\%, 25) - 20{,}000 = (-3000)(3.9849) - 20{,}000 = -\$31{,}954.70$

$P(B) = (-2500)(3.9849) - 25{,}000 = -\$34{,}962.25$

A is better.

(b) $\text{CC}(A) = 20{,}000 + \frac{3000}{0.25} = \$32{,}000$

$\text{CC}(B) = 25{,}000 + \frac{2500}{0.25} = \$35{,}000$

A is better.

(c) $\text{EUAC}(A) = (20{,}000)(A/P, 25\%, 25) + 3000 = (20{,}000)(0.2509) + 3000 = \$8018.00$

$\text{EUAC}(B) = (25{,}000)(0.2509) + 2500 = \$8772.50$

A is better.

15. (a) $\text{BV} = 18{,}000 - (5)\left(\frac{18{,}000 - 2000}{8}\right) = \boxed{\$8000}$

(b) From Eq. 2.13, the basis is

$(18{,}000 - 2000)(A/F, 8\%, 8) = (18{,}000 - 2000)(0.0940) = \$1504$

$D_1 = (1504)(1.000) = \boxed{\$1504}$

$D_2 = (1504)(1.0800) = \boxed{\$1624}$

$D_3 = (1504)(1.0800)^2 = (1504)(1.1664) = \boxed{\$1754}$

(c) $D_1 = \left(\frac{2}{8}\right)(18{,}000) = \boxed{\$4500}$

$D_2 = \left(\frac{2}{8}\right)(18{,}000 - 4500) = \boxed{\$3375}$

$D_3 = \left(\frac{2}{8}\right)(18{,}000 - 4500 - 3375) = \boxed{\$2531}$

$D_4 = \left(\frac{2}{8}\right)(18{,}000 - 4500 - 3375 - 2531)$
$= \boxed{\$1898}$

$D_5 = \left(\frac{2}{8}\right)(18{,}000 - 4500 - 3375 - 2531 - 1898)$
$= \boxed{\$1424}$

$\text{BV} = 18{,}000 - 4500 - 3375 - 2531 - 1898 - 1424$
$= \boxed{\$4272}$

16. $T = \left(\frac{1}{2}\right)(8)(9) = 36$

$D_1 = \left(\frac{8}{36}\right)(12{,}200) = \$2711$

$\Delta D = \left(\frac{1}{36}\right)(12{,}200) = \$339$

$D_2 = 2711 - 339 = \$2372$

$$\begin{aligned}\text{DR} &= (0.52)(2711)(P/A, 15\%, 8) \\ &\quad - (0.52)(339)(P/G, 15\%, 8) \\ &= (0.52)(2711)(4.4873) - (0.52)(339)(12.4807) \\ &= \boxed{\$4125.74}\end{aligned}$$

17.
$$\begin{aligned}\text{EUAC}_{5\text{ ft}} &= (600{,}000)(A/P, 12\%, 15) \\ &\quad + \left(\tfrac{14}{50}\right)(600{,}000) + \left(\tfrac{8}{50}\right)(650{,}000) \\ &\quad + \left(\tfrac{3}{50}\right)(700{,}000) + \left(\tfrac{1}{50}\right)(800{,}000) \\ &= \$418{,}080\end{aligned}$$

$$\begin{aligned}\text{EUAC}_{10\text{ ft}} &= (710{,}000)(A/P, 12\%, 15) \\ &\quad + \left(\tfrac{8}{50}\right)(650{,}000) + \left(\tfrac{3}{50}\right)(700{,}000) \\ &\quad + \left(\tfrac{1}{50}\right)(800{,}000) \\ &= \$266{,}228\end{aligned}$$

$$\begin{aligned}\text{EUAC}_{15\text{ ft}} &= (900{,}000)(A/P, 12\%, 15) \\ &\quad + \left(\tfrac{3}{50}\right)(700{,}000) + \left(\tfrac{1}{50}\right)(800{,}000) \\ &= \$190{,}120\end{aligned}$$

$$\begin{aligned}\text{EUAC}_{20\text{ ft}} &= (1{,}000{,}000)(A/P, 12\%, 15) \\ &\quad + \left(\tfrac{1}{50}\right)(800{,}000) \\ &= \$162{,}800\end{aligned}$$

$\boxed{\text{Build to 20 ft.}}$

18. Assume replacement after 1 year.

$$\begin{aligned}\text{EUAC}(1) &= (30{,}000)(A/P, 12\%, 1) + 18{,}700 \\ &= (30{,}000)(1.12) + 18{,}700 = \$52{,}300\end{aligned}$$

Assume replacement after 2 years.

$$\begin{aligned}\text{EUAC}(2) &= (30{,}000)(A/P, 12\%, 2) \\ &\quad + 18{,}700 + (1200)(A/G, 12\%, 2) \\ &= (30{,}000)(0.5917) + 18{,}700 \\ &\quad + (1200)(0.4717) = \$37{,}017\end{aligned}$$

Assume replacement after 3 years.

$$\begin{aligned}\text{EUAC}(3) &= (30{,}000)(A/P, 12\%, 3) \\ &\quad + 18{,}700 + (1200)(A/G, 12\%, 3) \\ &= (30{,}000)(0.4163) + 18{,}700 \\ &\quad + (1200)(0.9246) = \$32{,}299\end{aligned}$$

Similarly, calculate to obtain the numbers in the following table.

years in service	EUAC
1	\$52,300
2	\$37,017
3	\$32,299
4	\$30,207
5	\$29,152
6	\$28,602
7	\$28,335
8	\$28,234
9	\$28,240
10	\$28,312

$\boxed{\text{Replace after 8 years.}}$

19. Assume the head and horsepower data are already reflected in the hourly operating costs.

Let N = no. of hours operated each year.

$$\begin{aligned}\text{EUAC(A)} &= (3600 + 3050)(A/P, 12\%, 10) \\ &\quad - (200)(A/F, 12\%, 10) + 0.30N \\ &= (6650)(0.1770) - (200)(0.0570) + 0.30N \\ &= 1165.65 + 0.30N\end{aligned}$$

$$\begin{aligned}\text{EUAC(B)} &= (2800 + 5010)(A/P, 12\%, 10) \\ &\quad - (280)(A/F, 12\%, 10) + 0.10N \\ &= (7810)(0.1770) - (280)(0.0570) + 0.10N \\ &= 1366.41 + 0.10N\end{aligned}$$

$$\text{EUAC(A)} = \text{EUAC(B)}$$

$$1165.65 + 0.30N = 1366.41 + 0.10N$$

$$N = \boxed{1003.8 \text{ hours}}$$

20. From Eq. 2.31,

$$\frac{T_2}{T_1} = 0.88 = 2^{-b}$$
$$\log 0.88 = -b \log 2$$
$$-0.0555 = -(0.3010)b$$
$$b = 0.1843$$

$$T_4 = (6)(4)^{-0.1843} = \boxed{4.65 \text{ weeks}}$$

From Eq. 2.32,

$$\begin{aligned} T_{6\text{–}14} &= \left(\frac{6}{1-0.1843}\right) \\ &\quad \times \left[\left(14+\tfrac{1}{2}\right)^{1-0.1843} - \left(6-\tfrac{1}{2}\right)^{1-0.1843}\right] \\ &= \left(\frac{6}{0.8157}\right)(8.857-4.017) \\ &= \boxed{35.6 \text{ weeks}} \end{aligned}$$

Timed

1. First check that both alternatives have a ROR greater than the MARR. Work in thousands of dollars.

- Alternative A:

$$\begin{aligned} P(\text{A}) &= -120 + (15)(P/F, i\%, 5) \\ &\quad + (57)(P/A, i\%, 5)(1-0.45) \\ &\quad + \left(\frac{120-15}{5}\right)(P/A, i\%, 5)(0.45) \\ &= -120 + (15)(P/F, i\%, 5) \\ &\quad + (40.8)(P/A, i\%, 5) \end{aligned}$$

Try 15%.

$$\begin{aligned} P(\text{A}) &= -120 + (15)(0.4972) + (40.8)(3.3522) \\ &= \$24.23 \end{aligned}$$

Try 25%.

$$\begin{aligned} P(\text{A}) &= -120 + (15)(0.3277) + (40.8)(2.6893) \\ &= -\$5.36 \end{aligned}$$

Since $P(\text{A})$ goes through 0,

$$(\text{ROR})_\text{A} > \text{MARR} = 15\%$$

- Alternative B:

$$\begin{aligned} P(\text{B}) &= -170 + (20)(P/F, i\%, 5) \\ &\quad + (67)(P/A, i\%, 5)(1-0.45) \\ &\quad + \left(\frac{170-20}{5}\right)(P/A, i\%, 5)(0.45) \\ &= -170 + (20)(P/F, i\%, 5) \\ &\quad + (50.35)(P/A, i\%, 5) \end{aligned}$$

Try 15%.

$$\begin{aligned} P(\text{B}) &= -170 + (20)(0.4972) + (50.35)(3.352) \\ &= \$8.72 \end{aligned}$$

Since $P(\text{B}) > 0$ and will decrease as i increases,

$$(\text{ROR})_\text{B} > 15\%$$

ROR > MARR for both alternatives.

Do an incremental analysis to see if it is worthwhile to invest the extra 170 − 20 = \$50.

$$\begin{aligned} P(\text{B}-\text{A}) &= -50 + (5)(P/F, i\%, 5) \\ &\quad + (9.55)(P/A, i\%, 5) \end{aligned}$$

Try 15%.

$$\begin{aligned} P(\text{B}-\text{A}) &= -50 + (5)(0.4972) \\ &\quad + (9.55)(3.3522) \\ &= -\$15.50 \end{aligned}$$

Since $P(\text{B}-\text{A}) < 0$ and would become more negative as i increases, the ROR of the added investment is < 15%.

Select alternative A.

2. Use the year-end convention with the tax credit. The purchase is made at $t = 0$. However, the tax credit is received at $t = 1$ and must be multiplied by $(P/F, i\%, 1)$.

(Note that 0.0667 is actually $\frac{2}{3}$ of 10%.)

$$\begin{aligned} P &= -300{,}000 + (0.0667)(300{,}000)(P/F, i\%, 1) \\ &\quad + (90{,}000)(P/A, i\%, 5)(1-0.48) \\ &\quad + \left(\frac{300{,}000-50{,}000}{5}\right)(P/A, i\%, 5)(0.48) \\ &\quad + (50{,}000)(P/F, i\%, 5) \\ &= -300{,}000 + (20{,}000)(P/F, i\%, 1) \\ &\quad (50{,}000)(P/F, i\%, 5) \\ &\quad + (46{,}800)(P/A, i\%, 5) \\ &\quad + (24{,}000)(P/A, i\%, 5) \end{aligned}$$

By trial and error,

i	P
10%	\$17,616
15%	−\$20,412
12%	\$1448
13%	−\$6142
$12\frac{1}{4}$%	−\$479

$$\boxed{i \text{ is between } 12\% \text{ and } 12\tfrac{1}{4}\%.}$$

3. Assume loan payments are made at the end of each year. Find the annual payment.

$$\begin{aligned} \text{payment} &= (2{,}500{,}000)(A/P, 12\%, 25) \\ &= (2{,}500{,}000)(0.1275) \\ &= \$318{,}750 \\ \text{distributed profit} &= (0.15)(2{,}500{,}000) \\ &= \$375{,}000 \end{aligned}$$

After paying all expenses and distributing the 15% profit, the remainder should be 0.

$$\begin{aligned} 0 = \text{EUAC} &= 20{,}000 + 50{,}000 + 200{,}000 + 375{,}000 \\ &\quad + 318{,}750 - \text{annual receipts} \\ &\quad - (500{,}000)(A/F, 15\%, 25) \\ &= 963{,}750 - \text{annual receipts} \\ &\quad - (500{,}000)(0.0047) \end{aligned}$$

This calculation assumes $i = 15\%$, which equals the desired return. However, this assumption only affects the salvage calculation, and since the number is so small, the analysis is not sensitive to the assumption.

$$\text{annual receipts} = \$961{,}400$$

The average daily receipts are

$$\frac{\$961{,}400}{365} = \$2634$$

Use the expected value approach. The average occupancy is

$$\begin{aligned} &(0.40)(0.65) + (0.30)(0.70) + (0.20)(0.75) \\ &\quad + (0.10)(0.80) = 0.70 \end{aligned}$$

The average number of rooms occupied each night is

$$(0.70)(120) = 84 \text{ rooms}$$

The minimum required average rate is

$$\frac{\$2634}{84} = \boxed{\$31.36}$$

4. $$\text{annual savings} = \left(\frac{0.69 - 0.47}{1000}\right)(3{,}500{,}000) = \$770$$

$$\begin{aligned} P &= -7500 + (770 - 200 - 100) \\ &\quad \times (P/A, i\%, 25) = 0 \end{aligned}$$

$$(P/A, i\%, 25) = 15.957$$

Searching the tables and interpolating,

$$i \approx \boxed{3.75\%}$$

5. Work in millions of dollars.

$$\begin{aligned} P(\text{A}) &= -(1.3)(1 - 0.48)(P/A, 15\%, 6) \\ &= (1.3)(0.52)(3.7845) \\ &= -\$2.56 \quad \text{[millions]} \end{aligned}$$

Since this is an after-tax analysis and since the salvage value was mentioned, assume that the improvements can be depreciated.

Use straight-line depreciation.

- Alternative B:

$$\begin{aligned} D_j &= \tfrac{2}{6} = 0.333 \\ P(\text{B}) &= -2 - (0.20)(1.3)(1 - 0.48)(P/A, 15\%, 6) \\ &\quad - (0.15)(1 - 0.48)(P/A, 15\%, 6) \\ &\quad + (0.333)(0.48)(P/A, 15\%, 6) \\ &= -2 - (0.20)(1.3)(0.52)(3.7845) \\ &\quad - (0.15)(0.52)(3.7845) \\ &\quad + (0.333)(0.48)(3.7845) \\ &= -\$2.206 \quad \text{[millions]} \end{aligned}$$

- Alternative C:

$$\begin{aligned} D_j &= \frac{1.2}{3} = 0.4 \\ P(\text{C}) &= -(1.2)[1 + (P/F, 15\%, 3)] \\ &\quad - (0.20)(1.3)(1 - 0.48) \\ &\quad \times [(P/F, 15\%, 1) + (P/F, 15\%, 4)] \\ &\quad - (0.45)(1.3)(1 - 0.48) \\ &\quad \times [(P/F, 15\%, 2) + (P/F, 15\%, 5)] \\ &\quad - (0.80)(1.3)(1 - 0.48) \\ &\quad \times [(P/F, 15\%, 3) + (P/F, 15\%, 6)] \\ &\quad + (0.4)(0.48)(P/A, 15\%, 6) \\ &= -(1.2)(1.6575) \\ &\quad - (0.20)(1.3)(0.52)(0.8696 + 0.5718) \\ &\quad - (0.45)(1.3)(0.52)(0.7561 + 0.4972) \\ &\quad - (0.80)(1.3)(0.52)(0.6575 + 0.4323) \\ &\quad + (0.4)(0.48)(3.7845) \\ &= -\$2.436 \quad \text{[millions]} \end{aligned}$$

$$\boxed{\text{Choose B.}}$$

6. This is a replacement study. Since production capacity and efficiency are not a problem with the defender, the only question is when to bring in the challenger.

Since this is a before-tax problem, depreciation is not a factor, nor is book value.

The cost of keeping the defender one more year is

$$\begin{aligned}\text{EUAC(defender)} &= 200{,}000 + (0.15)(400{,}000)\\ &= \$260{,}000\end{aligned}$$

For the challenger,

$$\begin{aligned}\text{EUAC(challenger)} &= (800{,}000)(A/P, 15\%, 10)\\ &\quad + 40{,}000 + (30{,}000)\\ &\quad \times (A/G, 15\%, 10)\\ &\quad - (100{,}000)(A/F, 15\%, 10)\\ &= (800{,}000)(0.1993)\\ &\quad + 40{,}000 + (30{,}000)(3.3832)\\ &\quad - (100{,}000)(0.0493)\\ &= \$296{,}006\end{aligned}$$

Since the defender is cheaper, keep it. The same analysis next year will give identical answers. Therefore, keep the defender for the next 3 years, at which time the decision to buy the challenger will be automatic.

Having determined that it is less expensive to keep the defender than to maintain the challenger for 10 years, determine whether the challenger is less expensive if retired before 10 years.

If retired in 9 years,

$$\begin{aligned}\text{EUAC(challenger)} &= (800{,}000)(A/P, 15\%, 9) + 40{,}000\\ &\quad + (30{,}000)(A/G, 15\%, 9)\\ &\quad - (150{,}000)(A/F, 15\%, 9)\\ &= (800{,}000)(0.2096)\\ &\quad + 40{,}000 + (30{,}000)(3.0922)\\ &\quad - (150{,}000)(0.0596)\\ &= \$291{,}506\end{aligned}$$

Similar calculations yield the following results for all the retirement dates.

n	EUAC
10	$296,000
9	$291,506
8	$287,179
7	$283,214
6	$280,016
5	$278,419
4	$279,909
3	$288,013
2	$313,483
1	$360,000

Since none of these equivalent uniform annual costs is less than that of the defender, it is not economical to buy and keep the challenger for any length of time.

Keep the defender.

FLUID STATICS AND DYNAMICS

Untimed

1. Assume that flows from reservoirs A and B are toward D and then toward C. Then,

$$Q_{A\text{–}D} + Q_{B\text{–}D} = Q_{D\text{–}C}$$

or

$$A_A v_{A\text{–}D} + A_B v_{B\text{–}D} - A_C v_{D\text{–}C} = 0$$

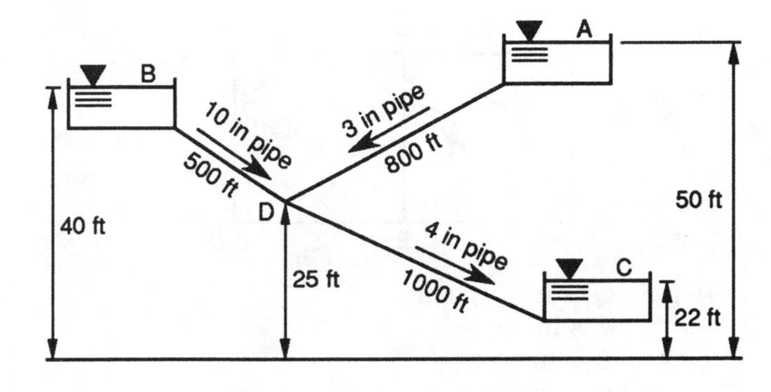

Assume a schedule-40 pipe so that

$$A_A = 0.05134 \text{ ft}^2 \qquad D_A = 0.2557 \text{ ft}$$
$$A_B = 0.5476 \text{ ft}^2 \qquad D_B = 0.8350 \text{ ft}$$
$$A_C = 0.08841 \text{ ft}^2 \qquad D_C = 0.3355 \text{ ft}$$

$$(0.05134)v_{A\text{–}D} + (0.5476)v_{B\text{–}D} - (0.08841)v_{D\text{–}C} = 0 \qquad \text{[Eq. 1]}$$

Ignoring the velocity heads, the conservation of energy equation between A and D is

$$z_A = \frac{p_D}{\gamma} + z_D + h_{f,A\text{–}D}$$

$$50 = \frac{p_D}{62.4} + 25 + \frac{(0.02)(800)(v_{A\text{–}D})^2}{(2)(0.2557)(32.2)}$$

or

$$v_{A\text{–}D} = \sqrt{25.73 - 0.0165 p_D} \qquad \text{[Eq. 2]}$$

Similarly, for B–D,

$$40 = \frac{p_D}{62.4} + 25 + \frac{(0.02)(500)(v_{B\text{–}D})^2}{(2)(0.8350)(32.2)}$$

or

$$v_{B\text{–}D} = \sqrt{80.66 - 0.0862 p_D} \qquad \text{[Eq. 3]}$$

For D–C,

$$22 = \frac{p_D}{62.4} + 25 - \frac{(0.02)(1000)(v_{D\text{–}C})^2}{(2)(0.3355)(32.2)}$$

or

$$v_{D\text{–}C} = \sqrt{3.24 + 0.0173 p_D} \qquad \text{[Eq. 4]}$$

Equations 1, 2, 3, and 4 must be solved simultaneously. To do this, assume a value for p_D. This value then determines all three velocities in Eqs. 2, 3, and 4. These velocities are substituted into Eq. 1. A trial and error solution yields

$$v_{A\text{–}D} = 3.21 \text{ ft/sec}$$
$$v_{B\text{–}D} = 0.418 \text{ ft/sec}$$
$$v_{D\text{–}C} = 4.40 \text{ ft/sec}$$

Flow is from B to D.

2. Assume 70°F water.

$$E_w = 320 \times 10^3 \text{ psi}$$
$$\gamma_w = 62.3 \text{ lbf/ft}^3$$

Assume class 20 cast iron pipe.

$$E_p = 20 \times 10^6 \text{ psi} \qquad \text{[Table 12.1]}$$

(a) From p. 22-1, the modulus of elasticity to be used is

$$E = \frac{(320 \times 10^3)(1.16)(20 \times 10^6)}{(1.16)(20 \times 10^6) + (24)(320 \times 10^3)} = 2.4 \times 10^5 \text{ psi}$$

From Eq. 3.20,

$$c = \sqrt{\frac{(2.4 \times 10^5)(144)(32.2)}{62.3}} = 4226 \text{ ft/sec}$$

From Eq. 3.199, the pressure increase is

$$\Delta p = \frac{(62.3)(4226)(6)}{32.2} = \boxed{49{,}058 \text{ lbf/ft}^2 \text{ (psf)}}$$

(b) From Eq. 3.198, the maximum closure time is

$$t = \frac{(2)(500)}{4226} = \boxed{0.237 \text{ sec}}$$

3. *steps 1, 2, and 3:*

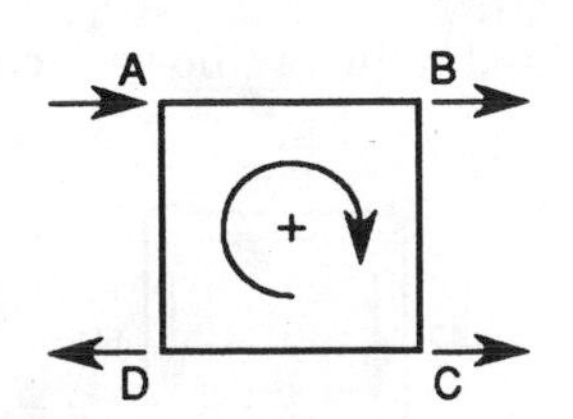

step 4: There is only one loop: ABCD.

step 5: From Eq. 3.133,

$$\text{pipe AB:}\quad K' = \frac{(10.44)(1000)}{(100)^{1.85}(8)^{4.87}} = 8.33 \times 10^{-5}$$

$$\text{pipe BC:}\quad K' = \frac{(10.44)(1000)}{(100)^{1.85}(6)^{4.87}} = 3.38 \times 10^{-4}$$

$$\text{pipe CD:}\quad K' = 8.33 \times 10^{-5} \quad \text{[same as AB]}$$

$$\text{pipe DA:}\quad K' = \frac{(10.44)(1000)}{(100)^{1.85}(12)^{4.87}} = 1.16 \times 10^{-5}$$

step 6: Assume the flows shown in the figure.

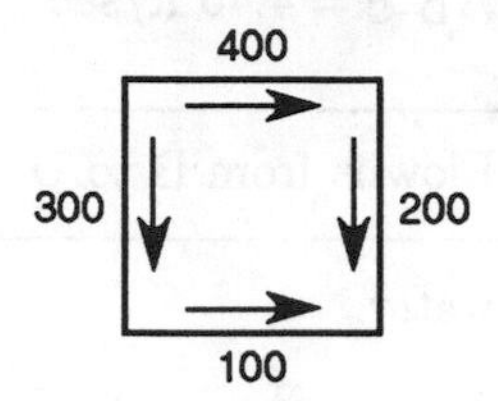

step 7:

$$\delta = \left(\frac{-1}{1.85}\right) \times \left[\frac{\begin{array}{c}(8.33\times10^{-5})(400)^{1.85}+(3.38\times10^{-4})(200)^{1.85}\\ -(8.33\times10^{-5})(100)^{1.85}-(1.16\times10^{-5})(300)^{1.85}\end{array}}{\begin{array}{c}(8.33\times10^{-5})(400)^{0.85}+(3.38\times10^{-4})(200)^{0.85}\\ +(8.33\times10^{-5})(100)^{0.85}+(1.16\times10^{-5})(300)^{0.85}\end{array}}\right]$$

$$= \left(\frac{-1}{1.85}\right)\left[\frac{10.67}{4.98 \times 10^{-2}}\right] = -116 \text{ gal/min}$$

step 8: The adjusted flows are shown.

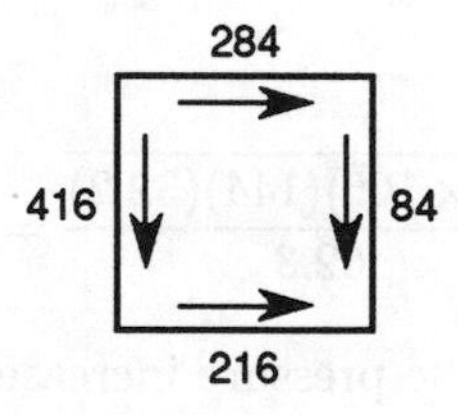

step 7: $\delta = -24$ gal/min

step 8: The adjusted flows are shown.

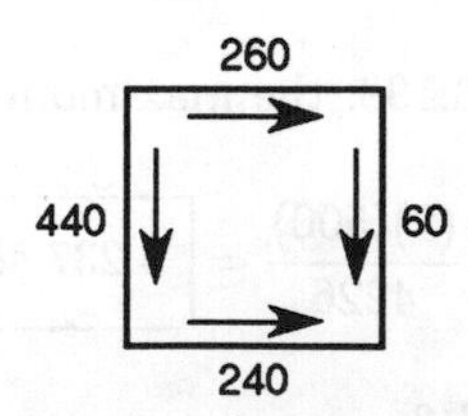

step 7: $\delta = -2$ gal/min [small enough]

step 8: The final adjusted flows are shown.

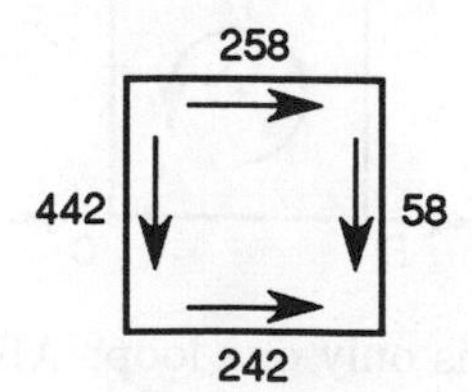

4. This is a Hardy Cross problem. The pressure at point C does not change the solution procedure. Refer to p. 3-28.

step 1: The Hazen-Williams roughness coefficient is given.

step 2: Choose clockwise as positive.

step 3: Nodes are already numbered.

step 4: Choose the loops as shown.

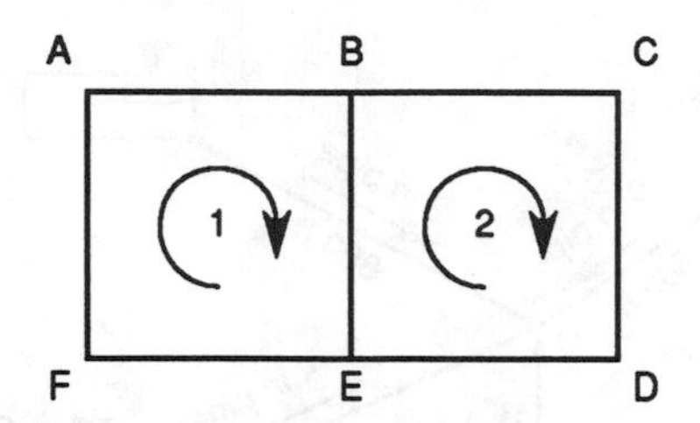

step 5: Q is in gal/min.
d is in inches.
L is in feet.

Use Eq. 3.133. Each pipe has the same length.

$$K'_{8\text{ in}} = \frac{(10.44)(1000)}{(100)^{1.85}(8)^{4.87}} = 8.33 \times 10^{-5}$$

$$K'_{6\text{ in}} = \frac{(10.44)(1000)}{(100)^{1.85}(6)^{4.87}} = 3.38 \times 10^{-4}$$

$$K'_{\text{CE}} = 2K'_{6\text{ in}} = 6.76 \times 10^{-4}$$

$$K'_{\text{EA}} = K'_{6\text{ in}} + K'_{8\text{ in}} = 4.21 \times 10^{-4}$$

step 6: Assume the flows shown.

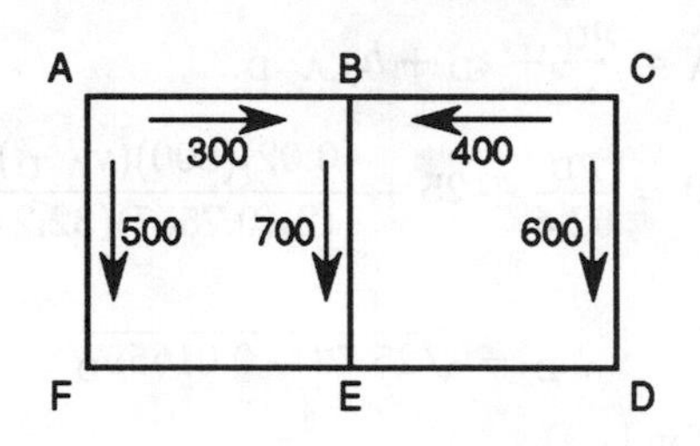

AB = 300 gal/min
BE = 700 gal/min
AE = 500 gal/min
CE = 600 gal/min
CB = 400 gal/min

step 7: If the elevations are included as part of the head loss, Eq. 3.137 becomes

$$\Sigma h = \Sigma K' Q_a^n + \delta \Sigma n K' Q_a^{n-1} + z_2 - z_1 = 0$$

However, since the loop closes on itself, $z_2 = z_1$, and the elevations can be omitted.

- First iteration:

Loop 1:

$$\delta_1 = \frac{-[(8.33\times10^{-5})(300)^{1.85}+(8.33\times10^{-5})(700)^{1.85}-(4.21\times10^{-4})(500)^{1.85}]}{(1.85)[(8.33\times10^{-5})(300)^{0.85}+(8.33\times10^{-5})(700)^{0.85}+(4.21\times10^{-4})(500)^{0.85}]}$$

$$= \frac{-(-22.97)}{0.213} = +108 \text{ gal/min}$$

Loop 2:

$$\delta_2 = \frac{-[(6.76\times10^{-4})(600)^{1.85}-(8.33\times10^{-5})(700)^{1.85}-(8.33\times10^{-5})(400)^{1.85}]}{(1.85)[(6.76\times10^{-4})(600)^{0.85}+(8.33\times10^{-5})(700)^{0.85}+(8.33\times10^{-5})(400)^{0.85}]}$$

$$= \frac{-(-72.5)}{0.353} = -205 \text{ gal/min}$$

- Second iteration:

$$\begin{aligned}
&\text{AB:} && 300 + 108 && = 408 \text{ gal/min}\\
&\text{BE:} && 700 + 108 - (-205) && = 1013 \text{ gal/min}\\
&\text{AE:} && 500 - 108 && = 392 \text{ gal/min}\\
&\text{CE:} && 600 + (-205) && = 395 \text{ gal/min}\\
&\text{CB:} && 400 - (-205) && = 605 \text{ gal/min}
\end{aligned}$$

Loop 1:

$$\delta_1 = \frac{-(9.48)}{0.205} = -46 \text{ gal/min}$$

Loop 2:

$$\delta_2 = \frac{-(1.08)}{0.292} = -3.7 \text{ gal/min} \quad \text{[round to –4]}$$

- Third iteration:

$$\begin{aligned}
&\text{AB:} && 408 + (-46) && = 362 \text{ gal/min}\\
&\text{BE:} && 1013 + (-46) - (-4) && = 971 \text{ gal/min}\\
&\text{AE:} && 392 - (-46) && = 438 \text{ gal/min}\\
&\text{CE:} && 395 + (-4) && = 391 \text{ gal/min}\\
&\text{CB:} && 605 - (-4) && = 609 \text{ gal/min}
\end{aligned}$$

Loop 1:

$$\delta_1 = \frac{-(0.066)}{0.213} = -0.31 \text{ gal/min} \quad \text{[round to 0]}$$

Loop 2:

$$\delta_2 = \frac{-(2.4)}{0.29} = -8.3 \text{ gal/min} \quad \text{[round to –8]}$$

Use the following flows.

$$\begin{aligned}
&\text{AB:} && 362 + 0 && = 362 \text{ gal/min}\\
&\text{BE:} && 971 + 0 - (-8) && = 979 \text{ gal/min}\\
&\text{AE:} && 438 - 0 && = 438 \text{ gal/min}\\
&\text{CE:} && 391 + (-8) && = 383 \text{ gal/min}\\
&\text{CB:} && 609 - (-8) && = 617 \text{ gal/min}
\end{aligned}$$

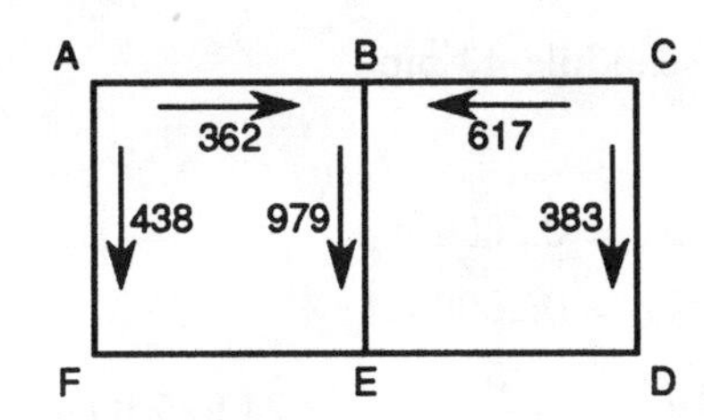

Using Eq. 3.136, the friction loss in each section is

$$h_{f,\text{AB}} = (8.33 \times 10^{-5})(362)^{1.85} = 4.5 \text{ ft}$$
$$h_{f,\text{BE}} = (8.33 \times 10^{-5})(979)^{1.85} = 28.4 \text{ ft}$$
$$h_{f,\text{AF}} = (8.33 \times 10^{-5})(438)^{1.85} = 6.4 \text{ ft}$$
$$h_{f,\text{FE}} = (3.38 \times 10^{-4})(438)^{1.85} = 26.0 \text{ ft}$$
$$h_{f,\text{CD}} = h_{f,\text{DE}} = (3.38 \times 10^{-4})(383)^{1.85} = 20.3 \text{ ft}$$
$$h_{f,\text{CB}} = (8.33 \times 10^{-5})(617)^{1.85} = 12.1 \text{ ft}$$

Assume 60°F water. From p. 3-41, $\gamma = 62.37 \text{ lbf/ft}^3$. The pressure at C is 40 psig. This gives

$$h_\text{C} = \frac{(40)(144)}{62.37} = 92.4 \text{ ft}$$

Next, use

$$h = h_\text{C} + z_\text{C} - z - \Sigma h_f$$

$$h_\text{D} = 92.4 + 300 - 150 - 20.3 = 222.1 \text{ ft}$$
$$h_\text{E} = 222.1 + 150 - 200 - 20.3 = 151.8 \text{ ft}$$
$$h_\text{F} = 151.8 + 200 - 150 + 26 = 227.8 \text{ ft}$$
$$h_\text{B} = 92.4 + 300 - 150 - 12.1 = 230.3 \text{ ft}$$
$$h_\text{A} = 230.3 + 150 - 200 + 4.5 = 184.8 \text{ ft}$$

Using $p = \gamma h$,

$$p_\text{A} = \frac{(62.37)(184.8)}{144} = \boxed{80.0 \text{ lbf/in}^2 \text{ (psig)}}$$

$$p_\text{B} = \frac{(62.37)(230.3)}{144} = \boxed{99.7 \text{ psig}}$$

$$p_\text{C} = \boxed{40 \text{ psig [given]}}$$

$$p_\text{D} = \frac{(62.37)(222.1)}{144} = \boxed{96.2 \text{ psig}}$$

$$p_\text{E} = \frac{(62.37)(151.8)}{144} = \boxed{65.7 \text{ psig}}$$

$$p_\text{F} = \frac{(62.37)(227.8)}{144} = \boxed{98.7 \text{ psig}}$$

The pressure increase across the pump is $80 - 20 = 60$ psig = 8640 psfg.

From Table 4.2, the hydraulic horsepower is

$$P = \frac{(8640)(800)}{2.468 \times 10^5} = \boxed{28 \text{ hp}}$$

5. Assume schedule-40 pipe.

- At A:

$$D_A = 0.5054 \text{ ft}$$
$$A_A = 0.2006 \text{ ft}^2$$
$$v_A = \frac{5 \text{ cfs}}{0.2006 \text{ ft}^2} = 24.925 \text{ ft/sec}$$
$$p_A = 10 \text{ psia} = 1440 \text{ psf}$$
$$z_A = 0$$

- At B:

$$D_B = 1.4063 \text{ ft}$$
$$A_B = 1.5533 \text{ ft}^2$$
$$v_B = \frac{5}{1.5533} = 3.219$$
$$p_B = 7 \text{ psia} = 1008 \text{ psf}$$
$$z_B = 15 \text{ ft}$$

Ignore the minor loss and assume 70°F water, so $\gamma = 62.3 \text{ lbf/ft}^3$. Using Eq. 3.56, the total head at A and B is

$$(\text{TH})_A = \frac{1440}{62.3} + \frac{(24.925)^2}{(2)(32.2)} + 0 = 32.76 \text{ ft}$$
$$(\text{TH})_B = \frac{1008}{62.3} + \frac{(3.219)^2}{(2)(32.2)} + 15 = 31.34 \text{ ft}$$

The flow is from high head to low head, so

$$\boxed{\text{Flow is from A to B.}}$$

6. From App. D (p. 3-42),

$$(750 \text{ gal/min})(0.002228) = 1.671 \text{ ft}^3/\text{sec (cfs)}$$

From p. 3-45, $D = 0.5054$ ft and $A = 0.2006 \text{ ft}^2$. Assume schedule-40 pipe.

$$v = \frac{1.671}{0.2006} = 8.33 \text{ ft/sec}$$

From p. 3-41 for 60°F water,

$$\nu = 1.217 \times 10^{-5} \text{ ft}^2/\text{sec}$$
$$N_{Re} = \frac{(8.33)(0.5054)}{1.217 \times 10^{-5}} = 3.46 \times 10^5$$

From p. 3-20, $\epsilon = 0.0002$.

$$\frac{\epsilon}{D} = \frac{0.0002 \text{ ft}}{0.5054 \text{ ft}} \approx 0.0004$$

From p. 3-21, $f = 0.0175$. From Eq. 3.71,

$$h_f = \frac{(0.0175)(3000)(8.33)^2}{(2)(0.5054)(32.2)} = 111.9 \text{ ft}$$

Use the Bernoulli equation (Eq. 3.70). Since velocity is the same at points A and B, it may be omitted.

$$\frac{144p_1}{62.37} = \frac{(50)(144)}{62.37} + 60 + 111.9$$
$$p_1 = \boxed{124.5 \text{ lbf/in}^2 \text{ (psig)}}$$

7. $$A_o = \left(\frac{\pi}{4}\right)\left(\frac{4}{12}\right)^2 = 0.08727 \text{ ft}^2$$
$$A_t = \left(\frac{\pi}{4}\right)(20)^2 = 314.16 \text{ ft}^2$$

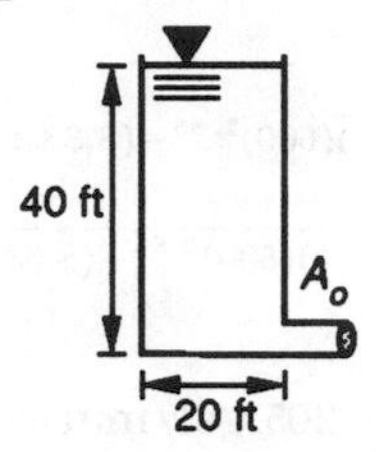

From Eq. 3.97, the time to drop from 40 ft to 20 ft is

$$t = \frac{(2)(314.16)\left(\sqrt{40} - \sqrt{20}\right)}{(0.98)(0.08727)\sqrt{(2)(32.2)}} = \boxed{1695.8 \text{ sec}}$$

8. $C_d = 1.00$ [given]

$$F_{va} = \frac{1}{\sqrt{1 - \left(\frac{D_2}{D_1}\right)^4}} = \frac{1}{\sqrt{1 - \left(\frac{8}{12}\right)^4}}$$
$$= 1.116$$
$$A_2 = \left(\frac{\pi}{4}\right)\left(\frac{8}{12}\right)^2 = 0.3491 \text{ ft}^2$$
$$p_1 - p_2 = [(0.491)(4) - (0.0361)(4)](144) = 262.0 \text{ psf}$$

From Eq. 3.164,

$$Q = (1.116)(1)(0.3491)\sqrt{\frac{(2)(32.2)(262)}{62.4}}$$
$$= \boxed{6.406 \text{ ft}^3/\text{sec (cfs)}}$$

9. From Eq. 3.161,

$$\Delta p = \left(\frac{\gamma}{2g}\right)\left(\frac{Q}{C_f A_o}\right)^2$$

For 70°F water,

$$\nu = 1.059 \times 10^{-5} \text{ ft}^2/\text{sec}$$
$$\gamma = 62.3 \text{ lbf/ft}^3$$
$$N_{Re} = \frac{Dv}{\nu} = \frac{(1)(2)}{1.059 \times 10^{-5}} = 1.89 \times 10^5$$
$$A_o = \left(\frac{\pi}{4}\right)(0.2)^2 = 0.0314 \text{ ft}^2$$
$$A_p = \left(\frac{\pi}{4}\right)(1)^2 = 0.7854 \text{ ft}^2$$
$$\frac{A_o}{A_p} = \frac{0.0314}{0.7854} = 0.040$$

From Fig. 3.26 (p. 3-32), $C_f \approx 0.60$.

$$Q = A\text{v} = (0.7854)(2) = 1.571 \text{ ft}^3/\text{sec}$$

$$\Delta p = \left[\frac{62.3}{(2)(32.2)}\right]\left[\frac{1.571}{(0.6)(0.0314)}\right]^2 = \boxed{6727 \text{ lbf/ft}^2 \text{ (psf)}}$$

10.

$$A_A = \left(\frac{\pi}{4}\right)\left(\frac{24}{12}\right)^2 = 3.142 \text{ ft}^2$$

$$A_B = \left(\frac{\pi}{4}\right)\left(\frac{12}{12}\right)^2 = 0.7854 \text{ ft}^2$$

$$\text{v}_A = \frac{Q}{A} = \frac{8}{3.142} = 2.546 \text{ ft/sec}$$

$$p_A = (20)(62.4) = 1248 \text{ lbf/ft}^2$$

Using the Bernoulli equation to solve for p_B,

$$\text{v}_B = \frac{8}{0.7854} = 10.19 \text{ ft/sec}$$

$$p_B = 1248 - \left[\frac{(10.19)^2 - (2.546)^2}{(2)(32.2)}\right](62.4) = 1153.67 \text{ psf}$$

From Eq. 3.196 with $\phi = 0$,

$$F_x = (1153.67)(0.7854) - (1248)(3.142) + \left[\frac{(8)(62.4)}{32.2}\right](10.19 - 2.546) = \boxed{-2897 \text{ lbf on the fluid (to the left)}}$$

$$F_y = \boxed{0}$$

Timed

1. For simplicity, assume nominal pipe sizes and ignore minor losses. Use Eq. 3.74.

$$h_{f,A-B} = \frac{(10.44)(20{,}000)(120)^{1.85}}{(150)^{1.85}(6)^{4.8655}} = 22.6 \text{ ft}$$

Investigate velocity head.

$$\text{v} = \frac{Q}{A} = \frac{(120)(0.002228)}{\left(\frac{\pi}{4}\right)\left(\frac{6}{12}\right)^2} = 1.36 \text{ ft/sec}$$

$$h_\text{v} = \frac{\text{v}^2}{2g} = \frac{(1.36)^2}{(2)(32.2)} = 0.029 \text{ ft}$$

Velocity heads are low and can be disregarded.

$$h_{f,B-C} = \frac{(10.44)(10{,}000)(160)^{1.85}}{(150)^{1.85}(6)^{4.8655}} = 19.25 \text{ ft}$$

$$h_{f,C-D} = \frac{(10.44)(30{,}000)(120)^{1.85}}{(150)^{1.85}(4)^{4.8655}} = 243.9 \text{ ft}$$

Assume a pressure of 20 psig at point A.

$$h_{p,A} = \frac{(20)(144)}{62.4} = 46.2 \text{ ft}$$

From the Bernoulli equation, ignoring velocity head,

$$h_{p,A} + z_A = h_{p,B} + z_B + h_{f,A-B}$$

$$46.2 + 620 = h_{p,B} + 460 + 22.6$$

$$h_{p,B} = 183.6 \text{ ft}$$

$$p_B = \gamma h = \frac{(62.4)(183.6)}{144} = 79.6 \text{ lbf/in}^2 \text{ (psig)}$$

For B to C,

$$183.6 + 460 = h_{p,C} + 540 + 19.25$$

$$h_{p,C} = 84.35 \text{ ft}$$

$$p_C = \frac{(84.35)(62.4)}{144} = 36.6 \text{ lbf/in}^2 \text{ (psig)}$$

For C to D,

$$84.35 + 540 = h_{p,D} + 360 + 243.9$$

$$h_{p,D} = 20.45 \text{ ft}$$

$$p_D = \frac{(20.45)(62.4)}{144} = 8.9 \text{ lbf/in}^2 \text{ (psig)} \quad \text{[too low]}$$

Since p_D is too low, add $20 - 8.9 = 11.1$ lbf/in^2 (psig) to each point.

(b)

$$p_A = 20.0 + 11.1 = \boxed{31.1 \text{ lbf/in}^2 \text{ (psig)}}$$

$$p_B = 79.6 + 11.1 = \boxed{90.7 \text{ lbf/in}^2 \text{ (psig)}}$$

$$p_C = 36.6 + 11.1 = \boxed{47.7 \text{ lbf/in}^2 \text{ (psig)}}$$

$$p_D = 8.9 + 11.1 = \boxed{20.0 \text{ lbf/in}^2 \text{ (psig)}}$$

The elevation of the hydraulic grade line above point D is

(a)

$$\Delta h_{A-D} = z_A - z_D + p_A - p_D = 620 - 360 + \frac{(31.1 - 20)(144)}{62.4} = \boxed{285.6 \text{ ft}}$$

HYDRAULIC MACHINES

Untimed

1. $$\left(2000\ \frac{\text{gal}}{\text{min}}\right)\left(0.002228\ \frac{\text{ft}^3/\text{sec}}{\text{gal/min}}\right) = 4.456\ \text{ft}^3/\text{sec}$$

Assume schedule-40 steel pipe.

$$D_1 = 0.9948\ \text{ft} \qquad A_1 = 0.7773\ \text{ft}^2$$
$$D_2 = 0.6651\ \text{ft} \qquad A_2 = 0.3473\ \text{ft}^2$$

$$p_1 = [14.7 - (8)(0.491)](144) = 1551.2\ \text{lbf/ft}^2\ \text{(psf)}$$
$$p_2 = (14.7 + 20)(144) + (4)(1.2)(62.4) = 5296.3\ \text{lbf/ft}^2\ \text{(psf)}$$
$$\text{v}_1 = \frac{4.456}{0.7773} = 5.73\ \text{ft/sec}$$
$$\text{v}_2 = \frac{4.456}{0.3473} = 12.83\ \text{ft/sec}$$

The total heads at points 1 and 2 are

$$(\text{TH})_1 = \frac{1551.2}{(62.4)(1.2)} + \frac{(5.73)^2}{(2)(32.2)} = 21.23\ \text{ft}$$
$$(\text{TH})_2 = \frac{5296.3}{(62.4)(1.2)} + \frac{(12.83)^2}{(2)(32.2)} = 73.29\ \text{ft}$$

The pump must add 73.29 – 21.23 = 52.06 ft of head.

The power required is

$$P = \Delta h\dot{w} = \frac{(52.06\ \text{ft})(1.2)\left(62.4\ \frac{\text{lbf}}{\text{ft}^3}\right)\left(4.456\ \frac{\text{ft}^3}{\text{sec}}\right)}{550\ \frac{\text{ft-lbf}}{\text{hp-sec}}} = 31.58\ \text{hp}$$

The input horsepower is

$$\frac{31.58}{0.85} = \boxed{37.15\ \text{hp}}$$

2. Assume schedule-40 pipe and 70°F water.

$$D_1 = 7.981\ \text{in}$$
$$D_2 = 6.065\ \text{in}$$

The flow rate is

$$A_2\text{v}_2\gamma_2 = \left(\frac{\pi}{4}\right)\left(\frac{6.065}{12}\right)^2(12)(62.4) = 150.2\ \text{lbf/sec}$$
$$\text{v}_1 = \left(\frac{6.065}{7.981}\right)^2\left(12\ \frac{\text{ft}}{\text{sec}}\right) = 6.930\ \text{ft/sec}$$

The inlet pressure is

$$(14.7 - 5)(144) = 1396.8\ \text{lbf/ft}^2\ \text{(psf)}$$

The total inlet head is

$$(\text{TH})_1 = \frac{1396.8}{62.4} + \frac{(6.930)^2}{(2)(32.2)} = 23.13\ \text{ft}$$

The head added by the pump is

$$H = \frac{(0.70)(20\ \text{hp})\left(550\ \frac{\text{ft-lbf}}{\text{hp-sec}}\right)}{150.2\ \frac{\text{lbf}}{\text{sec}}} = 51.26\ \text{ft}$$

The total available head is

$$23.13 + 51.26 = 74.39\ \text{ft}$$

The total head used is

$$\frac{(14.7)(144)}{62.4} + \frac{(12)^2}{(2)(32.2)} + z + 10 = z + 46.16$$

The maximum elevation is

$$z = 74.39 - 46.16 = \boxed{28.23\ \text{ft}}$$

3. The flow rate is

$$\dot{w} = \gamma\dot{V} = \left(62.4\ \frac{\text{lbf}}{\text{ft}^3}\right)\left(1000\ \frac{\text{ft}^3}{\text{sec}}\right) = 6.24\times 10^4\ \text{lbf/sec}$$

The head available for work is

$$\Delta H = 625 - 58 = 567\ \text{ft}$$

The work done is

$$W = \eta\dot{w}\Delta H = (0.89)\left(6.24\times 10^4\ \frac{\text{lbf}}{\text{sec}}\right)(567\ \text{ft}) = 3.149\times 10^7\ \text{ft-lbf/sec}$$

Using Eq. 4.26 to convert from hp to kW,

$$\frac{(3.149\times 10^7)(0.7457)}{550} = \boxed{4.27\times 10^4\ \text{kW}}$$

4. Since turbines are essentially pumps running backward, use Table 4.2.

$$\Delta p = (30\ \text{psi} - 5\ \text{psi})(144) = 3600\ \text{lbf/ft}^2\ \text{(psf)}$$
$$\text{hp} = \frac{(3600)(100)}{550} = \boxed{654.5\ \text{hp}}$$

5. From the Bernoulli equation, the total entering head is

(a) $H = 92.5 + \frac{(12)^2}{(2)(32.2)} + 5.26 = \boxed{100 \text{ ft}}$

(b) The water horsepower is

$$P = \frac{\dot{w}H}{550} = \frac{\left(25\ \frac{\text{ft}^3}{\text{sec}}\right)\left(62.4\ \frac{\text{lbf}}{\text{ft}^3}\right)(100\text{ ft})}{550\ \frac{\text{ft-lbf}}{\text{hp-sec}}}$$

$$= 283.6 \text{ hp}$$

$$\eta = \frac{250}{283.6} = \boxed{0.882\ (88.2\%)}$$

(c) From Eq. 4.38,

$$n_2 = n_1\sqrt{\frac{H_2}{H_1}} = 610\sqrt{\frac{225}{100}} = \boxed{915 \text{ rpm}}$$

(d) From Eq. 4.40,

$$(\text{hp})_2 = (\text{hp})_1\left(\frac{H_2}{H_1}\right)^{1.5} = (250)\left(\frac{225}{100}\right)^{1.5} = \boxed{843.8 \text{ hp}}$$

(e) From Eq. 4.39,

$$Q_2 = Q_1\sqrt{\frac{H_2}{H_1}} = 25\sqrt{\frac{225}{100}} = \boxed{37.5 \text{ cfs}}$$

6. Assume schedule-40 pipe. From p. 3-44,

$$D_i = 0.1342 \text{ ft}$$
$$A_i = 0.01414 \text{ ft}^2$$

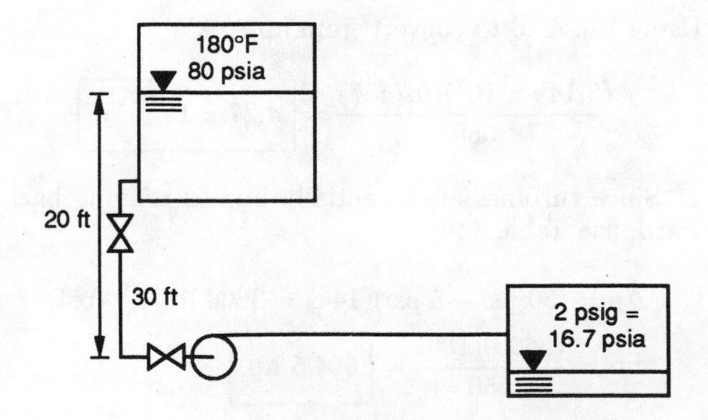

From App. D (p. 3-42),

$$\left(100\ \frac{\text{gal}}{\text{min}}\right)(0.002228) = 0.2228 \text{ ft}^3/\text{sec (cfs)}$$

$$\text{v} = \frac{Q}{A} = \frac{0.2228}{0.01414} = 15.76 \text{ ft/sec}$$

Assume regular screwed steel fittings. From p. 3-53, the equivalent lengths are

elbows: $2 \times 3.4 = 6.8$ ft
gate value: $2 \times 1.2 = 2.4$ ft
total: 9.2 ft

The total equivalent length is $30 + 9.2 = 39.2$ ft.

From p. 3-20, $\epsilon = 0.0002$.

$$\frac{\epsilon}{D} = \frac{0.0002}{0.1342} = 0.0015$$

At 180°F, $\nu = 0.385 \times 10^{-5}$ ft²/sec (p. 3-41).

$$N_{\text{Re}} = \frac{D\text{v}}{\nu} = \frac{(15.76)(0.1342)}{0.385 \times 10^{-5}} = 5.5 \times 10^5$$

From p. 3-21, $f = 0.022$.

$$h_f = \frac{(0.022)(39.2)(15.76)^2}{(2)(0.1342)(32.2)} = 24.78 \text{ ft}$$

At 180°F, $\gamma = 60.58$ (from p. 3-41), and $h_{\text{vapor}} = 17.87$ ft. From Eq. 4.12 (substituting 80 psia for p_a because the tank is pressurized),

$$\text{NPSHA} = \frac{(80)(144)}{60.58} + 20 - 24.78 - 17.87 = 167.5 \text{ ft}$$

Since the most the liquid can drop is 20 ft to the pump inlet, the minimim NPSHA is 147.5 ft. This exceeds 10 ft.

The pump will never cavitate as long as the 80 psia pressure is kept.

7. Assume each stage adds 150 ft of head. The suction lift is 10 ft. From p. 4-20 for a single-suction pump,

$$n_s = \boxed{2050 \text{ rpm}}$$

8. (a) At 70°F, the head dropped is

$$H = \frac{(500 \text{ psi} - 30 \text{ psi})(144)}{62.4 \ \frac{\text{lbf}}{\text{ft}^3}} = 1084.6 \text{ ft}$$

Using Eq. 4.52, the specific speed of the turbine is

$$n_s = \frac{1750\sqrt{250}}{(1084.6)^{1.25}} = 4.446 \text{ rpm}$$

Since the lowest suggested value of n_s for a reaction turbine is 10 (p. 4-18),

Recommend an impulse (Pelton) wheel.

(b) From p. 3-34,

$$Q = A\text{v} = \left(\frac{\pi}{4}\right)\left(\frac{4}{12}\right)^2 (35) = 3.054 \text{ cfs}$$

$$Q' = \frac{(35 - 10)(3.054)}{35} = 2.181 \text{ cfs}$$

Assuming $\gamma = 62.4$ lbf/ft^3,

$$F_x = \left[-\frac{(2.181)(62.4)}{32.2}\right](35 - 10)(1 - \cos 80°)$$
$$= -87.32 \text{ lbf}$$

$$F_y = \left[\frac{(2.181)(62.4)}{32.2}\right](35 - 10)(\sin 80°)$$
$$= 104.1 \text{ lbf}$$

$$R = \sqrt{(87.32)^2 + (104.1)^2} = \boxed{135.9 \text{ lbf}}$$

Note that parts (a) and (b) are independent, since a head of appproximately 1100 ft would produce a jet speed much higher than 35 ft/sec.

9. From Eq. 4.33,

$$(\text{hp})_2 = (0.5)\left(\frac{2000}{1750}\right)^3 = \boxed{0.746 \text{ hp}}$$

10. Assume screwed steel fittings. The equivalent lengths in the inlet and discharge lines (p. 3-53) are

inlet (essentially reentrant):	8.5 ft
check valve:	19.0 ft
elbows:	3 × 3.6 = 10.8 ft

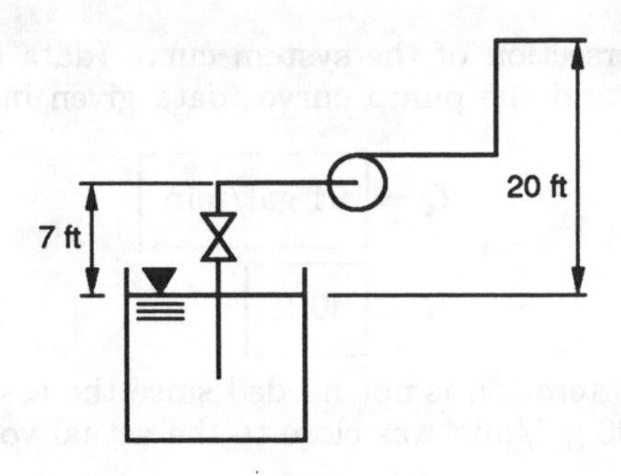

The total equivalent length of the 2-in line is

$$L_e = 12 + 8.5 + 19 + 10.8 + 80 = 130.3 \text{ ft}$$

Assume schedule-40 pipe. From p. 3-44,

$$D_1 = 0.1723 \text{ ft}$$
$$A_i = 0.0233 \text{ ft}^2$$

Since the flow rate is unknown, it must be assumed in order to find v. Assume 90 gal/min.

$$(90)(0.002228) = 0.20052 \text{ cfs}$$
$$\text{v} = \frac{Q}{A} = \frac{0.20052}{0.0233} = 8.606 \text{ ft/sec}$$
$$\nu = 1.059 \times 10^{-5} \text{ ft}^2\text{/sec} \quad [\text{p. 3-41}]$$
$$\epsilon = 0.0002 \text{ ft} \quad [\text{p. 3-20}]$$
$$\frac{\epsilon}{D} = \frac{0.0002}{0.1723} = 0.0012$$
$$N_{\text{Re}} = \frac{(0.1723)(8.606)}{1.059 \times 10^{-5}} = 1.4 \times 10^5$$

From p. 3-21, $f = 0.022$. At 90 gal/min, the friction loss in the line is

$$h_f = \frac{(0.022)(130.3)(8.606)^2}{(2)(0.1723)(32.2)} = 19.1 \text{ ft}$$

The velocity head is

$$h_\text{v} = \frac{(8.606)^2}{(2)(32.2)} = 1.2 \text{ ft}$$

For any other flow rate, the velocity head is

$$h_\text{v} = (1.2)\left(\frac{Q_2}{90}\right)^2$$

The total system head is

$$H = \Delta z + h_\text{v} + h_f$$
$$= 20 + (1.2 + 19.1)\left(\frac{Q_2}{90}\right)^2$$

Q_2	H	Q_2	H
0	20.0	60	29.0
10	20.3	70	32.3
20	21.0	80	36.0
30	22.3	90	40.3
40	24.0	100	45.0
50	26.3	110	50.3

The intersection of the system curve (data given previously) and the pump curve (data given in problem) is

$$Q = \boxed{94 \text{ gal/min}}$$

$$H = \boxed{40 \text{ ft}}$$

Another iteration is not needed since the assumed volume of 90 gal/min was close to the actual volume.

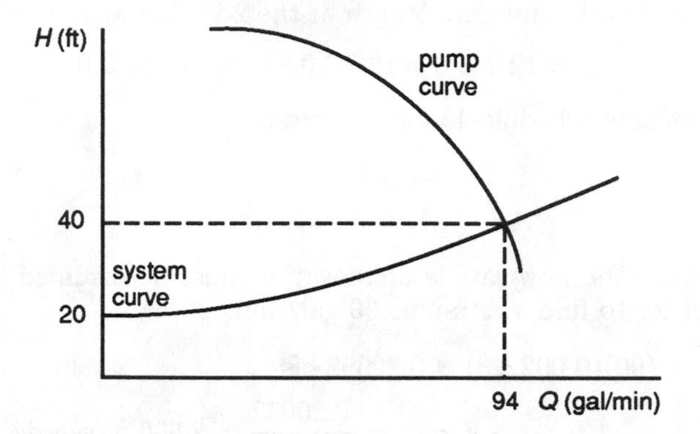

Timed

1. (a) $h_f = \Delta z = 48$ ft since $\Delta p = 0$ and $\Delta v = 0$ for open channel flow.

$$Q = \frac{(250)(10{,}000)}{(24)(60)} = 1736 \text{ gal/min}$$

Given $C = 130$, from Eq. 3.74,

$$d_{\text{in}}^{4.8655} = \frac{(10.44)(5000)(1736)^{1.85}}{(130)^{1.85}(48)} = 131{,}462$$

$$d = \boxed{11.27 \text{ in} \quad \text{[round to 12 in minimum]}}$$

Assumptions:

- 48 ft is true loss. Normally sewage friction loss is 10–25% more than water friction loss.
- Population growth is not considered.
- 12-in pipe is available.

(b) Without having a specific pump curve, the number of pumps can only be specified based on general rules. Use the *Ten-States' Standards*, which state:

- No station will have less than two identical pumps.
- Three pumps are required if the flow is greater than 1 mgd.
- There must be at least one backup pump.

$$\boxed{\text{Three pumps are required, plus one backup.}}$$

(c) With a variable speed pump, it will be possible to adjust to the wide variations in flow (100–250 gpcd). It may be possible to operate with two pumps and one backup.

(d) With three constant speed pumps,

$$Q = \frac{1736}{3} = 579 \text{ gal/min at maximum capacity}$$

From Table 4.2, assuming specific gravity ≈ 1.00, and using $\eta_{\text{pump}} = 0.80$,

$$\text{rated motor power} = \frac{(10)(579)}{(3960)(0.80)}$$

$$= \boxed{1.83 \text{ hp} \quad \text{[use 2.0 hp]}}$$

With two variable-speed pumps,

$$Q = \frac{1736}{2} = 868 \text{ gal/min}$$

$$\text{rated motor power} = \frac{(10)(868)}{(3960)(0.80)}$$

$$= \boxed{2.74 \text{ hp} \quad \text{[use 3.0 hp]}}$$

(e) Assume the question refers to demand control rather than motor speed control. Methods of control are

- variable run times
- pressure change in sump
- flow rate incoming
- sump level
- flow rate outgoing
- manual on/off

(f) In the wet well, the pump (and perhaps motor) is submerged. Forced ventilation air will prevent a concentration of explosive methane. From the *Ten-States' Standards*,

$$\boxed{\begin{array}{l}\text{12 air changes per hour if continuous,}\\ \text{30 per hour if intermittent.}\end{array}}$$

In the dry well,

$$\boxed{\begin{array}{l}\text{6 air changes per hour if continuous,}\\ \text{30 per hour if intermittent.}\end{array}}$$

2. (a) $h_s = \boxed{20 \text{ ft}}$

(b) $h_{\text{sd}} = \boxed{15 \text{ ft}}$

(c) There is no pipe size specified, so h_v cannot be calculated. Even so, v is typically in the 5–10 ft/sec range, and $h_v \approx 0$.

Since pipe lengths are not given, assume $h_f \approx 0$ and gage pressures.

$$20 + 15 + \frac{(80)(144)}{62.4} + 10 = \boxed{229.6 \text{ ft of water}}$$

(d) The flow rate is

$$\frac{(3.5 \text{ mgd})\left(62.4 \ \frac{\text{lbf}}{\text{ft}^3}\right)}{0.64632 \ \frac{\text{mgd}}{\text{cfs}}} = 337.9 \text{ lbf/sec}$$

From Table 4.2, the rated motor output power does not depend on the motor efficiency.

$$P = \frac{(229.6)(337.9)}{(550)(0.85)} = 160.0 \text{ hp}$$

$$\boxed{\text{Use a 200 hp motor.}}$$

(e) The chosen motor has more power than needed.

$$\dot{m} = \frac{(200)(550)(0.85)}{229.6} = 239.6 \text{ lbm/sec}$$

$$\frac{(239.6)(0.64632)}{62.4} = 2.48 \text{ mgd}$$

The excess required storage is

$$\left(\frac{8 \text{ hr}}{24}\right)[(1.25)(3.5) - 2.48] = \boxed{0.63 \text{ MG}}$$

3. This could be solved as a parallel pipe problem or as a pipe network problem. Follow the procedure on p. 3-28.

(a) Pipe network solution:

step 1: Use the Darcy equation since Hazen-Williams coefficients are not given (or assume C values based on the age of the pipe).

step 2: Clockwise is positive.

step 3: All nodes are lettered.

step 4: There is only one loop.

step 5: From p. 3-20, $\epsilon \approx 0.0008$ ft. Assume full turbulence and 10-in pipes.

$$\frac{\epsilon}{D} = \frac{0.0008}{\frac{10}{12}} \approx 0.001$$

From Fig. 3.13, $f \approx 0.020$ for full turbulence.

From Eq. 3.135,

$$K'_{\text{AB}} = (1.251 \times 10^{-7})\left[\frac{(0.02)(2000)}{\left(\frac{10}{12}\right)^5}\right] = 12.5 \times 10^{-6}$$

$$K'_{\text{BD}} = (1.251 \times 10^{-7})\left[\frac{(0.02)(1000)}{\left(\frac{10}{12}\right)^5}\right] = 6.23 \times 10^{-6}$$

$$K'_{\text{DC}} = (1.251 \times 10^{-7})\left[\frac{(0.02)(1500)}{\left(\frac{8}{12}\right)^5}\right] = 28.5 \times 10^{-6}$$

$$K'_{\text{CA}} = (1.251 \times 10^{-7})\left[\frac{(0.02)(1000)}{\left(\frac{12}{12}\right)^5}\right] = 2.50 \times 10^{-6}$$

step 6: Assume $Q_{\text{AB}} = 1$ mgd.

$$Q_{\text{ACDB}} = 0.5 \text{ mgd}$$

Convert Q to gal/min.

$$Q_{\text{AB}} = \frac{(1)(1 \times 10^6)}{(24)(60)} = 694 \text{ gal/min}$$

$$Q_{\text{ACDB}} = \frac{(0.5)(1 \times 10^6)}{(24)(60)} = 347 \text{ gal/min}$$

step 7: There is only one loop.

$$694 = (2)(347)$$

$$(694)^2 = (4)(347)^2$$

$$\delta = \frac{(-1)(1 \times 10^{-6})(347)^2[(12.5)(4) - 6.23 - 28.5 - 2.5]}{(2)(1 \times 10^{-6})(347)[(12.5)(2) + 6.23 + 28.5 + 2.5]}$$

$$= -35.6 \text{ gal/min} \quad [\text{use } -36 \text{ gal/min}]$$

step 8:

$$Q_{\text{AB}} = 694 + (-36) = 658 \text{ gal/min}$$

$$Q_{\text{ACDB}} = 347 - (-36) = 383 \text{ gal/min}$$

Repeat step 7.

$$658 = (1.72)(383)$$

$$(658)^2 = (2.95)(383)^2$$

$$\delta = \frac{(-1)(1 \times 10^{-6})(383)^2\,[(12.5)(2.95) - 6.23 - 28.5 - 2.5]}{(2)(1 \times 10^{-6})(383)[(12.5)(1.72) + 6.23 + 28.5 + 2.5]}$$

$$\approx 0$$

$$Q_{AB} = \boxed{658 \text{ gal/min}}$$

$$Q_{ACDB} = \boxed{383 \text{ gal/min}}$$

Check the Reynolds number in leg AB to verify that $f = 0.02$ was a good choice.

$$A_{\text{10-in pipe}} = \left(\frac{\pi}{4}\right)\left(\frac{10}{12}\right)^2 = 0.5454 \quad \text{[cast iron pipe]}$$

Flow rate = (658)(0.002228) = 1.47 cfs.

$$v = \frac{Q}{A} = \frac{1.47}{0.5454} = 2.7 \text{ ft/sec} \quad \text{[reasonable]}$$

Assume 40°F water.

$$\nu = 1.664 \times 10^{-5} \text{ ft}^2\text{/sec}$$

$$N_{Re} = \frac{\left(\frac{10}{12}\right)(2.7)}{1.664 \times 10^{-5}} = 1.35 \times 10^5$$

- Alternative closed-form solution:

Use the Darcy equation. Assume $\epsilon = 0.0008$. The relative roughness is

$$\frac{\epsilon}{D} = \frac{0.0008}{\left(\frac{10.1}{12}\right)} = 0.00095 \quad \text{[use 0.001]}$$

Assume $v_{max} = 5$ ft/sec and 50°F temperature. The Reynolds number is

$$N_{Re} = \frac{vD}{\nu} = \frac{(5)\left(\frac{10.1}{12}\right)}{1.41 \times 10^{-5}} = 2.98 \times 10^5$$

From the Moody diagram, $f \approx 0.0205$.

$$h_{f,AB} = \frac{fLv^2}{2Dg_c} = \frac{(0.0205)(2000)Q_{AB}^2}{(2)\left(\frac{10.1}{12}\right)(0.556)^2(32.2)}$$

$$= 2.446Q_{AB}^2$$

$$h_{f,ACDB} = \frac{(0.0205)(1000)Q_{ACDB}^2}{(2)\left(\frac{12.12}{12}\right)(0.801)^2(32.2)} + \frac{(0.0205)(1500)Q_{ACDB}^2}{(2)\left(\frac{8.13}{12}\right)(0.360)^2(32.2)} + \frac{(0.0205)(1000)Q_{ACDB}^2}{(2)\left(\frac{10.1}{12}\right)(0.556)^2(32.2)}$$

$$= 0.4912Q_{ACDB}^2 + 5.438Q_{ACDB}^2 + 1.223Q_{ACDB}^2$$

$$= 7.152Q_{ACDB}^2$$

$$h_{f,AB} = h_{f,ACDB}$$

$$2.446Q_{AB}^2 = 7.152Q_{ACDB}^2$$

$$Q_{AB} = \sqrt{\frac{7.152}{2.446}} = 1.71Q_{ACDB} \quad \text{[Eq. 1]}$$

$$Q_{AB} + Q_{ACDB} = 1041.7 \text{ gal/min} \quad \text{[Eq. 2]}$$

Solving Eqs. 1 and 2 simultaneously,

$$Q_{AB} = \boxed{657.3 \text{ gal/min}}$$

$$Q_{ACDB} = \boxed{384.4 \text{ gal/min}}$$

This answer is insensitive to the $v_{max} = 5$ ft/sec assumption. A second iteration using actual velocities from these flow rates does not change the answer.

The same technique can be used with the Hazen-Williams equation and an assumed value of C. If $C = 100$ is used, then

$$Q_{AB} = \boxed{725.4 \text{ gal/min}}$$

$$Q_{ACDB} = \boxed{316.3 \text{ gal/min}}$$

(b) From Fig. 3.13, $f \approx 0.022$. From Eq. 3.71,

$$h_{f,AB} = \frac{(0.022)(2000)(2.7)^2}{(2)\left(\frac{10}{12}\right)(32.2)} = 5.98 \text{ ft}$$

For leg BD, use $f = 0.022$ (assumed).

$$v = \frac{Q}{A} = \frac{(383)(0.002228)}{0.5454} = 1.56 \text{ ft/sec}$$

$$h_{f,DB} = \frac{(0.022)(1000)(1.56)^2}{(2)\left(\frac{10}{12}\right)(32.2)} = 1.0 \text{ ft}$$

At 40°F, $\gamma = 62.4$ lbf/ft^3. From the Bernoulli equation (omitting the velocity term),

$$\frac{p_B}{\gamma} + z_B + h_{f,DB} = \frac{p_D}{\gamma} + z_D$$

$$p_B = \left(\frac{62.4}{144}\right)\left[\frac{(40)(144)}{62.4} + 600 - 1 - 580\right]$$

$$= \boxed{48.2 \text{ lbf/in}^2 \text{ (psi)}}$$

$$\frac{p_A}{\gamma} + z_A = \frac{p_B}{\gamma} + z_B + h_{f,AB}$$

$$p_A = \left(\frac{62.4}{144}\right)\left[\frac{(48.2)(144)}{62.4} + 580 + 5.98 - 540\right]$$

$$= \boxed{68.1 \text{ lbf/in}^2 \text{ (psi)}}$$

OPEN CHANNEL FLOW

Untimed

1. From Eq. 5.9,

$$\text{(d)}\quad Q_o = \left(\frac{1.49}{0.013}\right)\left(\frac{\pi}{4}\right)\left(\frac{24}{12}\right)^2\left[\frac{24}{(12)(4)}\right]^{2/3}\sqrt{0.001}$$

$$= \boxed{7.173\ \text{ft}^3/\text{sec (cfs)}}$$

$$\text{(a)}\quad \text{v}_o = \frac{Q_o}{A} = \frac{7.173}{\left(\frac{\pi}{4}\right)(2)^2} = \boxed{2.28\ \text{ft/sec}}$$

$$\text{(b)}\quad \text{v}_{\text{current}} = \frac{6.00}{\left(\frac{\pi}{4}\right)(2)^2} = \boxed{1.91\ \text{ft/sec}}$$

Since Q is inversely proportional to n,

$$\text{(c)}\quad n_{\text{current}} = \left(\frac{Q_o}{Q_{\text{current}}}\right)n_o = \left(\frac{7.173}{6}\right)(0.013)$$

$$= \boxed{0.01554}$$

(e) To be self-cleansing (see Table 8.10),

$$\text{v} = \boxed{2\ \text{to}\ 2.5\ \text{ft/sec}}$$

2. (a) From Eq. 5.11,

$$D = (1.33)\left[\frac{(3.5)(0.013)}{\sqrt{0.01}}\right]^{3/8} = 0.99\ \text{ft}$$

$$\boxed{\text{Use 12-in pipe.}}$$

(b) From Eq. 5.9,

$$Q = \left(\frac{1.49}{0.013}\right)\left(\frac{\pi}{4}\right)(1)^2\left(\frac{1}{4}\right)^{2/3}\sqrt{0.01}$$

$$= \boxed{3.57\ \text{ft}^3/\text{sec (cfs)}}$$

$$\text{(c)}\quad \text{v} = \frac{Q}{A} = \frac{3.57}{\left(\frac{\pi}{4}\right)(1)^2} = \boxed{4.55\ \text{ft/sec}}$$

$$\text{(d)}\quad \frac{Q}{Q_{\text{full}}} = \frac{0.7}{3.57} \approx \boxed{0.2}$$

From App. C, letting n vary with depth,

$$\frac{d}{D} \approx 0.35$$

$$d = (0.35)(12) = \boxed{4.2\ \text{in}}$$

3. From Eq. 5.57,

$$\text{v}_1 = \sqrt{\left(\frac{32.2}{2}\right)\left(\frac{2.4}{1}\right)(1+2.4)} = 11.46\ \text{ft/sec}$$

$$Q = A\text{v} = (5)(1)(11.46) = \boxed{57.30\ \text{ft}^3/\text{sec (cfs)}}$$

4.

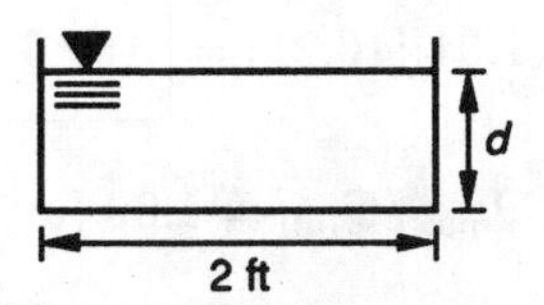

From Eq. 5.2,

$$r_H = \frac{2d}{2+2d} = \frac{d}{1+d}$$

From Eq. 5.9,

$$3 = \left(\frac{1.49}{0.012}\right)(2d)\left(\frac{d}{d+1}\right)^{2/3}\sqrt{0.01}$$

$$0.120805 = \left(\frac{d^{5/2}}{d+1}\right)^{2/3}$$

$$0.042 = \frac{d^{5/2}}{d+1}$$

By trial and error,

$$d = \boxed{0.314\ \text{ft}}$$

5.

- Full flowing:

$$r_H = \frac{D}{4} = \frac{4}{4} = 1\ \text{ft}$$

From Eq. 5.8,

$$\text{v} = \left(\frac{1.49}{0.013}\right)(1)^{2/3}\sqrt{0.02} = 16.21\ \text{ft/sec}$$

(Check this using App. B.)

$$Q = A\text{v} = \left(\frac{\pi}{4}\right)(4)^2(16.21)$$

$$= 203.7\ \text{ft}^3/\text{sec (cfs)}$$

• Actual:

$$\frac{d}{D} = \frac{1.5}{D} = 0.375$$

From App. C, assuming that n varies with depth,

$$\frac{v}{v_{full}} = 0.68$$

(If n is assumed to be constant, then $v/v_{full} \approx 0.85$.)

(a) $v = (0.68)(16.21) = \boxed{11.0 \text{ ft/sec}}$

(b) From App. C, the maximum value of v/v_{full} is 1.04 (at $d/D \approx 0.9$).

$$v_{max} = (1.04)(16.21) = \boxed{16.86 \text{ ft/sec}}$$

(c) Similarly, $Q_{max}/Q_{full} = 1.02$ (at $d/D \approx 0.96$).

$$Q_{max} = (1.02)(203.7) = \boxed{207.8 \text{ ft}^3\text{/sec (cfs)}}$$

6. From Eq. 5.5,

$$(0.2)(\text{width})(54.7) = (6)(\text{width})v_2$$
$$v_2 = 1.82 \text{ ft/sec}$$

From Eq. 5.60,

$$\Delta E = 0.2 + \frac{(54.7)^2}{(2)(32.2)} - 6 - \frac{(1.82)^2}{(2)(32.2)}$$
$$= \boxed{40.61 \text{ ft-lbf/lbm}}$$

7. Disregard the velocity of approach. Using $C_s = 3.5$ in Eq. 5.61,

(a) $Q = (3.5)(1)(2)^{3/2}$

$$= \boxed{9.9 \text{ ft}^3\text{/sec (cfs) per ft of width}}$$

(b) The upstream energy is

$$E_1 = 40 + 2 = 42 \text{ ft} \quad [v_1 \approx 0]$$

At the toe, $E_2 = E_1$. From Eq. 5.39,

$$d_2 + \frac{(9.9)^2}{(2)(32.2)(1)^2 d_2^2} = 42$$

By trial and error,

$$d_2 = \boxed{0.19 \text{ ft}}$$

8. From Eq. 5.61,

$$Q = (3.5)(1)(5)^{3/2} = \boxed{39.1 \text{ ft}^3\text{/sec-ft (cfs/ft)}}$$

9.

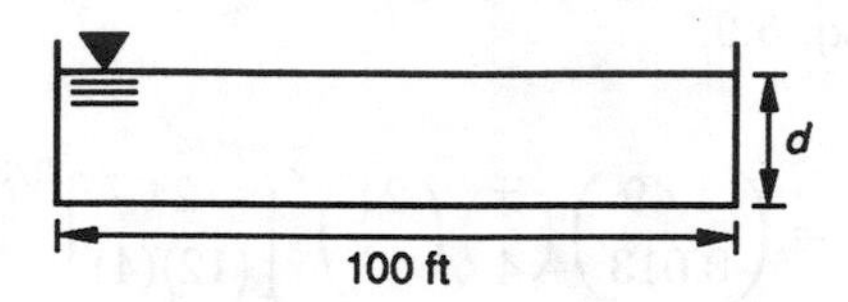

(a) $r_H = \frac{100d}{100 + 2d} = \frac{50d}{50 + d}$

From Eq. 5.9,

$$10{,}000 = \left(\frac{1.49}{0.012}\right)(100d)\left(\frac{50d}{50+d}\right)^{2/3}\sqrt{0.05}$$
$$0.26538 = \left(\frac{d^{5/2}}{50+d}\right)^{2/3}$$
$$0.1367 = \frac{d^{5/2}}{50+d}$$

By trial and error,

$$d = \boxed{2.2 \text{ ft}}$$

(b) From Eq. 5.43,

$$d_c = \sqrt[3]{\frac{(10{,}000)^2}{(32.2)(100)^2}} = \boxed{6.77 \text{ ft}}$$

(c) Since $d < d_c$,

$$\boxed{\text{Flow is rapid (shooting).}}$$

(d) $v = \frac{Q}{A} = \frac{10{,}000}{(100)(2.2)} = \boxed{45.45 \text{ ft/sec}}$

From Eq. 5.59,

$$d_2 = -\left(\frac{1}{2}\right)(2.2) + \sqrt{\frac{(2)(45.45)^2(2.2)}{32.2} + \frac{(2.2)^2}{4}}$$
$$= \boxed{15.74 \text{ ft}}$$

10. From Eq. 5.28,

$$b = 5 - (0.1)(2)(0.43) = 4.914 \text{ ft}$$

From Eq. 5.27,

$$C_1 = \left[0.6035 + (0.0813)\left(\frac{0.43}{6}\right) + \frac{0.000295}{6}\right] \times \left(1 + \frac{0.00361}{0.43}\right)^{3/2}$$
$$= (0.60938)(1.0126) = 0.617$$

From Eq. 5.26,

$$Q = \left(\tfrac{2}{3}\right)(0.617)(4.914)\sqrt{(2)(32.2)}(0.43)^{3/2}$$
$$= \boxed{4.574 \text{ ft}^3/\text{sec (cfs)}}$$

11.

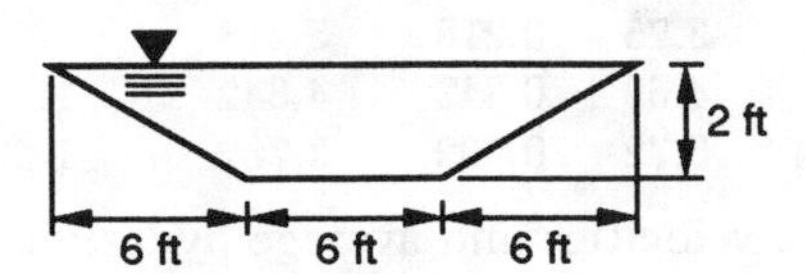

$$A = (6)(2) + \frac{(2)(6)(2)}{2} = 24 \text{ ft}^2$$
$$P = 6 + 2\sqrt{(6)^2 + (2)^2} = 18.65 \text{ ft}$$
$$r_H = \frac{24}{18.65} = 1.287$$
$$Q = (24)\left(\frac{1.49}{0.013}\right)(1.287)^{2/3}\sqrt{0.002}$$
$$= \boxed{145.6 \text{ ft}^3/\text{sec (cfs)}}$$

12.(a) From Eq. 5.61 using $C_s = 3.7$,

$$Q = (3.7)(60)(5)^{3/2} = \boxed{2482 \text{ ft}^3/\text{sec (cfs)}}$$

(b) Disregarding the velocity of approach, the initial energy is

$$E_1 = (\text{elev})_1 = 2425 \text{ ft}$$

At the toe (point 2),

$$E_1 = E_2 + h_f$$
$$2425 = (\text{elev})_2 + d_2 + \frac{Q^2}{2gw^2d_2^2} + (0.20)(2425 - 2345)$$
$$(1 - 0.20)(2425 - 2345) = d_2 + \frac{(2482)^2}{(2)(32.2)(60)^2 d_2^2}$$
$$64 = d_2 + \frac{26.57}{d_2^2}$$

By trial and error,

$$d_2 = \boxed{0.647 \text{ ft}}$$

(c) A submerged hydraulic jump will occur if $d_2 \geq d_c$. From Eq. 5.43,

$$d_c = \sqrt[3]{\frac{(2482)^2}{(32.2)(60)^2}} = \boxed{3.76 \text{ ft}}$$

(d) If there was a hydraulic jump, it would be between the conjugate depths, d_1 to d_2.

$$d_1 = 0.647 \text{ ft}$$
$$\text{v}_1 = \frac{Q}{A_1} = \frac{2482}{d_1 w} = \frac{2482}{(0.647)(60)}$$
$$= 63.94 \text{ ft/sec}$$

From Eq. 5.59,

$$d_2 = -\frac{1}{2}d_1 + \sqrt{\frac{2\text{v}_1^2 d_1}{g} + \frac{d_1^2}{4}}$$
$$= \left(-\frac{1}{2}\right)(0.647) + \sqrt{\frac{(2)(63.9)^2(0.647)}{32.2} + \frac{(0.647)^2}{4}}$$
$$= 12.49 \text{ ft}$$
$$\text{v}_2 = \frac{Q}{A_2} = \frac{2482}{d_2 w} = \frac{2482}{(12.49)(60)} = 3.31 \text{ ft/sec}$$

Use d_2 and v_2 in Eq. 5.60.

$$\Delta E = \left(d_1 + \frac{\text{v}_1^2}{2g}\right) - \left(d_2 + \frac{\text{v}_2^2}{2g}\right)$$
$$= \left[0.647 + \frac{(63.9)^2}{(2)(32.2)}\right] - \left[12.49 + \frac{(3.31)^2}{(2)(32.2)}\right]$$
$$= \boxed{51.4 \text{ ft (ft-lbf/lbm)}}$$

13.(a) From Eq. 5.34,

$$n = (1.522)(6)^{0.026} = 1.595$$

From Eq. 5.33,

$$Q = (4)(6)\left(\frac{18}{12}\right)^{1.595} = \boxed{45.8 \text{ ft}^3/\text{sec (cfs)}}$$

(b) The flume is self-cleansing, the head loss is small, and operation by unskilled personnel is possible.

14. This is a Cipoletti weir. From Eq. 5.32,

$$Q = (3.367)\left(\frac{18}{12}\right)\left(\frac{9}{12}\right)^{3/2} = \boxed{3.28 \text{ ft}^3/\text{sec (cfs)}}$$

15.

$$\text{v}_1 = \frac{Q}{A_1} = \frac{50}{(3)(6)} = 2.778 \text{ ft/sec}$$

From Eq. 5.43, the critical depth is

$$d_c = \sqrt[3]{\frac{(50)^2}{(32.2)(6)^2}} = 1.292 \text{ ft}$$
$$\text{v}_2 = \frac{Q}{A_2} = \frac{50}{(1.292)(6)}$$
$$= 6.450 \text{ ft/sec}$$

From Eq. 5.37 (using $z_1 = 0$ and $z_2 =$ height of hump),

$$\frac{v_1^2}{2g} + d_1 = \frac{v_2^2}{2g} + d_2 + z$$

$$z = \frac{(2.778)^2 - (6.450)^2}{(2)(32.2)} + 3.0 - 1.292$$

$$= \boxed{1.182 \text{ ft}}$$

16.

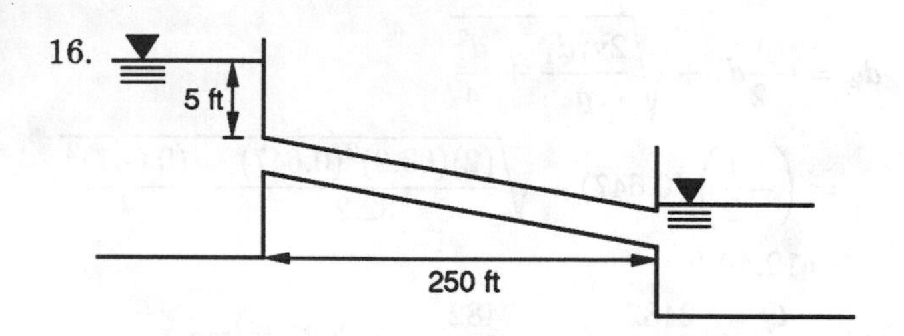

Since the inlet is submerged and $h_4/D = 1$, this is a type 6 flow. Either Eq. 5.74 or Eq. 5.67 can be used.

Assume $n = 0.013$ and use Eq. 5.67.

$$H = h_1 - h_3 = \text{inlet depth} + SL$$
$$= 5 + (0.006)(250)$$
$$= 6.5 \text{ ft}$$

$$r_H = \frac{D}{4} = \frac{3.5}{4} = 0.875 \text{ ft}$$

From Table 5.8, $K_e = 0.50$. From Eq. 5.67,

$$v = \sqrt{\frac{6.5}{\frac{1.5}{(2)(32.2)} + \frac{(0.013)^2(250)}{(2.21)(0.875)^{4/3}}}}$$
$$= 11.87 \text{ ft/sec}$$

$$Q = Av = \left(\frac{\pi}{4}\right)(3.5)^2(11.87)$$
$$= \boxed{114.2 \text{ ft}^3\text{/sec (cfs)}}$$

17. (a) The normal depth is d.

$$r_H = \frac{8d}{8 + 2d}$$

From Eq. 5.9,

$$150 = (8d)\left(\frac{1.49}{0.015}\right)\left(\frac{8d}{8+2d}\right)^{2/3}\sqrt{0.0015}$$

$$4.875 = d\left(\frac{8d}{8+2d}\right)^{2/3}$$

By trial and error,

$$d = \boxed{3.28 \text{ ft}}$$

(b) To get the backwater curve, choose depths and compute distances to those depths.

Preliminary parameters:

d	$A=8d$	$v=\frac{Q}{A}$	$\frac{v^2}{2g}$	$E=d+\frac{v^2}{2g}$	$P=2d+8$	r_H
6	48	3.12	0.151	6.151	20	2.40
5	40	3.75	0.218	5.218	18	2.22
4	32	4.69	0.342	4.342	16	2.00
3.28	26.24	5.72	0.502	3.782	14.56	1.80

The average velocities and average hydraulic radii are

d	v	v_{ave}	r_H	$r_{H,ave}$
6.00	3.12		2.40	
		3.435		2.31
5.00	3.75		2.22	
		4.22		2.11
4.00	4.69		2.00	
		5.21		1.90
3.28	5.72		1.80	

From Eq. 5.54, the average energy gradient is

$$S_{ave} = \left[\frac{nv_{ave}}{(1.486)(r_{H,ave})^{2/3}}\right]^2$$

d	S_{ave}	ΔE	$S_o - S_{ave}$
6.00			
	0.000394	0.933	0.001106
5.00			
	0.000671	0.876	0.000829
4.00			
	0.001175	0.552	0.000325
3.28			

From Eq. 5.56,

$$L_{6\text{–}5}: \quad \frac{\Delta E}{S_o - S_{ave}} = \frac{0.933}{0.001106} = 844 \text{ ft}$$

$$L_{5\text{–}4}: \quad \frac{0.876}{0.000829} = 1057 \text{ ft}$$

$$L_{4\text{–}3.28}: \quad \frac{0.552}{0.000325} = 1698 \text{ ft}$$

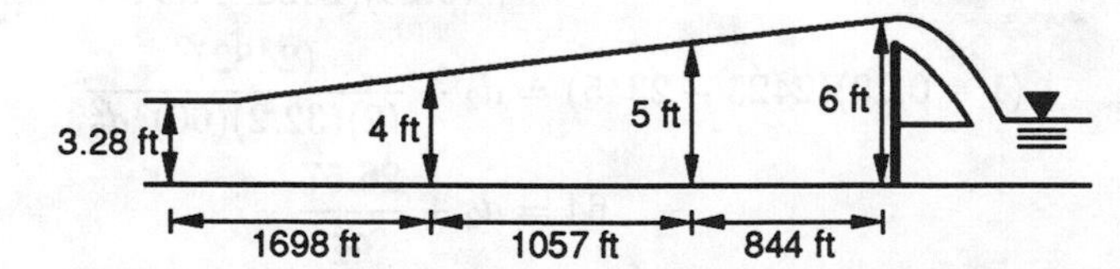

18. From App. E, p. 3-43,

$$A = 0.6318D^2$$
$$r_H = 0.3017D$$

From Eq. 5.9,

$$5 = (0.6318D^2)\left(\frac{1.49}{0.012}\right)(0.3017D)^{2/3}\sqrt{0.02}$$

$$1.0 = D^{8/3}$$

$$D = \boxed{1.0 \text{ ft}}$$

19. (a) From Eq. 5.57,

$$v_1 = \sqrt{\left[\frac{(32.2)(3.8)}{(2)(1.2)}\right](1.2 + 3.8)} = 15.97 \text{ ft/sec}$$

$$Q = Av = (1.2)(6)(15.97) = 115 \text{ ft}^3\text{/sec (cfs)}$$

$$v^2 = \frac{Q}{A} = \frac{115}{(3.8)(6)} = \boxed{5.04 \text{ ft/sec}}$$

(b) From Eq. 5.60,

$$\Delta E = \left[1.2 + \frac{(15.97)^2}{(2)(32.2)}\right] - \left[3.8 + \frac{(5.04)^2}{(2)(32.2)}\right]$$

$$= \boxed{0.97 \text{ ft (ft-lbf/lbm)}}$$

20.

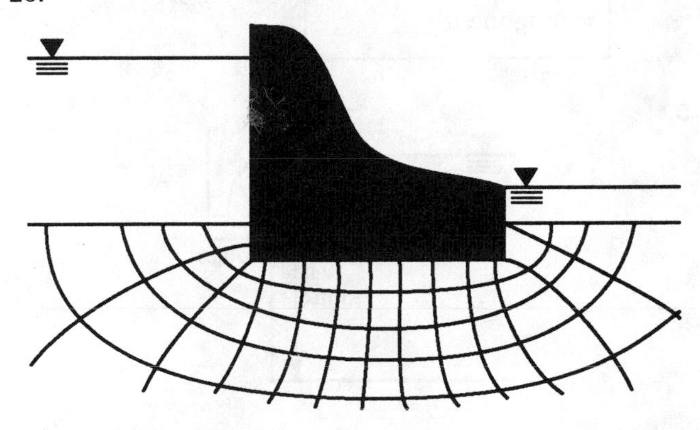

From Eq. 6.17, with 4 flow paths and 13 equipotential drops,

$$Q = (0.15)(9.0)\left(\frac{4}{13}\right)$$

$$= \boxed{0.42 \text{ ft}^3\text{/hr-ft width}}$$

Timed

1. From App. A, $n = 0.011$. From p. 5-6, $d = w/2$ for an optimum channel.

$$A = wd = \frac{w^2}{2}$$

$$P = w + (2)\left(\frac{w}{2}\right) = 2w$$

$$r_H = \frac{A}{P} = \frac{\left(\frac{w^2}{2}\right)}{2w} = \frac{w}{4}$$

(The 1.49 in Eq. 5.9 is an SI-to-English conversion. 1.0 should be used for SI.)

$$17 = \left(\frac{w^2}{2}\right)\left(\frac{1.0}{0.011}\right)\left(\frac{w}{4}\right)^{2/3}\sqrt{0.08}$$

$$w = \boxed{1.57 \text{ m}}$$

$$d = \boxed{0.79 \text{ m}}$$

2.

$$Q_o = C_d A_o \sqrt{2gh}$$

$$Q_1 = (0.7)(2)\sqrt{(2)(32.2)(1.5)}$$

$$= (11.235)\sqrt{1.5} \quad \text{[top inlet]}$$

$$Q_1 + Q_2 + Q_3 = (11.235)(\sqrt{1.5} + \sqrt{3.5} + \sqrt{5.5})$$

$$= 61.13 \text{ ft}^3\text{/sec (cfs)}$$

This is a type 6 culvert. Use Eq. 5.67 instead of Eq. 5.74. Since C_d is given, $K_e = 0$. Culvert size (and hence r_H) is unknown. To get a trial value, initially disregard friction.

$$H = 4 + (100)(0.05) = 9 \text{ ft}$$

$$v = \sqrt{\frac{9}{\frac{1+0}{(2)(32.2)}}} = 24.07 \text{ ft/sec}$$

$$A = \frac{Q}{V} = \frac{61.13}{\left(\frac{2}{3}\right)(24.07)} = 3.81 \text{ ft}^2$$

$$\text{box size} = \sqrt{3.81} = 1.95 \text{ ft}$$

(Round to $2\frac{1}{4}$ ft square to account for friction.)

Use Eq. 5.67.

$$r_H = \frac{(2.25)^2}{(4)(2.25)} = 0.5625 \text{ ft}$$

$$v = \sqrt{\frac{9}{\frac{1+0}{(2)(32.2)} + \frac{(0.013)^2(100)}{(2.21)(0.5625)^{4/3}}}}$$

$$= 16.77 \text{ ft/sec}$$

$$\text{box size} = \sqrt{\frac{61.13}{\left(\frac{2}{3}\right)(16.77)}} = \boxed{2.34 \text{ ft [28 in or more]}}$$

3. (a) Assume $C_s = 3.5$.

$$Q = (3.5)(500)H^{3/2} = 70{,}000 \text{ ft}^3\text{/sec (cfs)}$$

$$H = 11.7 \text{ ft}$$

$$v = \frac{70{,}000}{(500)(11.7)} = 11.97 \text{ ft/sec}$$

The specific energy at the toe acquired in the fall (if there is no friction) is

$$E = 100 + 11.7 + \frac{(11.97)^2}{(2)(32.2)} = 113.9 \text{ ft}$$

At the toe,

$$\text{v} = \frac{70{,}000}{(2)(500)} = 70.0 \text{ ft/sec}$$

Actual specific energy at the toe is

$$E = d + \frac{\text{v}^2}{2g} = 2 + \frac{(70.0)^2}{(2)(32.2)} = 78.1 \text{ ft}$$

$$\text{friction loss} = 113.9 - 78.1 = \boxed{35.8 \text{ ft}}$$

(b) From Eq. 5.59,

$$d = \frac{-2}{2} + \sqrt{\frac{(2)(70.0)^2(2)}{32.2} + \frac{(2)^2}{4}} = \boxed{23.7 \text{ ft}}$$

(c)

A 113.9 ft

B 78.1 ft

(d) The tailwater depth must be calculated. If the tailwater depth equals the critical depth at the toe, a hydraulic jump will occur at the toe.

4. (a)

$$\frac{A}{P} = \frac{\left(\frac{1}{2}\right)(\pi r^2)}{\left(\frac{1}{2}\right)(2\pi r)} = \frac{r}{2} = \frac{d}{4} = \frac{6}{4} = \boxed{1.5 \text{ ft}}$$

(b) Assume $n = 0.013$ for new concrete. From Eq. 5.9,

$$S = \left[\frac{nQ}{1.49A(r_H)^{2/3}}\right]^2 = \left[\frac{(0.013)(150)}{(1.49)\left(\frac{1}{2}\right)\pi(3)^2(1.5)^{2/3}}\right]^2 = \boxed{0.005}$$

(c) The normal depth is 3 ft. Equation 5.46 could be used, but it is easier to use App. D. This assumes n is constant.

$$d_c \approx 3.4 \text{ ft}$$

Since $d < d_c$, the flow is

$$\boxed{\text{supercritical}}$$

(d) When full, r_H is the same (1.5 ft). From Eq. 5.9, only A changes, and it doubles.

$$Q = (2)(150) = \boxed{300 \text{ ft}^3\text{/sec (cfs)}}$$

(e)
- headwater elevation
- tailwater elevation
- barrel slope
- length
- roughness

5.

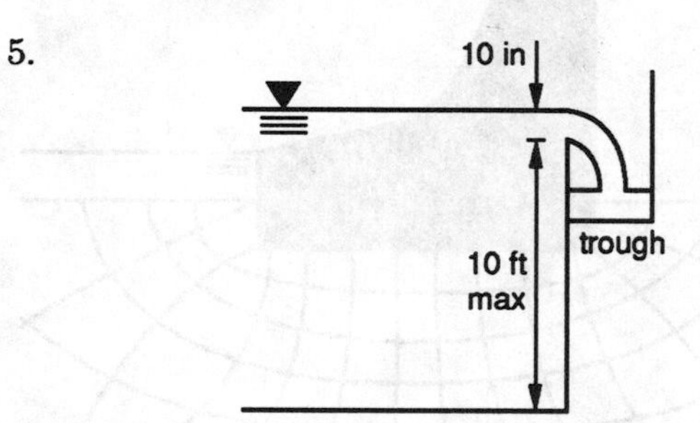

(a) Use Eq. 5.26 with $C_1 = 0.611$.

$$Q_{\text{mgd}} = \frac{Q_{\text{cfs}}}{1.547} = \frac{\left(\frac{2}{3}\right)(0.611)}{1.547} bH^{3/2}\sqrt{(2)(32.2)} = 2.11bH^{3/2}$$

At minimum flow, $Q = 2$ mgd.

$$H = \frac{10}{12} = 0.833 \text{ ft}$$

$$b_{\text{effective}} = \frac{2}{(2.11)(0.833)^{3/2}} = 1.25 \text{ ft}$$

From Eq. 5.28,

$$b_{\text{actual}} = 1.25 + (0.1)(2)(0.833) = \boxed{1.42 \text{ ft}}$$

(b) At maximum flow,

$$8 = (2.11)(1.25)H^{3/2}$$
$$H \approx 2.1 \text{ ft}$$

Use Eq. 5.28 to calculate $b_{\text{effective}}$. Continue iterating until H stabilizes.

$$\begin{aligned}b_{\text{effective}} &= 1.42 - (0.1)(2)(2.1) = 1.0 \text{ ft}\\ 8 &= (2.11)(1)H^{3/2}\\ H &= 2.4 \text{ ft}\\ b_{\text{effective}} &= 1.42 - (0.1)(2)(2.4) = 0.94 \text{ ft}\\ 8 &= (2.11)(0.94)H^{3/2}\\ H &= 2.53 \text{ ft}\\ b_{\text{effective}} &= 1.42 - (0.1)(2)(2.53) = 0.914 \text{ ft}\\ 8 &= (2.11)(0.914)H^{3/2}\\ H &= 2.58 \text{ ft}\\ b_{\text{effective}} &= 1.42 - (0.1)(2)(2.58) = 0.904 \text{ ft}\\ 8 &= (2.11)(0.904)H^{3/2}\\ H &= 2.60 \text{ ft} \quad \text{[acceptable]}\end{aligned}$$

Use a freeboard elevation of

$$580 + 10 + \frac{10}{12} + \frac{4}{12} = 591.17 \text{ ft}$$

The elevation at maximum flow is

$$591.17 - \text{freeboard height} = 591.17 - \frac{4}{12} = \boxed{590.83 \text{ ft}}$$

(c) The channel elevation is

$$590.83 - 2.60 = 588.23 \text{ ft}$$

The elevation at minimum flow is

$$588.23 + \frac{10}{12} = \boxed{589.07 \text{ ft}}$$

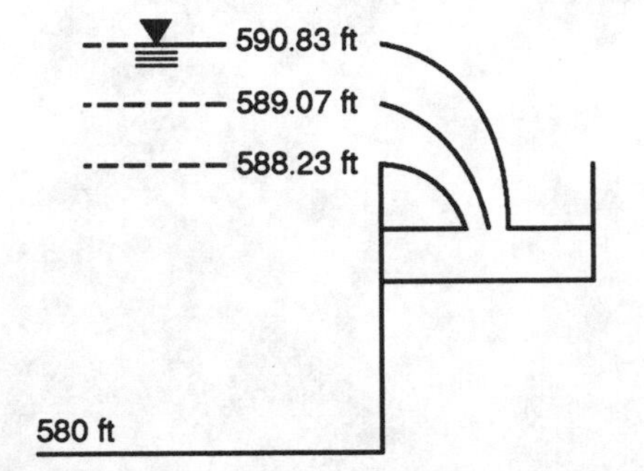

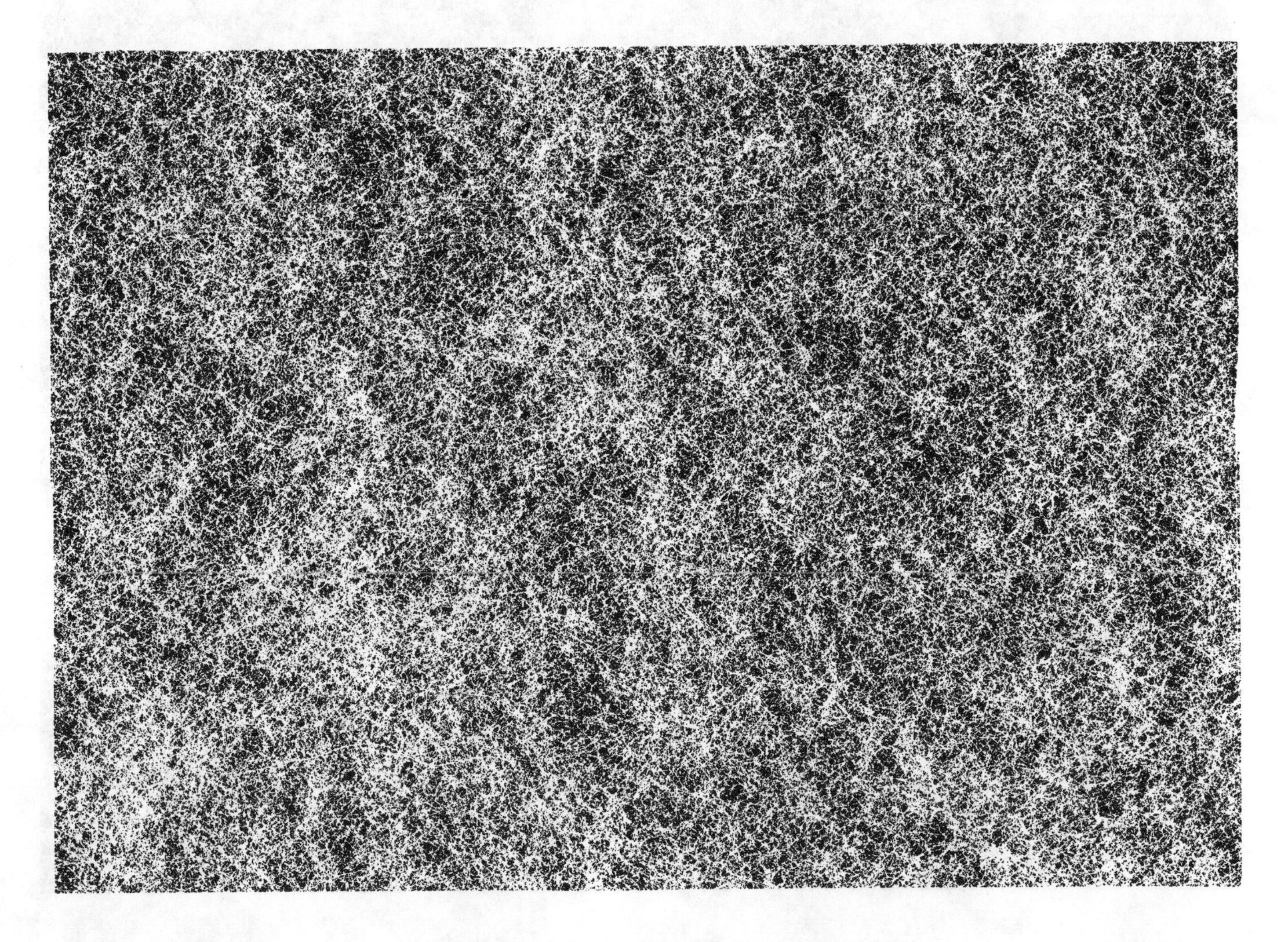

HYDROLOGY

Untimed

1. Assume that n is constant and there are four 5-acre areas.

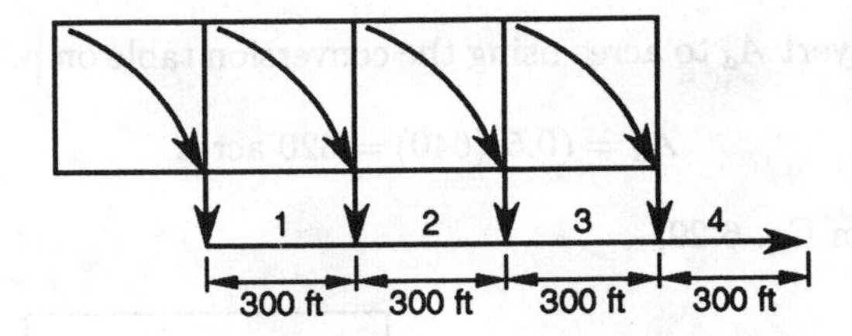

- First inlet:

Peak flow for the first inlet begins at $t = 15$ min.

$$I = \frac{100}{15+10} = 4 \text{ in/hr}$$

From Eq. 6.29,

$$Q = (0.55)(4)(5) = 11 \text{ ft}^3/\text{sec (cfs)}$$

From Eq. 5.11,

$$D = 1.33\left(\frac{nQ}{\sqrt{S}}\right)^{3/8} = (1.33)\left[\frac{(0.013)(11)}{\sqrt{0.005}}\right]^{3/8}$$
$$= 1.73 \text{ ft}$$
$$D \approx 21 \text{ in} \quad \text{[standard pipe size]}$$

$$\text{v}_{\text{full}} = \frac{Q}{A} = \frac{11 \frac{\text{ft}^3}{\text{sec}}}{\left(\frac{\pi}{4}\right)\left(\frac{21 \text{ in}}{12 \frac{\text{in}}{\text{ft}}}\right)^2}$$
$$= 4.6 \text{ ft/sec}$$

- Second inlet:

The flow time from inlet 1 to inlet 2 is

$$\frac{300}{4.6} = 65.2 \text{ sec} = 1.09 \text{ min}$$

The intensity at $t = 15.0 + 1.09 = 16.09$ is

$$I = \frac{100}{16.09+10} = 3.83 \text{ in/hr}$$

The runoff is

$$Q = (0.55)(3.83)(10) = 21.07 \text{ ft}^3/\text{sec (cfs)}$$

From Eq. 5.11,

$$D = (1.33)\left[\frac{(0.013)(21.07)}{\sqrt{0.005}}\right]^{3/8}$$
$$= 2.21 \text{ ft}$$
$$D \approx 27 \text{ in} \quad \text{[standard pipe size]}$$
$$\text{v}_{\text{full}} = \frac{21.07}{\left(\frac{\pi}{4}\right)\left(\frac{27}{12}\right)^2}$$
$$\approx 5.3 \text{ ft/sec}$$

- Third inlet:

The flow time from inlet 2 is

$$t = \frac{L}{\text{v}} = \frac{300}{(5.3)(60)} = 0.94 \text{ min}$$
$$t_c = 16.09 + 0.94 = 17.03 \text{ min}$$
$$I = \frac{100}{17.03+10} = 3.70 \text{ in/hr}$$
$$Q = (0.55)(3.70)(15) = 30.53 \text{ ft}^3/\text{sec (cfs)}$$

From Eq. 5.11,

$$D = (1.33)\left[\frac{(0.013)(30.53)}{\sqrt{0.005}}\right]^{3/8} = 2.54 \text{ ft}$$
$$D \approx 31 \text{ in}$$

Use a 33-in pipe (standard pipe size).

$$\text{v} = \frac{Q}{A} = \frac{30.53 \frac{\text{ft}^3}{\text{sec}}}{\left(\frac{\pi}{4}\right)\left(\frac{33}{12}\right)^2}$$
$$= 5.14 \text{ ft/sec}$$

- Fourth inlet:

The flow time from inlet 3 is

$$\frac{300}{(5.14)(60)} = 0.97 \text{ min}$$
$$t_c = 17.03 + 0.97 = 18.0 \text{ min}$$
$$I = \frac{100}{18.0+10} = 3.57 \text{ in/hr}$$
$$Q = (0.55)(3.57)(20) = 39.27 \text{ ft}^3/\text{sec (cfs)}$$

From Eq. 5.11,

$$D = (1.33)\left[\frac{(0.013)(39.27)}{\sqrt{0.005}}\right]^{3/8} = 2.79 \text{ ft}$$
$$= 33.5 \text{ in}$$

Use 36-in pipe.

2. (a)

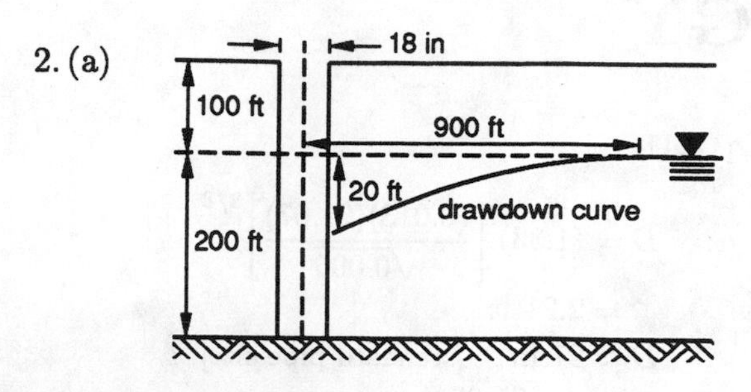

From Eq. 6.12,

$$K_p = \frac{T}{Y}$$

$$= \frac{10{,}000\ \frac{\text{gal}}{\text{day-ft}}}{200\text{ ft}} = 50\text{ gal/day-ft}^2$$

$$\left(50\ \frac{\text{gal}}{\text{day-ft}^2}\right)\left(0.1337\ \frac{\text{ft}^3}{\text{gal}}\right) = 6.685\text{ ft}^3\text{/day-ft}^2$$

From Eq. 6.15,

$$\text{drawdown} = 20\text{ ft at } r = \frac{18}{(2)(12)} = 0.75\text{ ft}$$

$$y_2 = 200 - 20 = 180\text{ ft at } r_2 = 0.75\text{ ft}$$

$$y_1 = 200\text{ ft at } r_1 = 900\text{ ft}$$

$$Q = \frac{\pi(6.685)[(200)^2 - (180)^2]}{(86{,}400)\ln\left(\frac{900}{0.75}\right)}$$

$$= \boxed{0.261\text{ ft}^3\text{/sec (cfs)}}$$

(b) Assume $\eta_{\text{pump}} = 0.65$ (Fig. 4.13). From Table 4.2, the motor horsepower is

$$P = \frac{(120)(0.261)(1)}{(8.814)(0.65)} = 5.47\text{ hp}$$

From Table 4.4,

$$\boxed{\text{Choose a 7.5-hp motor.}}$$

3. $$\frac{3300\text{ acre-ft}}{(43\text{ sq mi})\left(53.3\ \frac{\text{acre-ft}}{\text{sq mi-in}}\right)} = 1.44\text{ in}$$

(a) $$\frac{9300}{1.44} = \boxed{6458\text{ ft}^3\text{/sec (cfs)}}$$

(b) $$\text{unit hydrograph volume} = \frac{3300}{1.44} = 2292\text{ acre-ft}$$

$$(2.5)(2292) = \boxed{5730\text{ acre-ft}}$$

(c) $$(2.5)(6458) = \boxed{16{,}145\text{ ft}^3\text{/sec (cfs)}}$$

4. Use Table 6.1.

$$K = 170$$

$$b = 23$$

From Eq. 6.4, the intensity is

$$I = \frac{170}{60 + 23} = 2.05\text{ in/hr}$$

Convert A_d to acres using the conversion table on p. 6-1.

$$A_d = (0.5)(640) = 320\text{ acres}$$

From Eq. 6.29,

$$Q_p = (0.6)(2.05)(320) = \boxed{393.6\text{ ft}^3\text{/sec (cfs)}}$$

5.

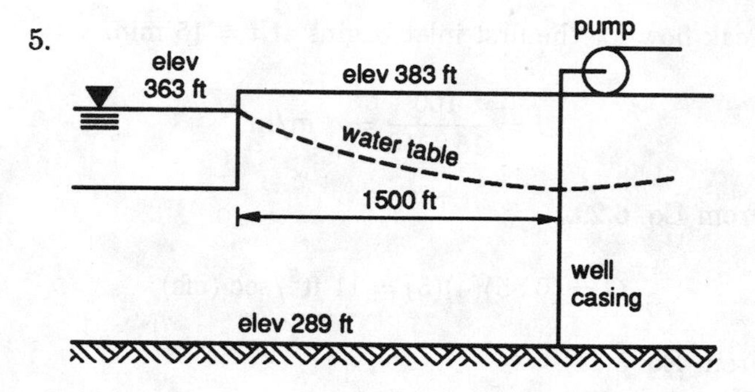

From p. 3-45,

$$D_i = 0.835\text{ ft}$$

$$A_i = 0.5476\text{ ft}^2$$

From p. 6-1,

$$Q = (120{,}000)(1.547 \times 10^{-6}) = 0.1856\text{ ft}^3\text{/sec (cfs)}$$

$$K_p = (1600)(0.1337) = 213.9\text{ ft}^3\text{/day-ft}^2$$

Use Eq. 6.15.

$$y_1 = 363 - 289 = 74\text{ ft}$$

$$r_1 = 1500\text{ ft}$$

$$r_2 = \frac{0.835}{2} = 0.4175\text{ ft}$$

$$0.1856 = \frac{\pi(213.9)[(74)^2 - y_2^2]}{(86{,}400)\ln\left(\frac{1500}{0.4175}\right)}$$

$$195.36 = (74)^2 - y_2^2$$

$$y_2 = 72.67\text{ ft}$$

The drawdown is

$$d = 74 - 72.67 = 1.33\text{ ft}$$

d is small compared to y, so Eq. 6.15 was an acceptable equation to use.

The velocity of the water in the pipe is

$$v = \frac{Q}{A} = \frac{0.1856}{0.5476} = 0.34 \text{ ft/sec}$$

This velocity is too small to include the velocity head.

The suction lift is $383 - 363 + 1.33 = 21.33$.

$$H = 21.33 + 100 = 121.33 \text{ ft}$$

The weight flow of the water is

$$\dot{w} = (0.1856)(62.4) = 11.58 \text{ lbf/sec}$$

From Table 4.2, the net water horsepower is

$$P = \frac{(121.33)(11.58)}{550} = \boxed{2.55 \text{ hp}}$$

6. (c) The actual hydrograph is plotted at the right. From the graph,

$$N \approx \boxed{6 \text{ hr}}$$

hour	runoff	ground water	surface water	surface water / 2.43
0	102	100	0	0
1	99	100	0	0
2	101	100	0	0
3	215	100	115	47.3
4	507	100	407	167.5
5	625	100	525	216.0
6	455	93	362	149.0
7	325	87	238	97.9
8	205	80	125	51.4
9	145	74	71	29.2
10	100	67	33	13.6
11	70	60	10	4.1
12	55	55	0	0
13	49	49	0	0
14	43	43	0	0
15	38	38	0	0
		total =	1886 ft³/sec	

The total volume of surface water is

$$\left(1886 \ \frac{\text{ft}^3}{\text{sec}}\right)(1 \text{ hr})\left(3600 \ \frac{\text{sec}}{\text{hr}}\right) = 6.79 \times 10^6 \text{ ft}^3$$

Using a conversion from p. 6-1, the basin area is

$$(1.2 \text{ mi}^2)(2.788 \times 10^7) = 3.35 \times 10^7 \text{ ft}^2$$

The average precipitation is

$$\frac{(6.79 \times 10^6 \text{ ft}^3)\left(12 \ \frac{\text{in}}{\text{ft}}\right)}{3.35 \times 10^7 \text{ ft}^2} = 2.43 \text{ in}$$

(a)

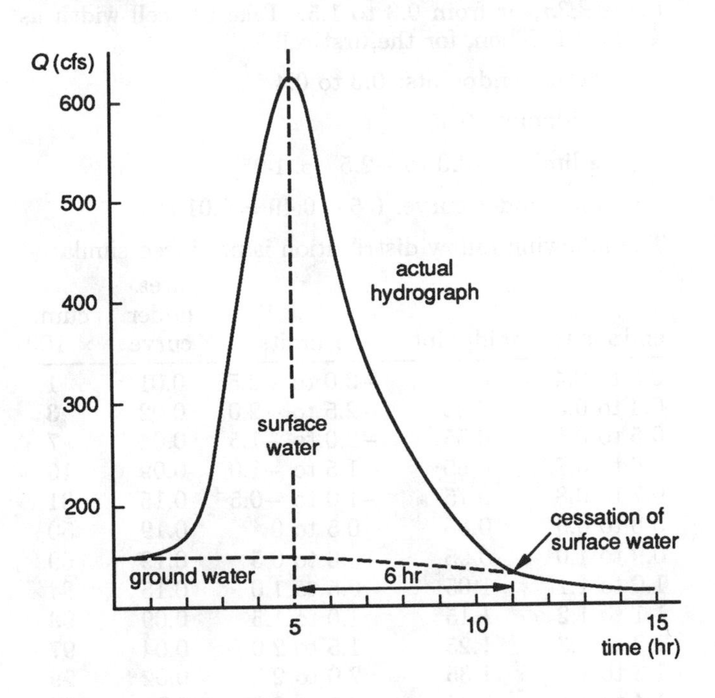

(b) and (d)

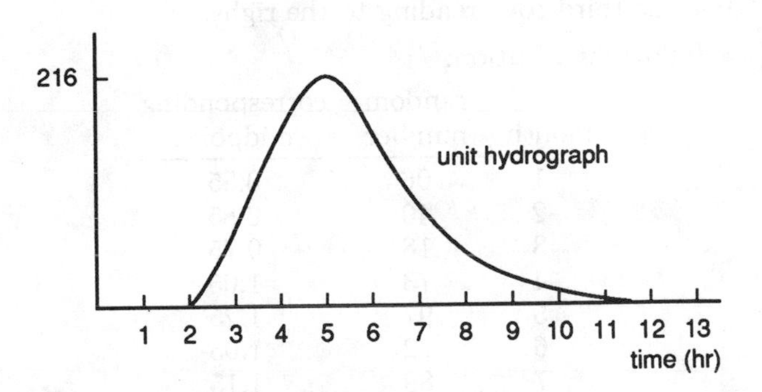

7. (a) From Eq. 6.20, the lag is

$$t_p = (1.8)[(20)(11)]^{0.3} = \boxed{9.08 \text{ hr}}$$

(b) From Eq. 6.22,

$$N = 72 + (3)(9.08) = \boxed{99.24 \text{ hr}}$$

(c) Assume $C_p = 0.6$.

$$A_d = (100)(640) = 64{,}000 \text{ acres}$$

From Eq. 6.21,

$$Q_p = \frac{(0.6)(64{,}000)}{9.08} = \boxed{4229 \text{ ft}^3\text{/sec (cfs)}}$$

8. For practical purposes, the inflow distribution extends $\pm 3\sigma$, or from 0.3 to 1.5. Take the cell width as $\frac{1}{2}\sigma$ or 0.1. Then, for the first cell,

actual endpoints: 0.3 to 0.4

midpoint: 0.35

z limits: -3.0 to -2.5 [p. 1-25]

area under curve: $0.5 - 0.49 = 0.01$

The following inflow distribution is produced similarly.

endpoints	midpoint	z limits	area under curve	cum. × 100
0.3 to 0.4	0.35	−3.0 to −2.5	0.01	1
0.4 to 0.5	0.45	−2.5 to −2.0	0.02	3
0.5 to 0.6	0.55	−2.0 to −1.5	0.04	7
0.6 to 0.7	0.65	−1.5 to −1.0	0.09	16
0.7 to 0.8	0.75	−1.0 to −0.5	0.15	31
0.8 to 0.9	0.85	−0.5 to 0	0.19	50
0.9 to 1.0	0.95	0 to 0.5	0.19	69
1.0 to 1.1	1.05	0.5 to 1.0	0.15	84
1.1 to 1.2	1.15	1.0 to 1.5	0.09	93
1.2 to 1.3	1.25	1.5 to 2.0	0.04	97
1.3 to 1.4	1.35	2.0 to 2.5	0.02	99
1.4 to 1.5	1.45	2.5 to 3.0	0.01	100

Choose 12 random numbers less than 100 from App. B. Use the third row, reading to the right.

- Inflow distribution:

month	random number	corresponding midpoint
1	06	0.55
2	40	0.85
3	18	0.75
4	73	1.05
5	97	1.25
6	72	1.05
7	89	1.15
8	83	1.05
9	24	0.75
10	41	0.85
11	88	1.15
12	86	1.15

Simulate the reservoir operation.

month	starting volume +	inflow −	constant use	= ending volume +	spill
1	5.5	0.55	0.7	5.35	
2	5.35	0.85	0.7	5.5	
3	5.5	0.75	0.7	5.55	
4	5.55	1.05	0.7	5.9	
5	5.9	1.25	0.7	6.45	
6	6.45	1.05	0.7	6.8	
7	6.8	1.15	0.7	7.0	0.25
8	7.0	1.05	0.7	7.0	0.35
9	7.0	0.75	0.7	7.0	0.05
10	7.0	0.85	0.7	7.0	0.15
11	7.0	1.15	0.7	7.0	0.45
12	7.0	1.15	0.7	7.0	0.45

9. Use the same simulation procedure for inflow as in Prob. 8. Proceed similarly.

- Demand distribution:

endpoints	midpoint	z limits	cum. × 100
0.1 to 0.2	0.15	−3.0 to −2.5	1
0.2 to 0.3	0.25	−2.5 to −2.0	3
0.3 to 0.4	0.35	−2.0 to −1.5	7
0.4 to 0.5	0.45	−1.5 to −1.0	16
0.5 to 0.6	0.55	−1.0 to −0.5	31
0.6 to 0.7	0.65	−0.5 to 0	50
0.7 to 0.8	0.75	0 to 0.5	69
0.8 to 0.9	0.85	0.5 to 1.0	84
0.9 to 1.0	0.95	1.0 to 1.5	93
1.0 to 1.1	1.05	1.5 to 2.0	97
1.1 to 1.2	1.15	2.0 to 2.5	99
1.2 to 1.3	1.25	2.5 to 3.0	100

Choose 12 random numbers. Use the fourth row, reading to the right.

- Demand distribution:

month	random number	use
1	04	0.35
2	75	0.85
3	41	0.65
4	44	0.65
5	89	0.95
6	39	0.65
7	42	0.65
8	09	0.45
9	42	0.65
10	11	0.45
11	58	0.75
12	04	0.35

Simulate reservoir operation.

month	starting volume +	inflow −	constant use	= ending volume +	spill
1	5.5	0.55	0.35	5.7	
2	5.7	0.85	0.85	5.7	
3	5.7	0.75	0.65	5.8	
4	5.8	1.05	0.65	6.2	
5	6.2	1.25	0.95	6.5	
6	6.5	1.05	0.65	6.9	
7	6.9	1.15	0.65	7.0	0.4
8	7.0	1.05	0.45	7.0	0.6
9	7.0	0.75	0.65	7.0	0.1
10	7.0	0.85	0.45	7.0	0.4
11	7.0	1.15	0.75	7.0	0.4
12	7.0	1.15	0.35	7.0	0.8

10. From Eq. 6.42,

$$E_{\text{reservoir}} = (0.7)(0.8) = \boxed{0.56 \text{ in}}$$

Timed

1.

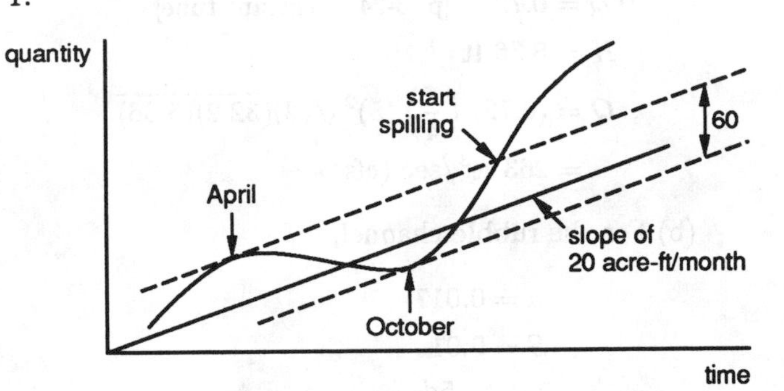

If the reservoir had been full in May, the maximum shortfall would have been 60 acre-ft, and the reservoir would be empty in October.

$$\boxed{60 \text{ acre-ft}}$$

2. (a)

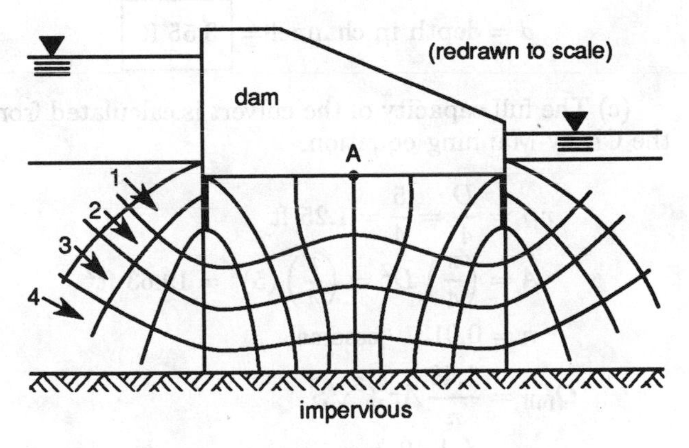

(b) no. of flow channels = 4
no. of equipotential drops = 16

$$H = 30 - 5 = 25 \text{ ft}$$

From Eq. 6.17,

$$Q = (0.05)(25)\left(\frac{4}{16}\right)$$

$$= \boxed{0.3125 \text{ ft}^3/\text{min-ft width}}$$

(c) The pressure at 35-ft depth (the depth of point A) is

$$p = \gamma h = \frac{(62.4)(35)}{144} = 15.17 \text{ lbf/in}^2 \text{ (psi)}$$

Each equipotential drop is

$$\Delta p = \frac{(62.4)(25)}{(144)(16)} = 0.677 \text{ lbf/in}^2 \text{ (psi)}$$

At point A (8 equipotential drops),

$$p = 15.17 - (8)(0.677) = \boxed{9.75 \text{ lbf/in}^2 \text{ (psi)}}$$

3. Since the first storm produces 1 inch net and the standard storm produces 2 inches, divide all runoffs by two. Offset the second storm by 4 hours.

hour	first storm	second storm	total
0	0		0
2	50		50
4	175	0	175
6	300	100	400
8	210	350	560
10	150	600	750
12	125	420	545
14	75	300	375
16	50	250	300
18	≈ 38	150	188
20	25	100	125
22	≈ 12	≈ 75	87
24	0	50	50
26		≈ 25	25
28		0	0

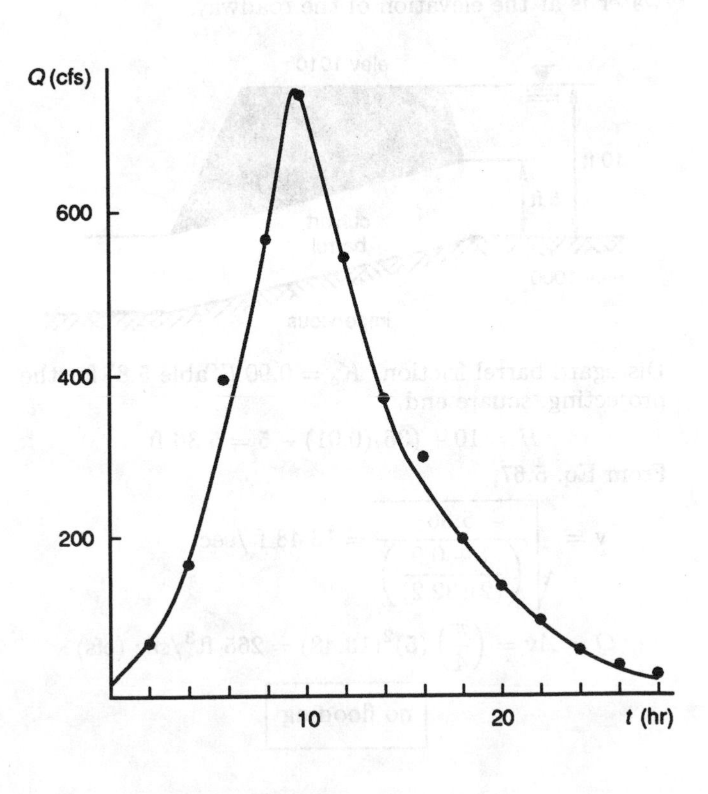

4. Draw the mass diagram.

$$\text{capacity} = \boxed{25\ \text{MG/mi}^2}$$

The spill starts near the end of September.

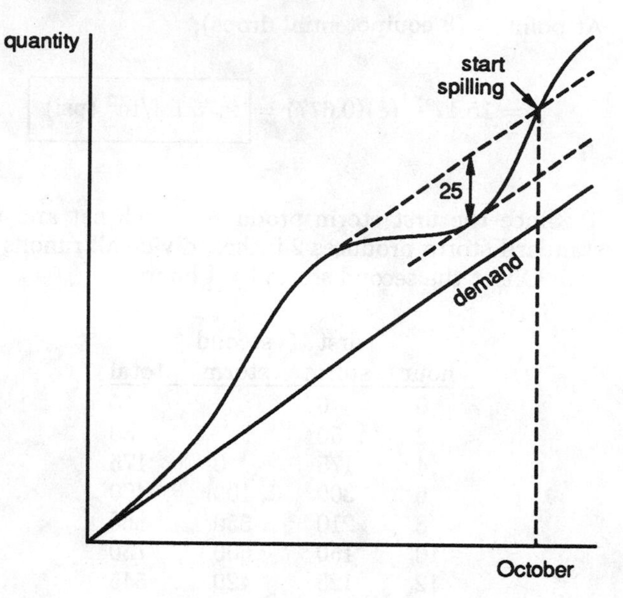

5. $I = 5.3$ in/hr

$$Q = CIA$$
$$= (0.5)(5.3)(75) = 198.8 \approx 200\ \text{ft}^3/\text{sec (cfs)}$$

(a) Take a worst-case approach. See if the culvert has a capacity equal to or greater than 200 ft^3/sec when water is at the elevation of the roadway.

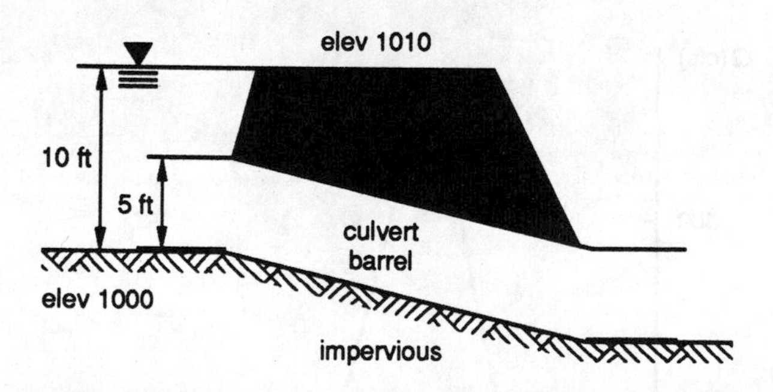

Disregard barrel friction. $K_e = 0.90$ (Table 5.8) for the projecting, square end.

$$H = 10 + (36)(0.01) - 5 = 5.36\ \text{ft}$$

From Eq. 5.67,

$$\text{v} = \sqrt{\frac{5.36}{\left(\frac{1+0.9}{(2)(32.2)}\right)}} = 13.48\ \text{ft/sec}$$

$$Q = A\text{v} = \left(\frac{\pi}{4}\right)(5)^2(13.48) = 265\ \text{ft}^3/\text{sec (cfs)}$$

$$\boxed{\text{no flooding}}$$

• Alternate method:

Use Eq. 5.74.

$$C_d = 0.72 \quad \text{[p. 3-24, reentrant tube]}$$
$$H = 5.36\ \text{ft}$$
$$Q = (0.72)\left(\frac{\pi}{4}\right)(5)^2\sqrt{(2)(32.2)(5.36)}$$
$$= 263\ \text{ft}^3/\text{sec (cfs)}$$

(b) For the rubble channel,

$$n = 0.017$$
$$S = 0.01$$
$$r_H = \frac{5d}{2d+5}$$
$$C = \left(\frac{1.49}{0.017}\right)\left(\frac{5d}{2d+5}\right)^{1/6}$$

$$Q = 200 = A\text{v} = (5d)\left(\frac{1.49}{0.017}\right)\left(\frac{5d}{2d+5}\right)^{0.667}\sqrt{0.01}$$
$$200 = (43.8d)\left(\frac{5d}{2d+5}\right)^{0.667}$$

By trial and error,

$$d = \text{depth in channel} = \boxed{3.55\ \text{ft}}$$

(c) The full capacity of the culvert is calculated from the Chezy-Manning equation.

$$r_H = \frac{D}{4} = \frac{5}{4} = 1.25\ \text{ft}$$
$$A = \left(\frac{\pi}{4}\right)D^2 = \left(\frac{\pi}{4}\right)(5)^2 = 19.63\ \text{ft}^2$$
$$n = 0.013 \quad \text{[assume]}$$
$$Q_{\text{full}} = \frac{1.49}{n}Ar_H^{2/3}\sqrt{S}$$
$$= \left(\frac{1.49}{0.013}\right)(19.63)(1.25)^{2/3}\sqrt{0.01}$$
$$= 261\ \text{ft}^3/\text{sec (cfs)}$$

This is close to (but not the same as) the 265 ft^3/sec calculated for the pressure flow.

$$\frac{Q}{Q_{\text{full}}} = \frac{200\ \frac{\text{ft}^3}{\text{sec}}}{261\ \frac{\text{ft}^3}{\text{sec}}} = 0.77$$

From p. 5-25 (assuming n varies),

$$\frac{d}{D} \approx 0.72$$

$$d_{\text{barrel}} = (0.72)(5\ \text{ft}) = \boxed{3.6\ \text{ft}}$$

6.

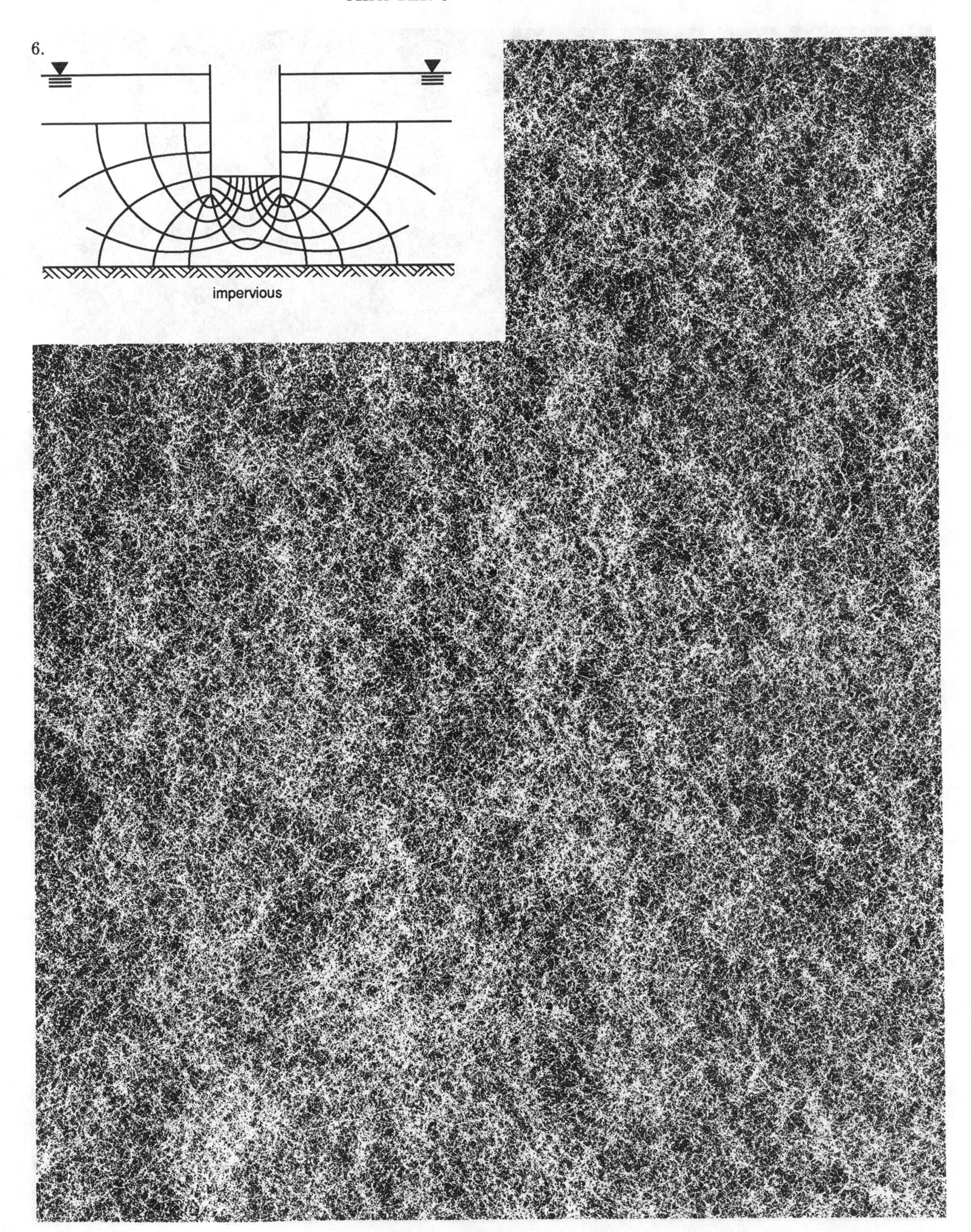

WATER SUPPLY ENGINEERING

Untimed

1. (a) Assume a flow rate of 4 gal/min-ft^2. The area is

$$A = \frac{600{,}000 \ \frac{\text{gal}}{\text{day}}}{\left(4 \ \frac{\text{gal}}{\text{min-ft}^2}\right)\left(24 \ \frac{\text{hr}}{\text{day}}\right)\left(60 \ \frac{\text{min}}{\text{hr}}\right)} = 104.2 \text{ ft}^2$$

$$\boxed{\text{use 10 ft} \times \text{10 ft}}$$

(b) Assume a rise of 24 in per minute. From Eq. 7.52, the volume is

$$V = (8 \text{ min})(100 \text{ ft}^2)\left(2 \ \frac{\text{ft}}{\text{min}}\right) = 1600 \text{ ft}^3$$

$$(1600 \text{ ft}^3)\left(7.481 \ \frac{\text{gal}}{\text{ft}^3}\right) = 11{,}970 \text{ gal}$$

$$\frac{11{,}970}{600{,}000} = \boxed{0.02 \ (2\%)}$$

2. (a) From Eq. 7.40, the surface area is

$$A_{\text{surface}} = \frac{2.8 \times 10^6 \ \frac{\text{gal}}{\text{day}}}{700 \ \frac{\text{gal}}{\text{ft}^2\text{-day}}} = 4000 \text{ ft}^2$$

Since $A = \left(\frac{\pi}{4}\right) D^2$,

$$D = \sqrt{\frac{(4)(4000)}{\pi}} = \boxed{71.4 \text{ ft}}$$

From Eq. 7.42, the volume is

$$V = \frac{\left(2.8 \times 10^6 \ \frac{\text{gal}}{\text{day}}\right)(2 \text{ hr})}{24 \ \frac{\text{hr}}{\text{day}}} = 2.333 \times 10^5 \text{ gal}$$

The depth is

$$d = \frac{V}{A_{\text{surface}}} = \frac{(2.333 \times 10^5 \text{ gal})\left(0.1337 \ \frac{\text{ft}^3}{\text{gal}}\right)}{4000 \text{ ft}^2}$$

$$= \boxed{7.8 \text{ ft}}$$

(b) From Eq. 7.40,

$$v^* = \frac{1.1 \times 10^6 \ \frac{\text{gal}}{\text{day}}}{4000 \text{ ft}^2} = \boxed{275 \text{ gal/day-ft}^2}$$

From Eq. 7.42,

$$t = \frac{(2.333 \times 10^5 \text{ gal})\left(24 \ \frac{\text{hr}}{\text{day}}\right)}{1.1 \times 10^6 \ \frac{\text{gal}}{\text{day}}} = \boxed{5.09 \text{ hr}}$$

3. (a) From Table 7.11, use 3 as the peak multiplier.

$$\frac{\left(110 \ \frac{\text{gal}}{\text{day}}\right)(15{,}000)(3)}{\left(24 \ \frac{\text{hr}}{\text{day}}\right)\left(60 \ \frac{\text{min}}{\text{hr}}\right)} = 3437.5 \text{ gal/min}$$

The fire fighting requirements are given by Eq. 7.26.

$$Q = (1020)\sqrt{15}\left[1 - (0.01)\sqrt{15}\right] = 3797.4 \text{ gal/min}$$

The total maximum demand for which the distribution system should be designed is

$$3437.5 + 3797.4 = \boxed{7234.9 \text{ gal/min}}$$

(b) The filter area should not be based on the maximum hourly rate since some of the demand during the peak hours can come from storage (clearwell or tanks). Also, fire requirements can bypass the filters if necessary. Therefore, base the design on a multiplier of 2 and disregard the fire requirements.

$$(110)(15{,}000)(2) = 3.3 \times 10^6 \text{ gal/day}$$

Using a flow rate of 4 gal per min/ft^2, the required filter area is

$$A = \frac{3.3 \times 10^6 \ \frac{\text{gal}}{\text{day}}}{\left(4 \ \frac{\text{gal}}{\text{min-ft}^2}\right)\left(24 \ \frac{\text{hr}}{\text{day}}\right)\left(60 \ \frac{\text{min}}{\text{day}}\right)} = \boxed{572.9 \text{ ft}^2}$$

(c) $\boxed{\text{No—60 ppm is soft water.}}$

(d) Disregarding the fire flow, the required avarage (not peak) daily chlorine mass is

$$\frac{\left(110 \ \frac{\text{gal}}{\text{day}}\right)(15{,}000)\left(2 \ \frac{\text{mg}}{l}\right)\left(8.345 \ \frac{\text{lbm-}l}{\text{MG-mg}}\right)}{10^6 \ \frac{\text{gal}}{\text{MG}}}$$

$$= \boxed{27.54 \text{ lbm/day}}$$

4. (a)

	as substance		factor from App. A		
Ca^{++}:	80.2	×	2.5	=	200.5
Mg^{++}:	24.3	×	4.1	=	99.63
Fe^{++}:	1	×	1.79	=	1.79
Al^{+++}:	0.5	×	5.56	=	2.78
			hardness	=	**304.7 mg/l**

(b) To remove the carbonate hardness,

$$CO_2: \quad 19 \times 2.27 = 43.13 \text{ mg}/l \text{ as } CaCO_3$$

Add lime to remove the carbonate hardness. It does not matter whether the HCO_3^- comes from Mg^{++}, Ca^{++}, or Fe^{++}; adding lime will remove it.

There may be Mg^{++}, Ca^{++}, or Fe^{++} ions left over in the form of noncarbonate hardness, but the problem asked for carbonate hardness.

$$HCO_3^-: \quad 185 \times 0.82 = 151.7 \text{ mg}/l$$

The total equivalents to be neutralized are

$$43.13 + 151.7 = 194.83 \text{ mg}/l$$

Convert $Ca(OH)_2$ using App. A.

$$\text{mg}/l \text{ of } Ca(OH)_2 = \frac{194.83}{1.35} = \boxed{144.3 \text{ mg}/l \text{ as substance}}$$

No soda ash is required since it is used to remove noncarbonate hardness.

5. (a) $Ca(HCO_3)_2$ and $MgSO_4$ both contribute to hardness. Since 100 mg/l of hardness is the goal, leave all $MgSO_4$ in the water. No soda ash is required. Take out $137 + 72 - 100 = 109$ mg/l of $Ca(HCO_3)_2$. From App. A (including the excess even though the reaction is not complete),

$$\text{pure } Ca(OH)_2 = 30 + \frac{109}{1.35} = 110.74 \text{ mg}/l$$

$$(110.74)(8.345) = \boxed{924 \text{ lbm/MG}}$$

(b) hardness removed $= 137 + 72 - 100 = 109$ mg/l

$$\left(109 \ \frac{\text{mg}}{l}\right)(8.345) = 909.6 \text{ lbm hardness/MG}$$

$$\left(\frac{0.5}{1000} \ \frac{\text{lbm}}{\text{grain}}\right)\left(909.6 \ \frac{\text{lbm}}{\text{MG}}\right)\left(7000 \ \frac{\text{grain}}{\text{lbm}}\right) = \boxed{3.18 \times 10^3 \text{ lbm/MG}}$$

6. (a) Assume a depth of 10 ft (p. 7-25) and a flow rate of 4 gal/min-ft².

$$A = \frac{1 \times 10^6 \ \frac{\text{gal}}{\text{day}}}{\left(24 \ \frac{\text{hr}}{\text{day}}\right)\left(60 \ \frac{\text{min}}{\text{hr}}\right)\left(4 \ \frac{\text{gal}}{\text{min-ft}^2}\right)} = 173.6 \text{ ft}^2$$

$$\text{width} = \sqrt{173.6} = 13.2 \text{ ft}$$

$$\boxed{13.2 \text{ ft} \times 13.2 \text{ ft} \times 10 \text{ ft}}$$

(b) Assume a rise rate of 2 ft/min. From Eq. 7.52, the water volume is

$$V = (5 \text{ min})(173.6 \text{ ft}^2)\left(2 \ \frac{\text{ft}}{\text{min}}\right)(5 \text{ filters}) = 8680 \text{ ft}^3$$

Using a conversion from p. 7-1,

$$(8680 \text{ ft}^3)(7.481) = 64{,}935 \text{ gal}$$

$$\frac{64{,}935}{5 \times 10^6} = \boxed{0.013 \ (1.3\%)}$$

7. (a) A bypass process is required.

$$\text{fraction bypassed: } \frac{80}{245} = 0.326$$

$$\text{fraction processed: } 1 - 0.326 = 0.674$$

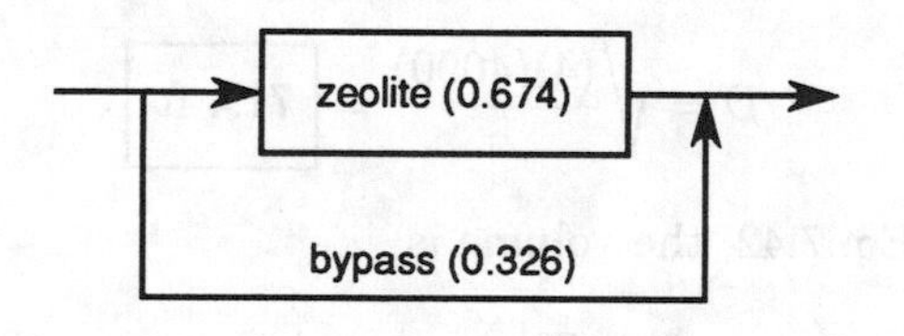

(b) The maximum hardness reduction is

$$\frac{(2 \text{ ft}^3)\left(20{,}000 \ \frac{\text{gr}}{\text{ft}^3}\right)}{7000 \ \frac{\text{gr}}{\text{lbm}}} = 5.71 \text{ lbm}$$

The hardness removal rate is

$$\frac{(0.674)\left(245 \ \frac{\text{mg}}{l}\right)\left(8.345 \times 10^{-6} \ \frac{\text{lbm}}{\text{gal}}\right)\left(20{,}000 \ \frac{\text{gal}}{\text{day}}\right)}{24 \ \frac{\text{hr}}{\text{day}}} = 1.15 \text{ lbm/hr}$$

$$t = \frac{5.71 \text{ lbm}}{1.15 \ \frac{\text{lbm}}{\text{hr}}} = \boxed{4.97 \text{ hr}}$$

8. From Eq. 7.35,

$$\text{crushing strength} = \left(1500\ \frac{\text{lbf}}{\text{ft}}\right)\left(\frac{12\ \text{in}}{12\ \frac{\text{in}}{\text{ft}}}\right) = 1500\ \text{lbf}$$

Assume the trench width is

$$B = \frac{\left(\frac{4}{3}\right)(12) + 8}{12\ \frac{\text{in}}{\text{ft}}} = 2\ \text{ft}$$

$$\frac{h}{B} = \frac{11}{2} = 5.5 \quad \text{[round to 6.0]}$$

From Table 7.15 for the $\gamma = 120$ lbf/ft^3 ($\rho = 120$ pcf) column, $C = 3.04$. (The load on the pipe is more important than the soil type.)

From Eq. 7.27, the actual load is

$$w = (3.04)(120)(2)^2 = 1459.2\ \text{lbf/ft}$$

(This disregards any live loads.)

Solve Eq. 7.36 for the load factor.

$$\text{load factor} = \frac{(1459.2)(1.5)}{1500} = 1.46$$

Select Class C or better from Fig. 7.5.

9.
$$v^* = \frac{\left(100{,}000\ \frac{\text{gal}}{\text{ft}^2\text{-day}}\right)\left(0.1337\ \frac{\text{ft}^3}{\text{gal}}\right)}{\left(24\ \frac{\text{hr}}{\text{day}}\right)\left(60\ \frac{\text{min}}{\text{hr}}\right)} = 9.28\ \text{ft/min}$$

Using interpolation,

$$0.45 + \left(\frac{9.28 - 5.0}{10.0 - 5.0}\right)(0.54 - 0.45) = 0.527 \text{ remains in flow}$$

$$1 - 0.527 = 0.473$$

47.3% removed

10. From Fig. 7.6,

$$v_s = 0.7\ \text{ft/sec}$$

Timed

1. For 60°F water,

$$\mu = 2.359 \times 10^{-5}\ \text{lbf-sec/ft}^2$$

From Eq. 7.50,

$$P = \left(2.359 \times 10^{-5}\ \frac{\text{lbf-sec}}{\text{ft}^2}\right)\left(45\ \frac{1}{\text{sec}}\right)^2 (200{,}000\ \text{ft}^3)$$
$$= 9554\ \text{ft-lbf/sec}$$

(c) $$\text{water horsepower} = \frac{9554}{550} = 17.4\ \text{hp}$$

(b) From Eq. 7.47,

$$D = \frac{P}{v} = \frac{9554\ \frac{\text{ft-lbf}}{\text{sec}}}{1.5\ \frac{\text{ft}}{\text{sec}}} = 6369\ \text{lbf}$$

(a) From Eq. 7.46,

$$A = \frac{(2)(32.2)(6369)}{(1.75)(62.4)(1.5)^2} = 1669\ \text{ft}^2$$

2. (a) The daily demand is

200 gal

(b)

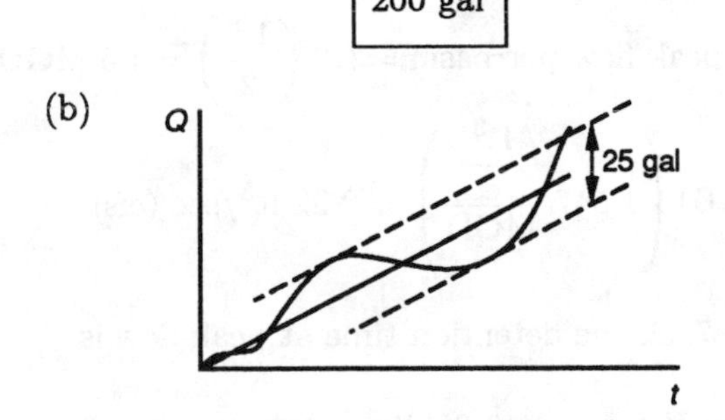

Population use = (25)(40,000) = 1,000,000 gal/day.

From Eq. 7.26, the fire fighting requirement is

$$Q = (1020)\sqrt{40}\left[1 - (0.01)\sqrt{40}\right] = 6043\ \text{gal/min}$$

The flow rate is 6043/1000 = 6 thousands of gallons, so maintain the flow for 6 hours.

$$\text{capacity} = 1{,}000{,}000 + (6043)(60)(6)$$
$$= 3{,}175{,}000\ \text{gal}$$

(c) The pump supplies demand from 4 a.m. until 8 a.m. The storage supplies demand from 8 a.m. until midnight and from midnight until 4 a.m.

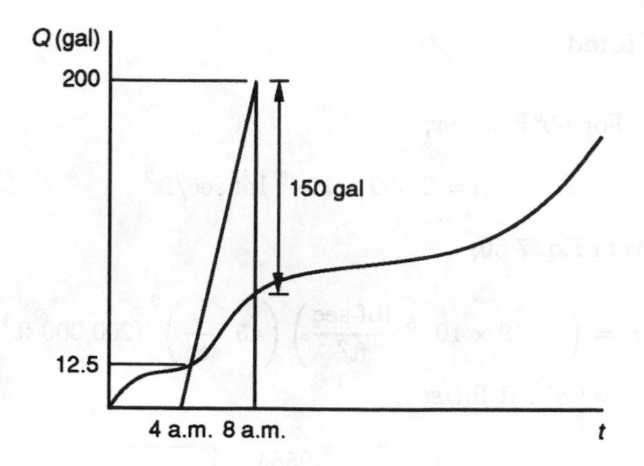

The population use is

$$(150 + 12.5)(40{,}000) = 6{,}500{,}000 \text{ gal}$$

Now add fire fighting.

$$6{,}500{,}000 + (6043)(60)(6) = \boxed{8{,}675{,}500}$$

3. Assume there are two basins in parallel.

$$\text{volume per basin} = (90)(16)(12) = 17{,}280 \text{ ft}^3$$

(The freeboard is not given.)

The area per basin is

$$A = (90)(16) = 1440 \text{ ft}^2$$

$$\text{3-month peak flow per basin} = (2)\left(\frac{1.5}{2}\right) = 1.5 \text{ MGD}$$

$$(1.5 \text{ MGD})\left(1.547 \frac{\frac{\text{ft}^3}{\text{sec}}}{\text{MGD}}\right) = 2.32 \text{ ft}^3/\text{sec (cfs)}$$

From Eq. 7.42, the detention time at peak flow is

$$t = \frac{V}{Q} = \frac{17{,}280 \text{ ft}^3}{(2.32)(60)\left(60 \frac{\text{ft}^3}{\text{hr}}\right)} = 2.07 \text{ hr}$$

Since $2.07 < 4$, this is not acceptable.

The weir loading is

$$\frac{1.5 \times 10^6 \frac{\text{gal}}{\text{day}}}{48 \text{ ft}} = 31{,}250 \text{ gal/day-ft}$$

Since $31{,}250 > 20{,}000$, this does not work either.

From Eq. 7.40, the overflow rate is

$$v^* = \frac{Q}{A} = \frac{(2.32)(60)}{1440} = 0.0967 \text{ ft/min}$$

Since $0.0967 < 0.5$, this is acceptable.

At low flow,

$$\text{flow} = \frac{0.7}{2.0} = 0.35 \quad \text{[35\% of high flow]}$$

$$t = \frac{2.07}{0.35} = 5.91 \text{ hr} \quad \text{[acceptable]}$$

$$\text{weir loading} = (0.35)(31{,}250) = 10{,}938 \quad \text{[acceptable]}$$

$$v^* = (0.35)(0.1) = 0.035 \quad \text{[acceptable]}$$

The basins have been correctly designed for low flow but not for peak flow. One or more basins should be used.

4. (a) The distance between points A and B is

$$\sqrt{(300)^2 + (1500)^2} = 1529.7 \text{ ft} \quad \text{[use 1530 ft]}$$

The difference in water table elevations is

$$229.75 - 223.25 = 6.5 \text{ ft}$$

$$\text{gradient} = \frac{6.5}{1530} = \boxed{4.25 \times 10^{-3} \text{ ft/ft}}$$

(b)

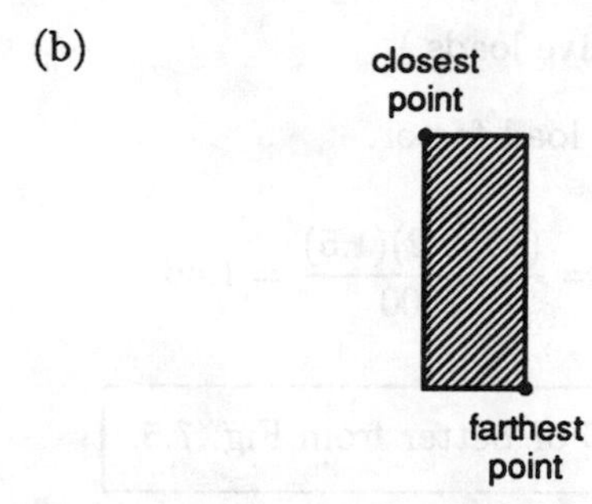

The distance to the borehole at the closest point is

$$\sqrt{(100)^2 + (500)^2} = 509.9 \quad \text{[use 510 ft]}$$

$$\text{minimum elevation} = 229.75 + 5 - (510)(4.25 \times 10^{-3})$$

$$= \boxed{232.6 \text{ ft}}$$

(c) From Eq. 9.23 (Darcy's law),

$$v = \frac{(3 \times 10^{-5})(4.25 \times 10^{-3})}{0.4} = 3.19 \times 10^{-7} \text{ ft/sec}$$

Assume the leak occurs at the farthest point.

$$\text{time} = \frac{510}{3.19 \times 10^{-7}} = 1.6 \times 10^9 \text{ sec}$$

$$\frac{1.6 \times 10^9}{(60)(60)(24)(365)} = \boxed{50.7 \text{ years}}$$

This assumes sandy, homogeneous soil.

WASTE-WATER ENGINEERING

Untimed

1. (a) From Eq. 8.12,

$$Q = \frac{(0.17)\left(\frac{20{,}000}{1000}\right)}{(8.345 \times 10^{-9})(300)} = \boxed{1.35 \times 10^6 \text{ gal/day}}$$

(*Ten-States' Standards* specify 100 gpcd in the absence of other information. In that case, $Q = (100)(20{,}000) = 2 \times 10^6$ gpcd.)

(b) The BOD load leaving the primary clarifier and entering the filter is

$$\text{BOD}_i = (1 - 0.30)(300) = 210 \text{ mg}/l$$

$$L_{\text{BOD}} = (1.35 \text{ MGD})\left(210 \ \frac{\text{mg}}{l}\right)\left(8.345 \ \frac{\text{lbm-}l}{\text{MG-mg}}\right) = 2366 \text{ lbm/day}$$

Equation 8.30 shows that if η is known, either L_{BOD} or F can be found. Since neither is known here, one must be assumed.

$$L_{\text{BOD}} = 60 \text{ lbm/1000 ft}^3\text{-day}$$

$$\frac{\left(2366 \ \frac{\text{lbm}}{\text{day}}\right)(1000)}{60 \ \frac{\text{lbm}}{\text{ft}^3\text{-day}}} = 39{,}433 \text{ ft}^3$$

$\boxed{\text{Assume a depth of 5 ft.}}$

The surface area is

$$A_s = \frac{39{,}433 \text{ ft}^3}{5 \text{ ft}} = 7887 \text{ ft}^2$$

The required diameter is

$$D = \sqrt{\frac{(4)(7887)}{\pi}} = \boxed{100.2 \text{ ft}}$$

(c) The efficiency of the filter and secondary clarifier is

$$\eta = \frac{210 - 50}{210} = 0.762 \ (76.2\%)$$

From Eq. 8.30,

$$0.762 = \frac{1}{1 + 0.0561\sqrt{\frac{60}{F}}}$$

$$F = 1.936$$

(Equation 8.34 also could have been used to find F.)

From Eq. 8.33,

$$1.936 = \frac{1 + R_R}{[1 + (0.1)(R_R)]^2}$$

$$R_R = 1.6$$

$$Q_R = (1.6)(1.35) = \boxed{2.16 \text{ MGD}}$$

(d)

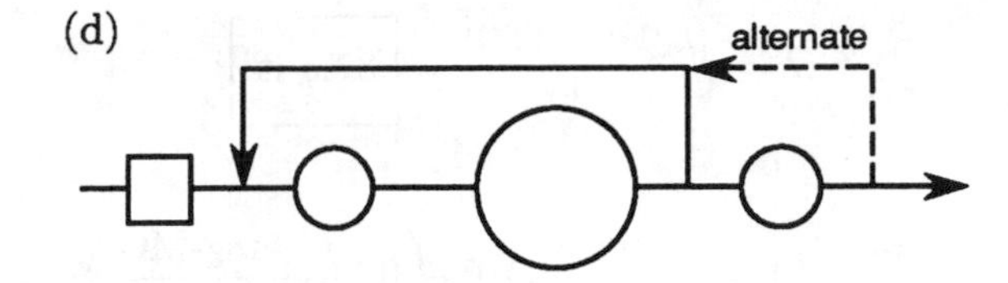

$$\text{(e) } \eta = \frac{300 - 50}{300} = \boxed{0.833 \ (83.3\%)}$$

2. Disregarding variations in peak flow, the average design volume is

$$\frac{(125 \text{ gpcd})(20{,}000)}{1{,}000{,}000} = 25 \text{ MGD}$$

(a) The settling tank surface area is

$$\frac{2.5 \times 10^6 \ \frac{\text{gal}}{\text{day}}}{1000 \ \frac{\text{gal}}{\text{day-ft}^2}} = \boxed{2500 \text{ ft}^2}$$

The required diameter when using two tanks in parallel is

$$D = \sqrt{\frac{(4)(2500)}{2\pi}} = \boxed{39.9 \text{ ft (use 40 ft) each}}$$

(b) Estimate 30%, or

$$(0.30)(250) = 75 \text{ mg}/l$$

BOD entering the filter is

$$250 - 75 = \boxed{175 \text{ mg}/l}$$

(c) From Eq. 8.32, the BOD loading is

$$L_{\text{BOD}} = (317.7)\left(\frac{50}{175 - 50}\right)^2 = 50.8 \text{ lbm/1000 ft}^3\text{-day}$$

The actual load is

$$(2.5 \text{ MGD})\left(175 \ \frac{\text{mg}}{l}\right)\left(8.345 \ \frac{\text{lbm-}l}{\text{MG-mg}}\right) = 3651 \text{ lbm/day}$$

The required volume is

$$\frac{\left(3651\ \frac{\text{lbm}}{\text{day}}\right)(1000)}{50.8\ \frac{\text{lbm}}{1000\ \text{ft}^3\text{-day}}} = 7.19\times 10^4\ \text{ft}^3$$

With a depth of 6 ft, the required surface area is

$$\frac{7.19\times 10^4}{6} = 11{,}980\ \text{ft}^2$$

If two filters are used in parallel, the required diameter is

$$D = \sqrt{\frac{(4)(11{,}980)}{2\pi}} = \boxed{87.3\ \text{ft}}$$

3. (a)

$$\frac{\left(0.191\ \frac{\text{lbm}}{\text{cd}}\right)(10^6)\left(0.1198\ \frac{\text{mg-MG}}{l\text{-lbm}}\right)}{122\ \frac{\text{gal}}{\text{cd}}} = \boxed{187.6\ \text{mg}/l \text{ at } 20°\text{C}}$$

(b)

$$\frac{(122)(10{,}000)(187.6) + (116{,}000)(0) + (180{,}000)(800) + (76{,}000)(1700)}{(122)(10{,}000) + 116{,}000 + 180{,}000 + 76{,}000} = \boxed{315.4\ \text{mg}/l \text{ at } 20°\text{C}}$$

(c)

$$\frac{(122)(10{,}000)(64) + (116{,}000)(51) + (180{,}000)(95) + (76{,}000)(84)}{(122)(10{,}000) + 116{,}000 + 180{,}000 + 76{,}000} = \boxed{67.5°\text{F}\ (19.7°\text{C})}$$

(e) The total discharge into the river is

$$[(122)(10{,}000) + 116{,}000 + 180{,}000 + 76{,}000] \times (1.547\times 10^{-6}) = \boxed{2.46\ \text{ft}^3/\text{sec (cfs)}}$$

step 1: Find the stream conditions immediately after mixing. Assume $\text{BOD}_{\text{stream}} = 0$.

$$\text{BOD}_5 = \frac{(2.46)(315.4) + (120)(0)}{2.46 + 120} = 6.34\ \text{mg}/l$$

Assume $\text{DO}_{\text{sewage}} = 0$.

$$\text{DO} = \frac{(2.46)(0) + (120)(7.5)}{2.46 + 120} = 7.35\ \text{mg}/l$$

$$T = \frac{(2.46)(19.7) + (120)(15)}{2.46 + 120} = 15.1°\text{C}$$

step 2: Calculate the rate constants at 15.1°C.

$$K_{D,15.1°\text{C}} = (1.047)^{15.1-20}(0.1) = 0.0798$$

$$K_{R,15.1°\text{C}} = (1.016)^{15.1-20}(0.2) = 0.185$$

step 3: Estimate BOD_u.

$$\text{BOD}_u = \frac{6.34}{1 - 10^{-(0.0798)(5)}} = 10.55\ \text{mg}/l$$

step 4: From App. B at 15°C, saturated DO = 10.2 mg/l. Since the actual is 7.35, the deficit is

$$D_o = 10.2 - 7.35 = 2.85\ \text{mg}/l$$

step 5: Calculate t_c from Eq. 8.24.

$$t_c = \left(\frac{1}{0.185 - 0.0798}\right) \times \log_{10}\left[\frac{(0.0798)(10.55) - (0.185)(2.85) + (0.0798)(2.85)}{(0.0798)(10.55)} \times \left(\frac{0.185}{0.0798}\right)\right] = 1.65\ \text{days}$$

step 6: The distance downstream is

$$(1.65\ \text{days})(3\ \text{mph})\left(24\ \frac{\text{hr}}{\text{day}}\right) = \boxed{118.8\ \text{mi}}$$

(d)

step 7: From Eq. 8.25,

$$D_c = \left[\frac{(0.0798)(10.55)}{0.185}\right](10)^{-(0.0798)(1.65)} = 3.36\ \text{mg}/l$$

step 8: $\text{DO}_{\text{min}} = 10.2 - 3.36 = \boxed{6.84\ \text{mg}/l}$

4. (a) Use Eq. 8.12 modified for the 0.18 lbm/cd standard. Do not apply the population expansion factor to the industrial effluents.

$$P_e = (20)(1.15) + \frac{(1100)(1.3\times 10^6)(8.345\times 10^{-9})}{0.18} + \frac{(500)(1.0\times 10^6)(8.345\times 10^{-9})}{0.18} = \boxed{112.5\ \text{[thousands of people]}}$$

(b) Since the plant loading is requested, the organic loading can be given in lbm/day. Compare to Eq. 8.29 for the filter loading.

$$L_{\text{BOD}} = (112{,}500)(0.18) = \boxed{20{,}250\ \text{lbm/day}}$$

(c) Eq. 8.27 cannot be used because the filter area is unknown.

$$\begin{aligned} L_H &= (20{,}000)(1.15)(100 \text{ gpcd}) \\ &\quad + 1.3\times 10^6 + 1.0\times 10^6 \\ &= \boxed{4.6\times 10^6 \text{ gal/day}} \end{aligned}$$

5. (a) and (b) From p. 8-17, there is nothing particularly special about a basin that removes 30% BOD. Choose two basins in parallel, each working with half of the total flow. Choose an overflow rate of 1000 gpd/ft². The area per basin is

$$A = \frac{4.4\times 10^6 \text{ gpd}}{(2)\left(1000\ \frac{\text{gpd}}{\text{ft}^2}\right)} = 2200 \text{ ft}^2$$

$$\text{diameter} = \sqrt{\left(\tfrac{4}{\pi}\right)(2200)} = 52.9 \text{ ft}$$

Assume an 8-ft depth. The detention time is

$$t = \frac{(2200)(8 \text{ ft}^3)}{\left(2.2\ \frac{\text{MGD}}{\text{tank}}\right)\left(1.547\ \frac{\frac{\text{ft}^3}{\text{sec}}}{\text{MGD}}\right)\left(3600\ \frac{\text{sec}}{\text{hr}}\right)} = 1.436 \text{ hr}$$

This is close to the 1.5–2.0 hr guideline given on p. 8-17. (If the detention time had not been satisfactory, the depth could have been changed.)

Check the weir loading.

$$\text{circumference} = \pi(52.9) = 166.2 \text{ ft}$$

$$\text{weir loading} = \frac{2.2\times 10^6}{166.2} = 13{,}240 \text{ gpd/ft}$$

This is within the 10,000–20,000 gpd/ft guideline given on p. 8-17.

> Choose two circular tanks that are 53 ft in diameter and 8 ft deep.

(c) With two basins (always the minimum number), the basin area is still 2200 ft² each.

$$\text{side length} = \sqrt{2200} = 46.9 \text{ ft}$$

Assume an 8-ft depth. The detention time is the same as was calculated in part (a).

$$t = \frac{(2200)(8)}{(2.2)(1.547)(3600)} = 1.436 \text{ hr} \quad \text{[acceptable]}$$

The weir loading is

$$\frac{2.2\times 10^6}{(4)(46.9)} = 11{,}730 \text{ gpd/ft} \quad \text{[acceptable]}$$

(d)

> Choose two square basins, each 47 ft × 47 ft × 8 ft deep.

(e) If the first stage of treatment removes 30% BOD, then the BOD entering the filter is

$$\begin{aligned} &(4.4 \text{ MGD})\left(160\ \frac{\text{mg}}{l}\right)\left(8.345\ \frac{\text{lbm-}l}{\text{MG-mg}}\right)(0.70) \\ &\quad = 4112.4 \text{ lbm/day} \end{aligned}$$

The allowable BOD per day is

$$(4.4)(20)(8.345) = 734.4 \text{ lbm/day}$$

The required efficiency of the second stage filter and basin combination is

$$1 - \frac{734.4}{4112.4} = 0.82\ (82\%)$$

$$\text{BOD}_{\text{in}} = (160)(1 - 0.30) = 112 \text{ mg}/l$$

From Eq. 8.32,

$$L_{\text{BOD}} = (317.7)\left(\frac{20}{112 - 20}\right)^2 = 15 \text{ lbm/1000 ft}^3\text{-day}$$

Use Eq. 8.30 to solve for the effective number of passes through the filter.

$$0.82 = \frac{1}{1 + 0.0561\sqrt{\frac{15}{F}}}$$

$$F = 0.98 \quad \text{[no recirculation required]}$$

Try a depth of 6 ft (typical). The required surface area is

$$A = \frac{\left(4112.4\ \frac{\text{lbm}}{\text{day}}\right)\left(1000\ \frac{\text{ft}^3}{1000 \text{ ft}^3}\right)}{\left(15\ \frac{\text{lbm}}{1000 \text{ ft}^3\text{-day}}\right)(6 \text{ ft})} = 45{,}690 \text{ ft}^2$$

Using two filters in parallel, the required diameter is

$$D = \sqrt{\left(\frac{4}{\pi}\right)\left(\frac{45{,}690}{2}\right)} = 170.5 \text{ ft}$$

(If this had been too large, the depth could have been increased.)

Check the hydraulic loading using Eq. 8.27.

$$L_H = \frac{4.4 \times 10^6 \text{ gpd}}{45{,}690 \text{ ft}} = 96.3 \text{ gpd/ft}$$

Since this is less than 100, it is acceptable.

filter	
depth:	6 ft
diameter:	171 ft
recirculation:	0

The design steps for the sedimentation basin are the same as for the primary basin, so

basin	
depth:	8 ft
diameter:	53 ft

6. (a) The total BOD is

$$\left(1000 \ \frac{\text{mg}}{l}\right)\left(8.345 \ \frac{\text{lbm-}l}{\text{MG-mg}}\right)\left(\frac{10{,}000 \text{ MG}}{1{,}000{,}000 \text{ day}}\right)$$
$$+ (250)(8.345)\left(\frac{25{,}000}{1{,}000{,}000}\right) = 135.6 \text{ lbm/day}$$

From p. 8-21, choose 50 lbm BOD/acre. The required area is

$$\frac{135.6}{50} = \boxed{2.7 \text{ acres}}$$

(b) From Eq. 7.42,

$$t = \frac{(2.7 \text{ acres})\left(43{,}560 \ \frac{\text{ft}^2}{\text{acre}}\right)(4 \text{ ft})}{(35{,}000 \text{ gpd})\left(1.547 \times 10^{-6} \ \frac{\text{ft}^3}{\text{sec-gpd}}\right)\left(3600 \ \frac{\text{sec}}{\text{hr}}\right)}$$
$$= \boxed{2414 \text{ hr} \ (14.4 \text{ weeks})}$$

7. This problem is more complex than it first appears because the BOD and SS are not mutually exclusive. Some of the SS is organic in nature, and this shows up as BOD. Make the following assumptions.

- 30% of the SS is organic material already in the BOD, and should be treated as BOD.
- 70% of the SS is nonorganic material.

$$SS' = (0.70)(225) \approx 158 \text{ mg}/l$$

- The final SS concentration of 20 mg/l is all inorganic.
- Aerobic digestion in the primary clarifier is insignificant. BOD reduction in the primary settling tank does not contribute to sludge production.

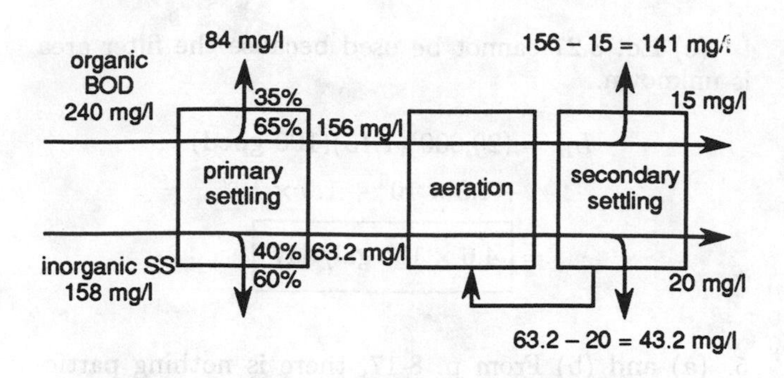

(a) The total dry weight of SS removed in all processes is

$$\left(158 - 20 \ \frac{\text{mg}}{l}\right)\left(8.345 \ \frac{\text{lbm-}l}{\text{MG-mg}}\right)\left(10 \ \frac{\text{MG}}{\text{day}}\right)$$
$$= 11{,}516 \text{ lbm/day}$$

The BOD entering the secondary process is

$$(1 - 0.35)\left(240 \ \frac{\text{mg}}{l}\right) = 156 \text{ mg}/l$$

Solids from BOD reduction in the secondary process are given by Eq. 8.57.

$$k(\Delta\text{BOD})Q = (0.60)\left(156 - 15 \ \frac{\text{mg}}{l}\right) \times \left(10 \ \frac{\text{MG}}{\text{day}}\right)\left(8.345 \ \frac{\text{lbm-}l}{\text{MG-mg}}\right)$$
$$= 7060 \text{ lbm/day}$$

The total dry sludge mass is

$$11{,}516 + 7060 = \boxed{18{,}576 \text{ lbm/day}}$$

(b) From Eq. 8.59,

$$\frac{1}{\text{SG}} = \frac{1-s}{1} + \frac{s}{(\text{SG})_{\text{solids}}}$$
$$\frac{1}{1.02} = \frac{1-0.06}{1} + \frac{0.06}{(\text{SG})_{\text{solids}}}$$
$$(\text{SG})_{\text{solids}} = 1.485$$

The density of the solids is

$$\rho_{\text{solids}} = (1.485)\left(62.4 \ \frac{\text{lbm}}{\text{ft}^3}\right) = 92.66 \text{ lbm/ft}^3$$

The solid dry sludge volume is

$$V_{\text{solid}} = \frac{m}{\rho} = \frac{18{,}576 \ \frac{\text{lbm}}{\text{day}}}{92.66 \ \frac{\text{lbm}}{\text{ft}^3}} = \boxed{200.5 \text{ ft}^3/\text{day}}$$

Although the sludge is completely dry, it would not be compressed solid. Furthermore, commercial drying processes would leave 60–80% (estimate 70%) moisture content. Finally, air voids (approximately 10%) would be present in the dried sludge. In this case, the disposal volume would be

$$V_t = V_{\text{solid}} + V_{\text{water}}$$

$$V_{\text{solid}} = \frac{m_{\text{solid}}}{\rho}$$

$$V_{\text{water}} = \frac{m_{\text{water}}}{62.4} = \frac{0.70 m_t}{62.4} = \frac{0.70 m_{\text{solids}}}{(62.4)(1-0.70)}$$

$$V \approx (1.10)\left[\frac{18{,}576}{92.7} + \frac{(18{,}576)(0.70)}{(62.4)(1-0.70)}\right] = \boxed{984.5 \text{ ft}^3/\text{day}}$$

Timed

1. (a)

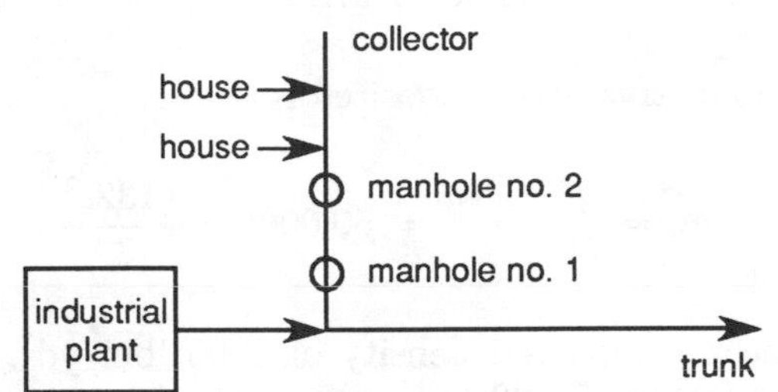

If the first manhole overflows, the piezometric head at the rim must be less than the head in the the trunk. Further up at manhole no. 2, the increase in elevation is sufficient to raise rim no. 2, so the head in the trunk must be lowered.

Alternate solutions:

- relief storage (surge chambers) between trunk and manhole no. 1
- storage at plant and gradual release
- private trunk for plant
- private treatment and discharge for plant
- larger trunk capacity using larger or parallel pipes
- backflow preventors

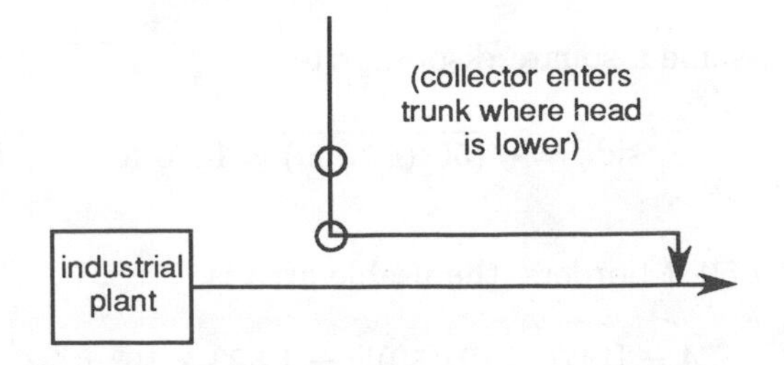

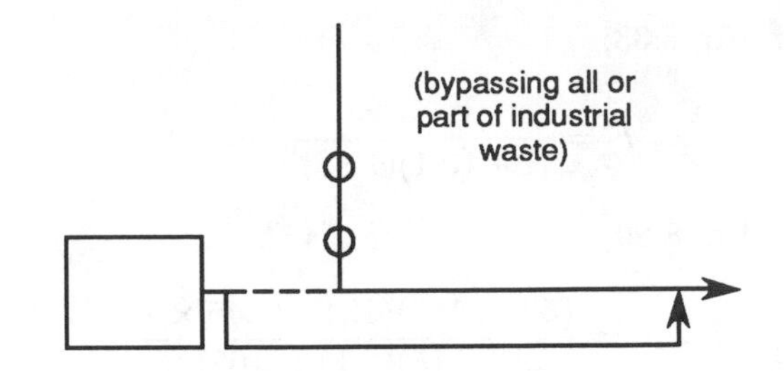

2. (NRC is the acronym for the National Research Council.)

The total efficiency required is

$$\frac{320-28}{320} = 0.9125 \ (91.25\%)$$

The primary clarifier removes 30% BOD.

$$\text{BOD}_{\text{out}} = (1-0.3)(320) = 224 \text{ mg}/l$$

Assume the first trickle filter has an efficiency of η = 65%. (This will be the only assumption for this problem.)

$$\text{BOD}_{\text{out}} = (1-0.65)(224) = 78.4 \text{ mg}/l$$

From Eq. 8.32,

$$L_{\text{BOD}} = (317.7)\left(\frac{78.4}{224-78.4}\right)^2 = 92.1 \text{ lbm}/1000 \text{ ft}^3\text{-day}$$

From Eq. 8.29, using volume = depth × area,

$$92.1 = \frac{(5.2\times10^6)(224)(8.345\times10^{-3})}{(5)(\text{area})}$$

$$\text{area} = 2.11\times10^4 \text{ ft}^2$$

$$\text{diameter} = \sqrt{\frac{4A}{\pi}} = \sqrt{\frac{(4)(2.11\times10^4 \text{ ft}^2)}{\pi}} = 163.9 \text{ ft} \quad [\text{use 165 ft}]$$

Check the efficiency assumption. Use Eq. 8.27 to find R_R.

$$L_H = (32 \text{ mgad})\left(22.96 \ \frac{\frac{\text{gpd}}{\text{ft}^2}}{\text{mgad}}\right) = 734.72 \text{ gpd/ft}^2$$

From Eq. 8.27,

$$734.72 = \frac{5.2\times10^6 + Q_R}{2.11\times10^4}$$

$$Q_R = 1.03\times10^7 \text{ MGD}$$

From Eq. 8.28,

$$R_R = \frac{Q_R}{Q_W} = \frac{1.03\times10^7}{5.2\times10^6} = 1.98$$

From Eq. 8.33,

$$F = \frac{1 + 1.98}{[1 + (0.1)(1.98)]^2} = 2.08$$

From Eq. 8.29,

$$L_{\text{BOD}} = \frac{(5.2 \times 10^6)(224)(8.345 \times 10^{-3})}{(5)(2.11 \times 10^4)} = 92.1 \text{ lbm/1000 ft}^3\text{-day}$$

From Eq. 8.30,

$$\eta = \frac{1}{1 + 0.0561\sqrt{\frac{92.1}{2.08}}} = 0.728 \ (72.8\%) \quad \text{[acceptable]}$$

Continue using $\eta = 0.73$. The BOD out of the first filter is

$$\text{BOD}_{\text{out}} = (1 - 0.728)(224) = 60.9 \text{ mg}/l$$

The true BOD loading for the second filter is given by Eq. 8.29.

$$L_{\text{BOD},2} = \frac{(5.2 \times 10^6)(60.9)(8.345 \times 10^{-3})}{(5)(2.11 \times 10^4)} = 25.0 \text{ lbm/1000 ft}^3\text{-day}$$

This assumes food is abundant (which it is not). From Eq. 8.36,

$$L_{\text{BOD},2} = \frac{25.0}{(1 - 0.728)^2} = 337.9 \text{ lbm/1000 ft}^3\text{-day}$$

From Eq. 8.30, using the same value of F (since R_R, L_H, and A are unchanged),

$$\eta = \frac{1}{1 + 0.0561\sqrt{\frac{337.9}{2.08}}} = 0.583 \ (58.3\%)$$

$$\text{final BOD} = (1 - 0.583)(60.9) = \boxed{25.4 \text{ mg}/l}$$

two basins	
depth:	5 ft
diameter:	165 ft

3. (a) and (b) Refer to p. 8-30.

- availability of land
- ease of access
- availability of cover soil if used
- wind speed and direction
- flat topology
- dry climate
- low risk of aquifer contamination
- impermeable lower strata
- temperate climate

(c) Assume a 20-year useful life and linear population growth. Find the rate of increase of waste production.

The mass of waste deposited in the first day will be

$$(10{,}000 \text{ people})\left(5 \ \frac{\text{lbm}}{\text{person-day}}\right) = 50{,}000 \text{ lbm/day}$$

The mass of waste deposited on the last day will be

$$(20{,}000)(5) = 100{,}000 \text{ lbm/day}$$

The increase rate is

$$\frac{\Delta m}{\Delta t} = \frac{100{,}000 - 50{,}000}{(15 \text{ years})\left(365 \ \frac{\text{days}}{\text{year}}\right)} = 9.132 \text{ lbm/day}$$

The mass deposited on day D is

$$m_D = 50{,}000 + (9.132)(D - 1) \approx 50{,}000 + 9.132D$$

The cumulative mass deposited is

$$m_t = \int_0^t m_D \, dt = 50{,}000t + \frac{9.132t^2}{2}$$

Assume a compacted density of 1000 lbm/yd^3 and a loading factor of 1.00 (no soil cover). The capacity of the site with a 6 ft lift is

$$m_{\text{max}} = \frac{(30 \text{ acres})\left(43{,}560 \ \frac{\text{ft}^2}{\text{acre}}\right)(6 \text{ ft})\left(1000 \ \frac{\text{lbm}}{\text{yd}^3}\right)}{27 \ \frac{\text{ft}^3}{\text{yd}}} = 2.9 \times 10^8 \text{ lbm}$$

The time to fill is found by solving the quadratic equation.

$$2.9 \times 10^8 = 50{,}000t + \frac{9.132t^2}{2}$$

$$t^2 + 10{,}951t = 6.351 \times 10^7$$

$$t = \boxed{4193 \text{ days}}$$

4. Assume a square disposal site.

$$\text{side} = \sqrt{(50)(43{,}560)} = 1476 \text{ ft}$$

With 50-ft borders, the usable area is

$$A = [1476 - (2)(50)]^2 = 1.893 \times 10^6 \text{ ft}^2$$

(c) If the site is excavated 20 ft, 10 ft of soil is used as cover, and the maximum above-ground height is 20 ft, the service capacity of compacted waste is

$$(1.893 \times 10^6)(30) = \boxed{5.68 \times 10^7 \text{ ft}^3}$$

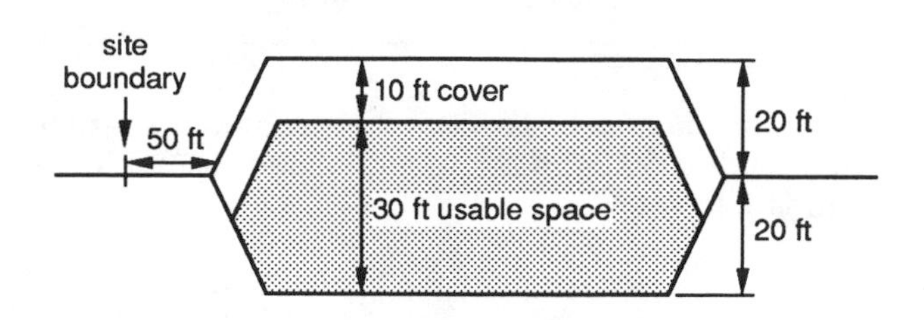

(not to scale or representative of actual construction)

(b) Assume 5 lbm/cd, compaction to 1000 lbm/yd^3, and a 1.25 loading factor. The volume of landfill per day is given by Eq. 8.62.

$$\frac{(10{,}000)(5)(1.25)}{1000} = \boxed{62.5 \text{ yd}^3/\text{day}}$$

(a) The service life is

$$\frac{5.68 \times 10^7 \text{ ft}^3}{\left(27 \frac{\text{ft}^3}{\text{yd}^3}\right)\left(62.5 \frac{\text{yd}^3}{\text{day}}\right)\left(365 \frac{\text{days}}{\text{yr}}\right)} = \boxed{92.2 \text{ yr}}$$

(d)

- availability of land
- ease of access
- availability of cover soil if used
- wind speed and direction
- flat topology
- dry climate
- low risk of aquifer contamination
- impermeable lower strata
- temperate climate

5. Assume: 125 gpcd for average flow (p. 8-8)
2:1 peak flow factor (Table 8.8)
800 mg/l total solids (Table 8.5)

(a)

$$\frac{(10{,}000)(125 \text{ gpcd})\left(800 \frac{\text{mg}}{l}\right)(8.345)}{1 \times 10^6 \frac{\text{gal}}{\text{MG}}}$$

$$= \boxed{8345 \text{ lbm/day solids}}$$

(b)

$$S = \frac{400}{(4)(5280)} = 0.01894$$

$$Q = \frac{(125)(2)(10{,}000)\left(0.1337 \frac{\text{ft}^3}{\text{gal}}\right)}{(24)(60)\left(60 \frac{\text{sec}}{\text{day}}\right)}$$

$$= 3.87 \text{ ft}^3/\text{sec (cfs)}$$

Assume $n = 0.013$. From Eq. 5.11,

$$D = (1.33)\left[\frac{(0.013)(3.87)}{\sqrt{0.01894}}\right]^{3/8} = 0.912 \text{ ft}$$

$$\boxed{\text{12-in pipe}}$$

6. (a) [Note that the *Ten-States' Standards* (p. 8-35) would have required two grit chambers in parallel.]

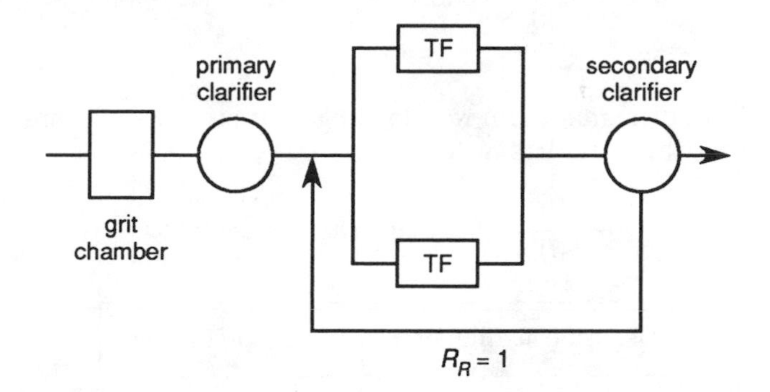

Assume a peak flow multiplier of 2. The flow rate per minute is

$$Q = \frac{(2)(1 \text{ MGD})\left(10^6 \frac{\text{gal}}{\text{MG}}\right)\left(0.1337 \frac{\text{ft}^3}{\text{gal}}\right)}{(24)(60)\left(60 \frac{\text{sec}}{\text{day}}\right)}$$

$$= 3.095 \text{ ft}^3/\text{sec (cfs)}$$

With a detention time of 3 min, the volume of the grit chamber would be

$$(3.095)(60)(3) = 557.1 \text{ ft}^3$$

Choose a length of 20 ft and assume width = 1.25 × depth.

$$557.1 = (20)(1.25)(\text{depth})^2$$

$$\text{depth} = 4.72 \text{ ft} \quad [\text{round to } 4\tfrac{3}{4} \text{ ft}]$$

$$\text{width} = (1.25)(4.72) = 5.9 \text{ ft} \quad [\text{round to 6 ft}]$$

$$\boxed{\text{6-ft} \times 4\tfrac{3}{4}\text{-ft} \times \text{20-ft grit chamber}}$$

(b) This is a shallow chamber. From p. 8-16, use

$$\left(3\ \frac{\text{cfm}}{\text{ft}}\right)(20\ \text{ft}) = \boxed{60\ \text{cfm of air}}$$

(c) (Note that the *Ten-States' Standards* require two basins.) Choose a detention time of 2 hr (p. 8-17). The volume is

$$(3.095\ \text{cfs})(2\ \text{hr})\left(3600\ \frac{\text{sec}}{\text{hr}}\right) \approx 22{,}300\ \text{ft}^3$$

Choose a 12-ft depth and a round shape.

$$D = \sqrt{\frac{(4)(22{,}300)}{\pi(12)}} = 48.64\ \text{ft} \quad [\text{round to 50 ft}]$$

Check the overflow rate.

$$v^* = \frac{10^6\ \frac{\text{gal}}{\text{day}}}{\frac{22{,}300}{12}} = 538.1\ \text{gal/day-ft}^2 \quad [\text{acceptable}]$$

Check the minimum weir loading assuming that the entire rim is used as a weir.

$$\frac{1\times 10^6}{\pi(50)} = 6366\ \text{gpd/ft} \quad [\text{acceptable}]$$

$$\boxed{\text{Use a 50-ft-diameter, 12-ft-deep clarifier.}}$$

(d) Assume the primary sedimentation basin removes 30% of the BOD. BOD incoming to the trickle filter process is

$$(1-0.30)(250) = 175\ \text{mg}/l$$

$$\left(175\ \frac{\text{mg}}{l}\right)\left(8.345\ \frac{\text{lbm-}l}{\text{MG-mg}}\right)(1\ \text{MGD}) = 1460\ \text{lbm/day}$$

Assume the primary basin damps out variations in flow, and peak multipliers do not apply. The required process efficiency is

$$\eta = \frac{\text{BOD}_{\text{in}} - \text{BOD}_{\text{out}}}{\text{BOD}_{\text{in}}} = \frac{175-30}{175} = 0.829\ (82.9\%)$$

From Fig. 8.10 (or Eqs. 8.30 and 8.33) with $\eta \approx 83\%$ and $R_R = 1$,

$$L_{\text{BOD}} \approx 22\ \text{lbm/1000 ft}^3\text{-day}$$

$$\text{filter volume} = \frac{\left(1460\ \frac{\text{lbm}}{\text{day}}\right)\left(1000\ \frac{\text{ft}^3}{1000\ \text{ft}^3}\right)}{\left(22\ \frac{\text{lbm}}{\text{day-1000 ft}^3}\right)(2\ \text{filters})} = 33{,}180\ \text{ft}^3$$

Assume a 6-ft-deep rock bed (p. 8-17) and a circular basin.

$$D = \sqrt{\frac{(4)(33{,}180)}{\pi(6)}} = 83.9\ \text{ft} \quad [\text{round to 85 ft}]$$

$$\boxed{\text{Use two 85-ft-diameter, 6-ft-deep filters.}}$$

(e) For the final clarifier, the maximum overflow rate should be 800 gpd/ft^2 (p. 8-25), and the minimum depth is 7 ft.

$$A = \frac{1\times 10^6\ \frac{\text{gal}}{\text{day}}}{800} = 1250\ \text{ft}^2$$

$$D = \sqrt{\frac{(4)(1250)}{\pi}} = 39.9 \quad [\text{round to 40 ft}]$$

$$\boxed{\text{Use a 40-ft-diameter, 7-ft-deep basin.}}$$

7. (a) Assume the last clarifier removes 30% of the remaining BOD.

$$\text{BOD}_3 = \frac{30\ \frac{\text{mg}}{l}}{1-0.3} = 42.9 \approx 43\ \text{mg}/l$$

The BOD loading to the trickle filter is given by Eq. 8.29 and does not consider recirculation.

$$L_{\text{BOD}} = \frac{(1.5)(250)(8.345)\left(1000\ \frac{\text{ft}^3}{1000\ \text{ft}^3}\right)}{\pi\left(\frac{75}{2}\right)^2(6)} = 118\ \text{lbm/day-1000 ft}^3 \quad [\text{high rate}]$$

From Eq. 8.33, the effective number of filter passes is

$$F = \frac{1+0.5}{[1+(0.1)(0.5)]^2} = 1.36$$

From Eq. 8.30, the filter/clarifier efficiency is

$$\eta = \frac{1}{1+0.0561\sqrt{\frac{118}{1.36}}} = 0.657 \approx 0.66\ (66\%)$$

This assumes high rate operation at 20°C, with uniform loading over a rock filter.

$$\text{BOD}_2 = (250)(1 - 0.66) = 85 \text{ mg}/l$$

Note that this does not include the recirculation flow. The NRC formula places greater emphasis on organic loading than on hydraulic loading. The removal fraction in the RBC must be

$$\frac{85 - 43}{85} = 0.494$$

Solving the given performance equation,

$$0.494 = \frac{1}{\left(1 + \frac{2.45A}{1.5 \times 10^6}\right)^3}$$

$$A = \boxed{1.62 \times 10^5 \text{ ft}^2}$$

(b) Solve the problem backward to get $\text{BOD}_{\text{out}} = 30 \text{ mg}/l$.

$$\text{BOD}_3 = \frac{30}{1 - 0.3} \approx 42.9 \text{ mg}/l$$

$$\text{BOD}_2 = \frac{42.9}{1 - 0.49} \approx 84.1 \text{ mg}/l$$

$$\eta_{\text{trickle filter and clarifier}} = \frac{250 - 84.1}{250} \approx 0.66 \text{ (66\%)}$$

From Eq. 8.30, solving for F,

$$F = \frac{L_{\text{BOD}}}{\left(\frac{1-\eta}{0.0561\eta}\right)^2} = \frac{118}{\left[\frac{1 - 0.66}{(0.0561)(0.66)}\right]^2} = 1.4$$

From Eq. 8.33,

$$1.4 = \frac{1 + R_R}{(1 + 0.1R_R)^2}$$

$$R_R = \boxed{0.56}$$

(c) Approximate the BOD removal of the secondary (first inline) clarifier.

$$\text{BOD}_{\text{entering}} = \frac{\text{BOD}_2}{1 - \eta} = \frac{85}{1 - 0.30} = 121.4 \text{ mg}/l$$

[assuming 30% removal]

The sludge production is

$$(0.4)[(121.4 - 85) + (43 - 30)](8.345)(1.5) = 247.3 \text{ lbm/day}$$

The sludge volume is

$$V = \frac{247.3}{62.4} = \boxed{3.96 \text{ ft}^3/\text{day}}$$

(d) Assume organic loading has the units of lbm/1000 ft^2-day.

$$L_{\text{BOD}} = \frac{(85)(1.5)(8.345)\left(1000 \frac{\text{ft}^3}{1000 \text{ ft}^3}\right)}{1.62 \times 10^5}$$

$$= \boxed{6.57 \text{ lbm}/1000 \text{ ft}^3\text{-day}}$$

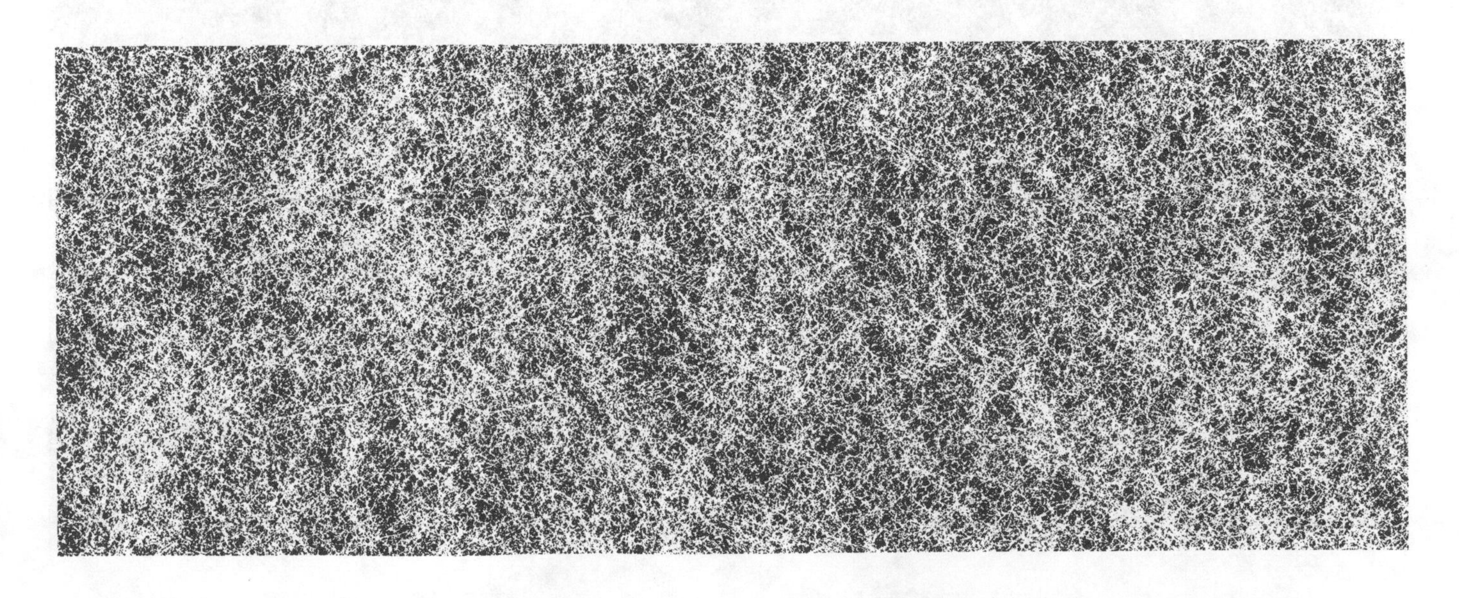

SOILS

Untimed

1. (a) $\rho = \dfrac{25.74 \text{ kg}}{0.01456 \text{ m}^3} = \boxed{1767.9 \text{ kg/m}^3}$

(b) $\gamma = \rho g = \left(1767.9 \ \dfrac{\text{kg}}{\text{m}^3}\right)\left(9.81 \ \dfrac{\text{m}}{\text{s}^2}\right)$

$= \boxed{17\,343 \text{ N/m}^3}$

(c) $W_w = 25.74 - 22.10 = 3.64 \text{ kg}$

$V_s = \dfrac{W_s}{\rho_s} = \dfrac{22.10 \text{ kg}}{(2.69)\left(1000 \ \dfrac{\text{kg}}{\text{m}^3}\right)} = 0.00822 \text{ m}^3$

$V_w = \dfrac{3.64 \text{ kg}}{1000 \ \dfrac{\text{kg}}{\text{m}^3}} = 0.00364 \text{ m}^3$

$V_g = 0.01456 - 0.00364 - 0.00822 = 0.0027 \text{ m}^3$

$e = \dfrac{0.00364 + 0.0027}{0.00822} = \boxed{0.771}$

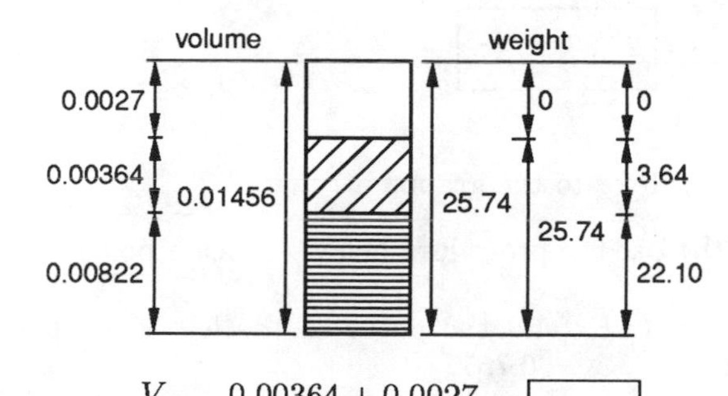

(d) $n = \dfrac{V_v}{V_t} = \dfrac{0.00364 + 0.0027}{0.01456} = \boxed{0.435}$

(e) $s = \dfrac{0.00364}{0.00364 + 0.0027} = \boxed{0.574}$

2. (a) Equation 9.10 should be used to correct the actual densities given. For the first sample,

$$\gamma_d = \frac{98}{1 + 0.10} = 89.1 \text{ lbf/ft}^3$$

Similarly, for the rest of the samples (see graph),

w	γ_d
10%	89.1
13%	93.8
16%	102.6
18%	105.9
20%	107.5
22%	104.9
25%	98.4

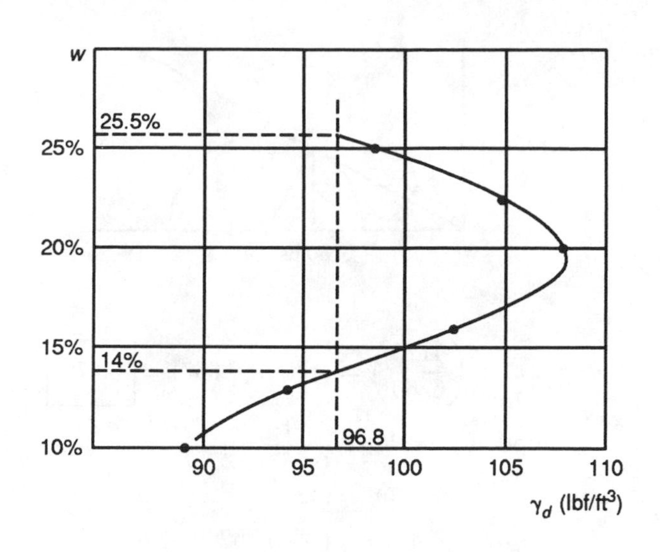

(b) $\gamma_{d,\text{max}} = \boxed{107.5 \text{ lbf/ft}^3}$

$w = \boxed{19\%}$

(c) $(0.90)(107.5) = 96.8 \text{ lbf/ft}^3$

$\boxed{14\% \leq w \leq 25.5\%}$

3. At maximum density, the weight of water and solids in 1 ft³ of soil is

$W_s = 107.5 \text{ lbf}$

$W_w = (0.19)(107.5) = 20.43 \text{ lbf}$

At 10% water content, $\gamma_d \approx 89.1 \text{ lbf/ft}^3$. For 1 ft³,

$W_s = 89.1 \text{ lbf}$

$W_w = (0.10)(89.1) = 8.91 \text{ lbf}$

To get 1 yd³ (27 ft³) of maximum density soil, the requirement is

$$\frac{(27)(107.5)}{89.1} = 32.58 \text{ ft}^3 \text{ of 10\% moisture soil}$$

The required water is

$$\frac{[(27)(20.43) - (32.58)(8.91)]\left(7.48 \ \dfrac{\text{gal}}{\text{ft}^3}\right)}{62.4 \ \dfrac{\text{lbf}}{\text{ft}^3}} = \boxed{31.33 \text{ gal}}$$

4. For a dry or drained sample, $c = 0$ (Eq. 9.35).

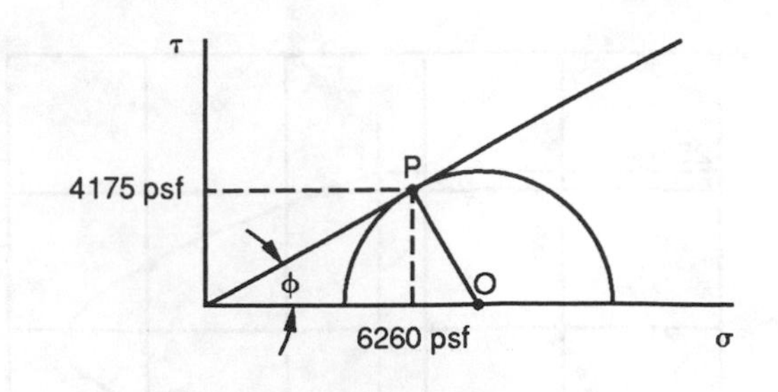

$$\phi = \arctan\left(\frac{4175}{6260}\right) = \arctan(0.667) = \boxed{33.7^\circ}$$

$$\text{slope of line PO} = \frac{-1}{0.667} = -1.5$$

$$y = mx + b$$

$$4175 = (-1.5)(6260) + b$$

$$b = 13{,}565 \text{ psf}$$

$$y = -1.5x + 13{,}565$$

When $y = 0$, $x = \dfrac{13{,}565}{1.5} = 9043$ psf.

$$\text{length PO} = \sqrt{(6260 - 9043)^2 + (4175)^2} = 5018$$

$$\text{principal stresses} = 9043 \pm 5018$$

$$= \boxed{14{,}061 \text{ psf}, \ 4025 \text{ psf}}$$

5. (a) Use Table 9.3.

	opening	% passing
$\frac{1}{2}$ in	12.70 mm	52
no. 4	4.76	37
no. 10	2.00	32
no. 20	0.85	23
no. 40	0.42	11
no. 60	0.25	7
no. 100	0.15	4

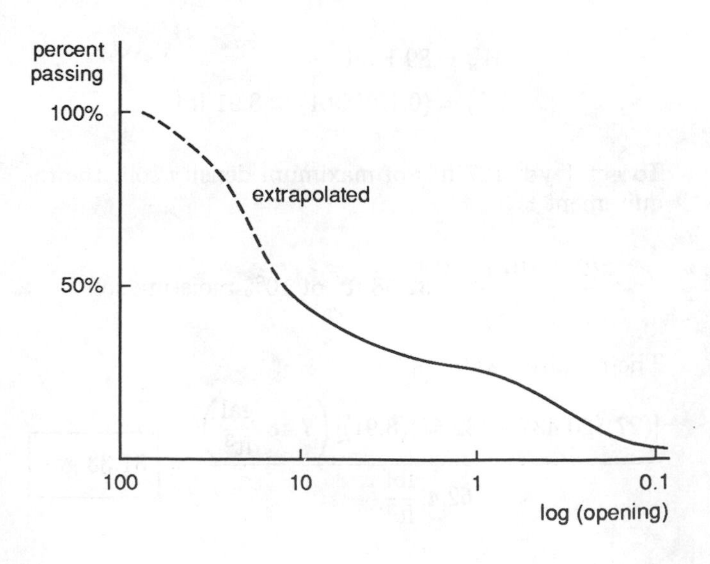

Use a 3-cycle semi-log graph. From it,

$$D_{10} = 0.4 \text{ mm}$$

$$D_{30} = 1.6 \text{ mm}$$

$$D_{60} \approx 17 \text{ mm}$$

(b) and (c) From Eqs. 9.2 and 9.3,

$$C_u = \frac{17}{0.4} = \boxed{42.5}$$

$$C_z = \frac{(1.6)^2}{(0.4)(17)} = \boxed{0.376}$$

6. From Eq. 9.26,

$$k = \frac{(1.5 \text{ ml})(120 \text{ mm})}{(225 \text{ mm})\left(\frac{\pi}{4}\right)(60 \text{ mm})^2(6.5 \text{ min})}$$

$$= 4.35 \times 10^{-5} \text{ ml/mm}^2\text{·min}$$

$$1 \text{ ml} = 1 \text{ cm}^3 = 1000 \text{ mm}^3$$

$$k = \frac{\left(4.35 \times 10^{-5} \frac{\text{ml}}{\text{mm}^2\text{·min}}\right)\left(1000 \frac{\text{mm}^3}{\text{ml}}\right)\left(525{,}600 \frac{\text{min}}{\text{yr}}\right)}{1000 \frac{\text{mm}}{\text{m}}}$$

$$= \boxed{22.86 \text{ m/yr}}$$

7. (a) Refer to the graph.

(b) Use the procedure below to locate point a.

step 1: Extend the curve to the left and estimate $e_o = 0.757$.

step 2: Draw a horizontal line from e_o (line H).

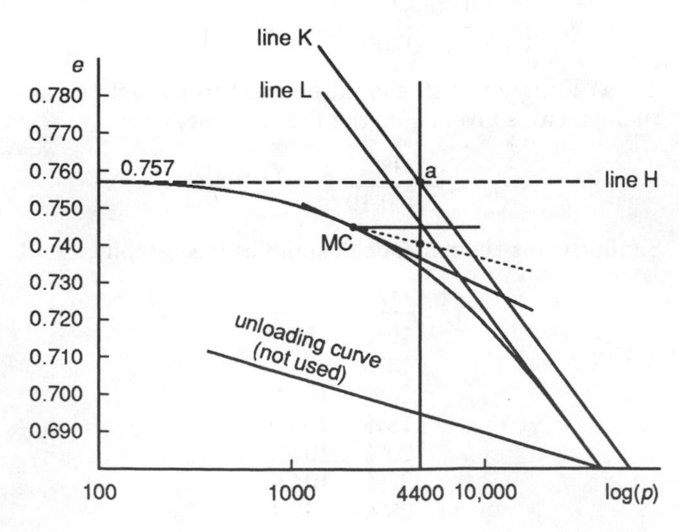

step 3: Extend the tail tangent upward (line L).

step 4: Find the point of maximum curvature by inspection (point MC). At point MC,

$$e = 0.746$$
$$p = 3000 \text{ psf}$$

step 5: Draw horizontal and tangent lines from point MC.

step 6: Bisect the angle (dotted line).

step 7: The intersection of the dotted line and line L gives p_o.

$$p_o = 4400 \text{ psf}$$

step 8: Draw a vertical line up from p_o to intersect line H. This identifies point a.

step 9: Since point a is close to the curve, the slope of line K should be estimated as the same as the slope of line L.

Draw a line passing through point a and parallel to line L. Choose two points on the line.

(4400 psf, 0.757)
(30,000 psf, 0.700)

From Eq. 9.31,

$$C_c = \frac{0.757 - 0.700}{\log\left(\frac{30{,}000}{4400}\right)} = \boxed{0.068}$$

(c) Construct the idealized consolidation curve.

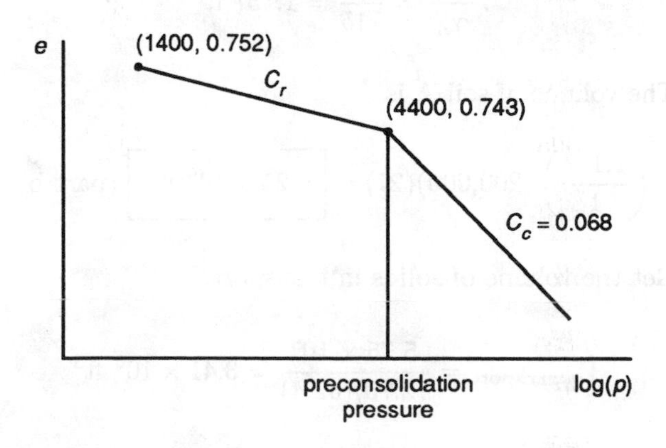

The consolidation below 4400 psi is given by Eq. 10.39. Equation 10.40 could also be used with the recompression index, C_r (similar to C_c, but used between 1400 and 4400 psi).

$$S_r = (10)\left(\frac{0.752 - 0.743}{1 + 0.752}\right) = 0.051 \text{ ft}$$

The remaining settlement allowed is

$$\tfrac{1}{12} - 0.051 = 0.032 \text{ ft}$$

From Eq. 10.40,

$$0.032 = \left(\frac{0.068}{1 + 0.743}\right) 10 \log\left(\frac{4400 + \Delta p}{4400}\right)$$
$$\Delta p = 914 \text{ psf}$$
$$p = 4400 + 914 = \boxed{5314 \text{ psf}}$$

8. Use Table 9.2, classification A-6.

$$I_p = w_l - w_p = 37 - 18 = 19$$

From Eq. 9.1,

$$\begin{aligned} I_g &= (57 - 35)[0.2 + (0.005)(37 - 40)] \\ &\quad + (0.01)(57 - 15)(19 - 10) \\ &= 7.85 \quad \text{[round to 8]} \end{aligned}$$

$$\text{classification} = \boxed{\text{A-6(8)}}$$

9. (a) From Eq. 9.15,

$$0.4 = \frac{0.97 - e}{0.97 - 0.45}$$
$$e = 0.76$$

Since $e = V_v/V_s$,

$$V_{\text{total}} = V_s + V_v = V_s + eV_s = (1 + e)V_s$$

Consider 1 ft^3 of saturated sand.

$$\text{weight of solids} = \frac{(2.65)(62.4)}{1 + 0.76} = 93.95 \text{ lbf/ft}^3$$
$$\text{weight of water} = \frac{(0.76)(62.4)}{1 + 0.76} = 26.95 \text{ lbf/ft}^3$$

Total weight (density) is

$$93.95 + 26.95 = \boxed{120.9 \text{ lbf/ft}^3}$$

(b)
$$\begin{aligned} \frac{\Delta V}{V_{t,40}} &= \frac{V_{t,40} - V_{t,65}}{V_{t,40}} \\ &= \frac{V_s(1 + e_{40}) - V_s(1 + e_{65})}{V_s(1 + e_{40})} \\ &= \frac{e_{40} - e_{65}}{1 + e_{40}} = \frac{0.76 - 0.63}{1.76} = 0.0739 \end{aligned}$$

$$\Delta t = (0.0739)(4) = \boxed{0.295 \text{ ft}}$$

10. Start with 1 ft^3 of fill in final form.

$$\gamma_{d,\text{required}} = (0.95)(124.0) = 117.8 \text{ pcf}$$

The weight of solids in 1 ft^3 of fill is

$$W_s = (117.8)(1) = 117.8 \text{ lbf}$$

Determine the borrow required.

$$\gamma_s = (62.4)(2.65) = 165.36 \text{ pcf}$$
$$W_s = 117.8 \text{ lbf} \quad \text{[same as fill]}$$
$$V_s = \frac{W_s}{\gamma_s} = \frac{117.8}{165.36} = 0.712 \text{ ft}^3$$
$$V_t = V_s(1+e) = (0.712)(1+0.6)$$
$$= \boxed{1.139 \text{ ft}^3}$$

The moisture content is irrelevant.

Timed

1.

- Soil A:

Calculate the required fill soil density.

$$\gamma_{F,\text{dry}} = (0.95)(112) = 106.4 \text{ lbf/ft}^3$$

This is also the weight of the solid in the fill.

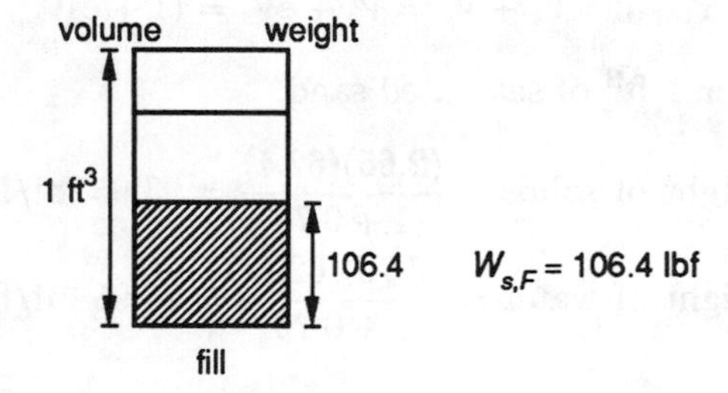

The weight of solid in the fill is the same as the weight of solid in the borrow.

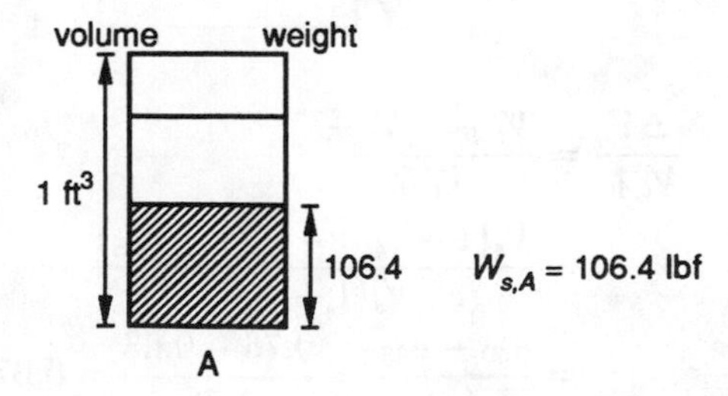

The weight of water in the fill is

$$W_{w,F} = wW_{s,F} = (0.10)(106.4) = 10.64 \text{ lbf}$$

Air has essentially no weight, so total fill density is

$$W_{t,F} = 106.4 + 10.64 = 117.04 \text{ lbf}$$

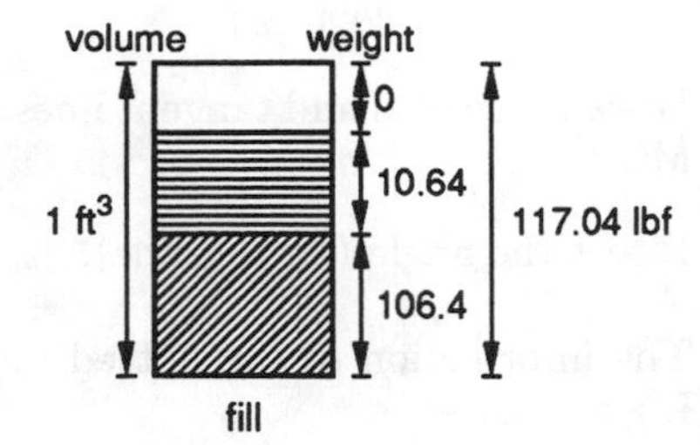

The volume is 200,000 yd^3 (given). The weight of borrow soil solids is

$$(200{,}000)(27)(106.4) = 5.75 \times 10^8 \text{ lbf}$$

Build the borrow soil weight from the requirements.

$$W_{s,A} = W_{s,F} = 106.4 \text{ lbf}$$
$$W_{w,A} = wW_{s,A} = (0.25)(106.4) = 26.6 \text{ lbf}$$
$$W_{a,A} = 0$$

The weight to be excavated to get 1 ft^3 fill is

$$W_{t,A} = 26.6 + 106.4 = 133 \text{ lbf}$$

The volume of borrow A per ft^3 of fill is

$$\frac{133}{\gamma_A} = \frac{133}{115} = 1.157 \text{ ft}^3$$

The volume of soil A is

$$\left(\frac{1.157}{1}\right)(200{,}000)(27) = \boxed{6.25 \times 10^6 \text{ ft}^3} \quad \text{(part b)}$$

Get the volume of solids in transport.

$$V_{s,\text{transport}} = \frac{5.75 \times 10^8}{(2.70)(62.4)} = 3.41 \times 10^6 \text{ ft}^3$$

The volume of voids in transport is

$$V_{\text{void}} = eV_s = (0.92)(3.41 \times 10^6) = 3.14 \times 10^6 \text{ ft}^3$$

The total transport volume is

$$V_t = 3.41 \times 10^6 + 3.14 \times 10^6 = 6.55 \times 10^6 \text{ ft}^3$$

The cost to excavate and transport A is

$$C_A = (0.20)\left(\frac{6.25\times10^6}{27}\right) + (0.30)\left(\frac{6.55\times10^6}{27}\right)$$
$$= \$1.19\times10^5$$

- Soil B:

$$\gamma_{dry} = (0.95)(110) = 104.5\ \text{lbf/ft}^3$$
$$W_{s,F} = 104.5\ \text{lbf}$$
$$W_{s,F} = W_{s,B} = 104.5\ \text{lbf}$$
$$W_{w,F} = wW_{s,F} = (0.10)(104.5) = 10.45\ \text{lbf}$$
$$W_{t,F} = 104.5 + 10.45 = 114.95\ \text{lbf}$$

The weight of borrow soil solids is

$$(200{,}000)(27)(104.5) = 5.64\times10^8\ \text{lbf}$$
$$W_{s,B} = 104.5\ \text{lbf}$$
$$W_{w,B} = (0.20)(104.5) = 20.9\ \text{lbf}$$
$$W_{t,B} = 20.9 + 104.5 = 125.4\ \text{lbf}$$

The volume of borrow B per ft^3 of fill is

$$\frac{125.4}{120} = 1.045\ \text{ft}^3$$

The volume of soil B is

$$\left(\frac{1.045}{1}\right)(200{,}000)(27) = \boxed{5.64\times10^6\ \text{ft}^3}\quad\text{(part b)}$$

Soil B has the minimum transport volume. (part c)

The volume in transport is

$$\frac{\text{weight}}{\gamma_{transport}} = \frac{(5.64\times10^6)(120)}{95} = 7.13\times10^6\ \text{ft}^3$$

The transport cost is

$$C_B = (0.10)\left(\frac{5.64\times10^6}{27}\right) + (0.40)\left(\frac{7.13\times10^6}{27}\right)$$
$$= \$1.27\times10^5$$

C_A is the minimum cost. (part a)

2. Graph the results.

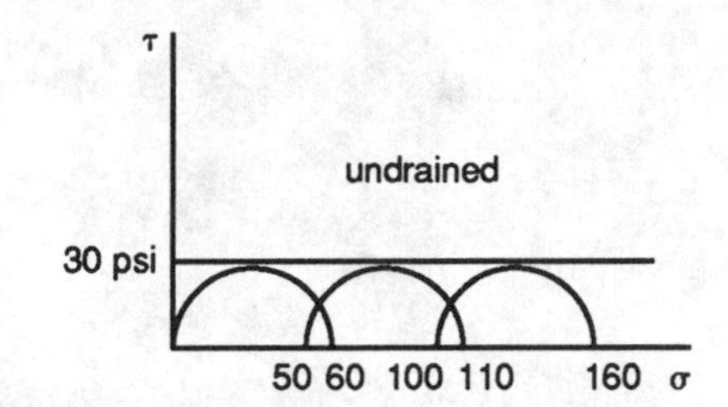

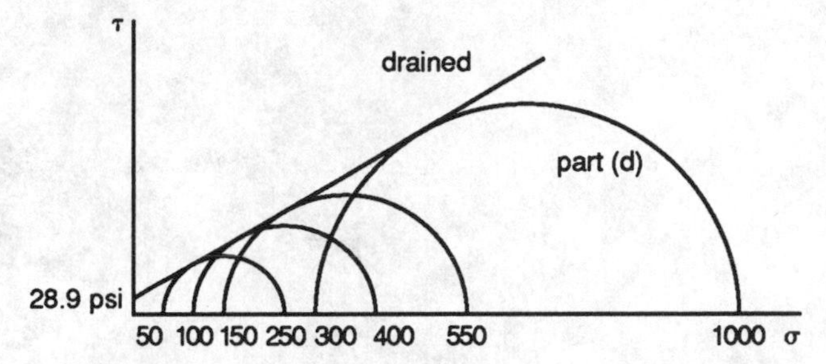

(a) undrained: $\phi = \boxed{0°}$

drained: $\phi = \boxed{30°}$ [found using a protractor]

(b) undrained: $c = \boxed{30\ \text{psi}}$

drained: $c = \boxed{28.9\ \text{psi}}$

(c) undrained: $\theta = 45° + \left(\frac{1}{2}\right)(0) = \boxed{45°}$ [Eq. 9.36]

drained: $\theta = 45° + \left(\frac{1}{2}\right)(30) = \boxed{60°}$

(d) By drawing circles to touch the rupture line and 300 psi by trial and error,

$$\sigma_A = \boxed{1000\ \text{psi}}$$

FOUNDATIONS AND RETAINING WALLS

Untimed

1. Follow the procedure on p. 10-12.

steps 1 and 2: The top 10 ft of soil will have settled under the 500 psf load as much as it is going to. The next 8 ft will settle due to the drop in the water table. The soil under the water table will also experience a reconsolidation. The midpoints of these two settling layers are at depths of 14 ft and 28 ft, respectively. The settlement will be due to the change in effective pressure at these midpoints. Because the raft is so large, its pressure load is assumed to not decrease with depth. Besides, its size was not given.

- Before the drop in the water table:

$$p_{14} = 500 + (10)(100) + (4)(120 - 62.4) = 1730 \text{ psf}$$
$$p_{28} = 500 + (10)(100) + (18)(120 - 62.4) = 2537 \text{ psf}$$

- After the drop in the water table:

$$p_{14} = 500 + (14)(100) = 1900 \text{ psf}$$
$$p_{28} = 500 + (18)(100) + (10)(120 - 62.4) = 2876 \text{ psf}$$

step 3: $e_o = 0.6$ [given]

step 4: $C_c = 0.02$ [given]

step 5: From Eq. 10.40, due to the 8 ft layer,

$$S_1 = \left(\frac{0.02}{1+0.6}\right)(8)\log_{10}\left(\frac{1900}{1730}\right) = 0.00407 \text{ ft}$$

Due to the submerged layer,

$$S_2 = \left(\frac{0.02}{1+0.6}\right)(20)\log_{10}\left(\frac{2876}{2537}\right) = 0.01362 \text{ ft}$$

The total settlement is

$$S_1 + S_2 = 0.00407 + 0.01362 = \boxed{0.0177 \text{ ft}}$$

2. (a) If the footing rests on the sand, then $D_f = 0$. From Table 10.3, for $\phi = 38°$ and a square footing,

$$N_\gamma \approx (0.85)(77) \approx 66$$
$$p_{\text{net}} = \left(\tfrac{1}{2}\right)(B)(121)(66) = 3993B \text{ psf}$$

From Eq. 10.10, using $F = 2$,

$$p_{\text{net}} = 2p_a = (2)\left(\frac{360{,}000}{B^2}\right) = \frac{720{,}000}{B^2}$$

Equating these two expressions for p_{net},

$$\frac{720{,}000}{B^2} = 3993B$$

$$B = \boxed{5.6 \text{ ft}}$$

(b) From Eq. 10.10,

$$p_{\text{net}} = Fp_a = (2)\left(\frac{360{,}000}{B^2}\right) = \frac{720{,}000}{B^2}$$

From Table 10.3, $N_q \approx 64$.

From Eq. 10.9,

$$\frac{720{,}000}{B^2} = \left(\tfrac{1}{2}\right)(B)(121)(66) + (121)(4)(64-1)$$
$$\frac{720{,}000}{B^2} = 3993B + 30{,}492$$

By trial and error,

$$B \approx \boxed{4.0 \text{ ft}}$$

3. The base must be at least 4 ft below the surface to be below the frost line. Therefore, the overall height of the wall is at least 18 ft.

step 1: Refer to p. 14-33.

choose $B = 0.65H = (0.65)(18) = 11.7$ ft [use 11.5 ft]

choose $d = \frac{1}{10}H = \left(\frac{1}{10}\right)(18) = 1.8$ ft [use 1.75 ft]

choose $t = d = 1.75$ ft at base

choose $t = 1.75 - \left(\frac{0.5}{12}\right)(18) = 1$ ft at top [not shown]

choose heel extension $= \left(\frac{2}{3}\right)(11.5 - 1.75) = 6.5$ ft

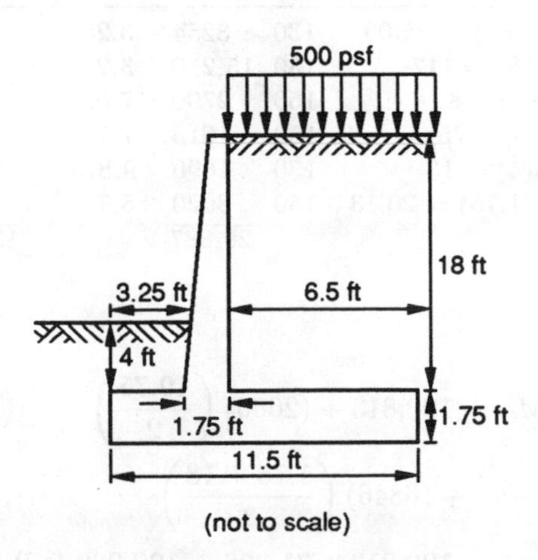

step 2: Convert the surcharge to an equivalent weight (depth) of soil.

$$\frac{500}{130} = 3.85 \text{ ft}$$

From Eq. 10.54,

$$k_A = \frac{1 - \sin 35^\circ}{1 + \sin 35^\circ} = 0.27$$

The surcharge pressure is

$$p_{\text{surcharge}} = (0.27)(130)(3.85) = 135 \text{ psf}$$

The surcharge load is

$$R_{\text{surcharge}} = (135)(19.75) = 2666 \text{ lbf}$$

From Eq. 10.60, the active soil resultant is

$$R_A = \left(\tfrac{1}{2}\right)(0.27)(130)(19.75)^2 = 6846 \text{ lbf}$$

Since the soil above the heel is horizontal, R_A is horizontal and there is no vertical component of R_A. (Verify this using App. A.)

step 3: Ignore the passive pressure.

step 4:

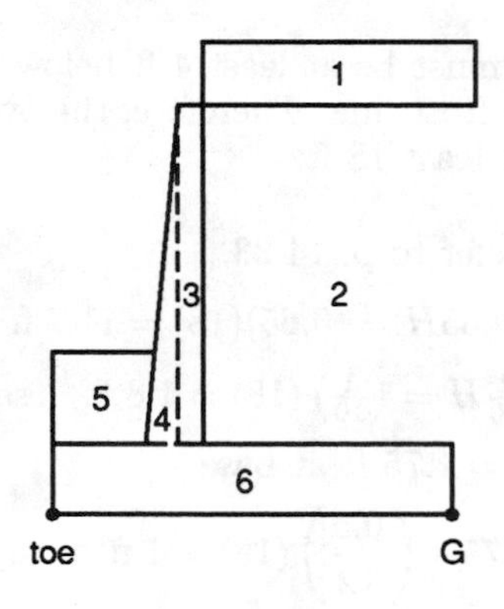

i	area	γ	W_i	r_i from G	moment$_G$
1	(6.5)(3.85) = 25.03	130	3254	3.25	10,575
2	(6.5)(18) = 117	130	15,210	3.25	49,433
3	(1)(18) = 18	150	2700	7.00	18,900
4	$\left(\frac{1}{2}\right)$(18)(0.75) = 6.75	150	1013	7.75	7851
5	(3.25)(4) = 13	130	1690	9.875	16,689
6	(11.5)(1.75) = 20.13	150	3020	5.75	17,365
			26,887		120,813

step 5:

$$M_G = 120{,}813 + (2666)\left(\frac{19.75}{2}\right) + (6846)\left(\frac{1.75 + 18}{3}\right)$$
$$= 120{,}813 + 71{,}396 = 192{,}209 \text{ ft-lbf}$$

step 6:

$$r = \frac{192{,}209}{26{,}887} = 7.15 \text{ ft}$$

$$e^* = 7.15 - \left(\tfrac{1}{2}\right)(11.5) = 1.40 \text{ ft}$$

This e^* is less than $11.5/6 = 1.92$ ft, so it is acceptable.

step 7:

$$p_{\text{max,toe}} = \left(\frac{26{,}887}{11.5}\right)\left[1 + \frac{(6)(1.40)}{11.5}\right] = 4046 \text{ psf} \quad \text{[acceptable]}$$

$$p_{\text{min,heel}} = \left(\frac{26{,}887}{11.5}\right)\left[1 - \frac{(6)(1.40)}{11.5}\right] = 630 \text{ psf}$$

The upward earth pressure distribution on the base is

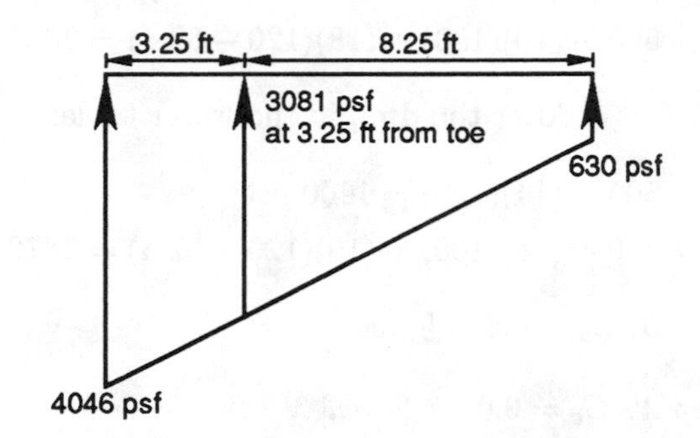

step 8: Since there is no key, the friction between concrete and soil resists sliding.

$$R_s = (\Sigma W)(0.5) = (26{,}887)(0.5) = 13{,}444 \text{ lbf}$$

step 9:

$$F_{\text{sliding}} = \frac{13{,}444}{2666 + 6846} = 1.41$$

(This is less than 1.5, so it is not acceptable.)

A key is needed for this design, so extend the stem down about 1 ft.

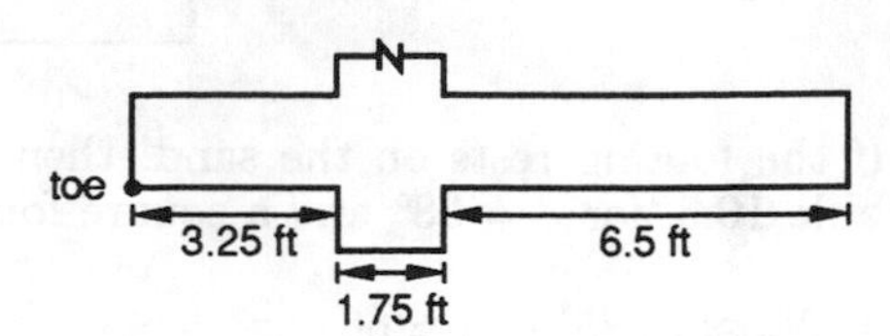

Check the resistance to sliding again. From the toe to the key, the soil must shear.

$$R_{s_1} = \left[(3.25)(3081) + \left(\tfrac{1}{2}\right)(3.25)(4046 - 3081)\right](\tan 35^\circ)$$
$$= (11{,}581)(\tan 35^\circ) = 8109 \text{ lbf}$$

From the key to the heel, the base slides on the soil.

$$R_{s_2} = (26{,}887 - 11{,}581)(0.5) = 7653 \text{ lbf}$$

$$F_{\text{sliding}} = \frac{8109 + 7653}{2666 + 6846} = 1.66 \quad \text{[acceptable]}$$

step 10: To calculate the factor of safety against overturning, it is necessary to take moments about the pivot point (which, in this case, is the toe).

i	W_i	r_i from toe	M_{toe}
1	3254	8.25	26,846
2	15,210	8.25	125,483
3	2700	4.50	2,150
4	1013	3.75	3799
5	1690	1.63	2755
6	3020	5.75	17,365
			188,398

(The weight of part 6 disregards the key, which could be included.)

From Eq. 10.83, including the surcharge effect,

$$F_{\text{overturning}} = \frac{188{,}398}{(6846)\left(\frac{19.75}{3}\right) + (2666)\left(\frac{19.75}{2}\right)} = 2.64 \quad \text{[acceptable]}$$

4. (a) From Eq. 10.54,

$$k_A = \frac{1 - \sin 36^\circ}{1 + \sin 36^\circ} = 0.26$$

At $H = 10$ ft,

$$\begin{aligned} p_h &= k_A p_v = k_A \gamma H \\ &= (0.26)(96)(10) = 249.6 \text{ psf} \\ p &= 249.6 + (0.26)(121 - 62.4)(H - 10) \\ &\quad + (62.4)(H - 10) \\ &= 249.6 + (77.6)(H - 10) \end{aligned}$$

At $H = 26$ ft, $p = 249.6$ psf $+ 1241.6$ psf.

The pressure distributions on the wall are as shown.

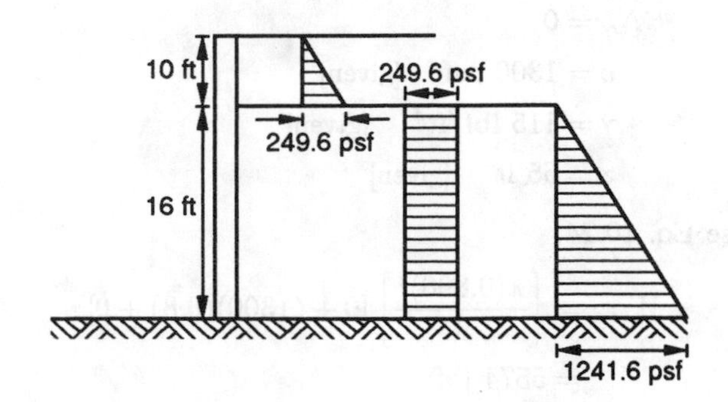

The resultants of each of the preceding three distributions are

$(\frac{1}{2})(10)(249.6) = 1248$ lbf, located $(\frac{2}{3})(10) = 6.67$ ft from the top (19.33 ft from the bottom)

$(16)(249.6) = 3994$ lbf, located 8 ft from the bottom

$(\frac{1}{2})(16)(1241.6) = 9933$ lbf, located $(\frac{1}{3})(16) = 5.33$ ft from the bottom

The active resultant is

$$1248 + 3994 + 9933 = \boxed{15{,}175 \text{ lbf}}$$

(b) Taking moments about the base,

$$\text{moment arm} = \frac{(1248)(19.33) + (3994)(8) + (9933)(5.33)}{15{,}175} = \frac{109{,}019}{15{,}175} = \boxed{7.18 \text{ ft from the bottom}}$$

(c) The original overturning moment is

$$M_{\text{original}} = 109{,}019 \text{ ft-lbf}$$

From Eq. 10.60, if the water table was lowered, the resultant would be

$$\begin{aligned} R_A &= (\tfrac{1}{2})(0.26)(96)(26)^2 \\ &= 8434 \text{ lbf, located } (\tfrac{1}{3})(26) = 8.67 \text{ ft from the bottom} \end{aligned}$$

$$\begin{aligned} M_{\text{drained}} &= (8434)(8.67) = 73{,}123 \text{ ft-lbf} \\ M_{\text{reduction}} &= 109{,}019 - 73{,}123 \\ &= \boxed{35{,}896 \text{ ft-lbf}} \end{aligned}$$

5. (a) From Eq. 10.35,

$$0.31 = \frac{e_2 - 1.04}{\log\left(\frac{2600}{3900}\right)}$$

$$e_2 = 0.985$$

$$\Delta e = 0.985 - 1.04 = \boxed{-0.055}$$

(b) From Eq. 10.40,

$$S = (16)\left(\frac{-0.055}{1 + 1.04}\right) = \boxed{-0.43 \text{ ft}}$$

(c) From Eq. 10.43,

$$a_v = \frac{1.04 - 0.985}{3900 - 2600} = 4.23 \times 10^{-5} \text{ ft}^2/\text{lbf}$$

The permeability is

$$\frac{4\times10^{-7}\ \frac{\text{mm}}{\text{sec}}}{\left(10\ \frac{\text{mm}}{\text{cm}}\right)\left(2.54\ \frac{\text{cm}}{\text{in}}\right)\left(12\ \frac{\text{in}}{\text{ft}}\right)} = 1.31\times10^{-9}\ \text{ft/sec}$$

From Eq. 10.42,

$$C_v = \frac{(1.31\times10^{-9})(1+1.04)}{(62.4)(4.23\times10^{-5})} = 1.01\times10^{-6}\ \text{ft}^2/\text{sec}$$

From Table 10.7, for $U_z = 75\%$, $T_v = 0.48$.

From Eq. 10.41,

$$t = \frac{(0.48)\left(\frac{16\ \text{ft}}{2}\right)^2}{1.01\times10^{-6}\ \frac{\text{ft}^2}{\text{sec}}} = 3.04\times10^7\ \text{sec} = 352\ \text{days}$$

6. (a) Refer to Fig. 10.26. The distribution is rectangular. From Eq. 10.89,

$$\begin{aligned} p_{max} &= (0.65)(121)(30)\tan^2\left(45° - \frac{40°}{2}\right) \\ &= 513\ \text{psf} \end{aligned}$$

(b) Assume simple supports of each horizontal lagging timber.

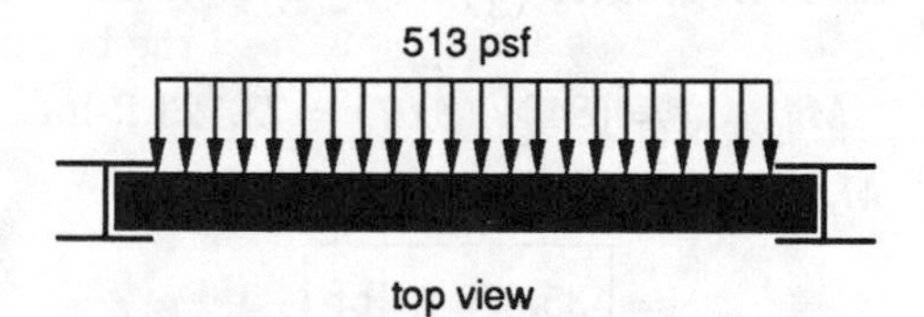

From beam equations, the maximum moment per foot of timber width is

$$M = \frac{(513)(8)^2}{8} = \boxed{4104\ \text{ft-lbf}}$$

7. From Eq. 9.35, for $\phi = 0°$,

$$\begin{aligned} c &= S_{us} = 1100\ \text{psf} \\ \beta &= \arctan\left(\tfrac{1}{2}\right) = 26.6° \\ d &= \frac{15}{43} = 0.35 \end{aligned}$$

From Fig. 10.13, the stability number is $N_o \approx 6.5$. The cohesion safety factor is

$$F = \frac{N_o c}{\gamma H} = \frac{(6.5)(1100)}{(112)(43)} = \boxed{1.48} \quad \text{[acceptable]}$$

8. *method 1:* Use Fig. 10.11. About 25 squares are covered. From Eq. 10.31,

$$p = (25)(0.005)(3000) = \boxed{375\ \text{psf}}$$

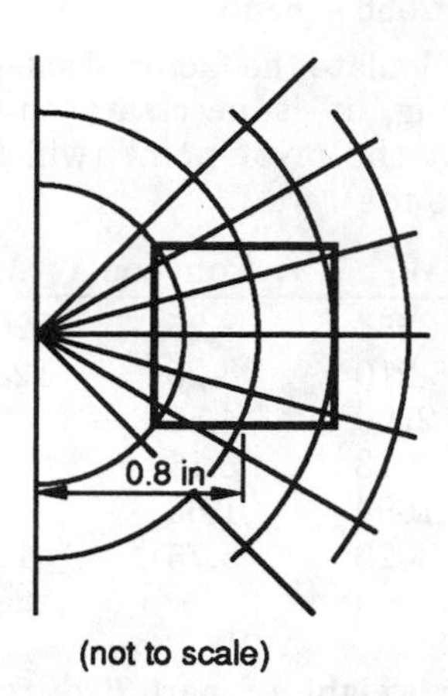

(not to scale)

method 2: Use App. C. At (−B, +0.88), read ≈ 0.1p.

$$p = (0.1)(3000) = \boxed{300\ \text{psf}}$$

9. Use Eq. 10.30.

$$\begin{aligned} p_v &= \frac{(3)(6000)}{(2\pi)(3.5)^2}\left[\frac{1}{1+\left(\frac{4}{3.5}\right)^2}\right]^{5/2} \\ &= \boxed{28.96\ \text{psf}} \end{aligned}$$

10.

- End capacity:

$$B = \frac{10.75\ \text{in}}{12\ \frac{\text{in}}{\text{ft}}} = 0.896\ \text{ft}$$

For saturated clay $\phi = 0°$, so from Table 10.3,

$$\begin{aligned} N_c &= (5.7)(1.20) = 6.8 \quad \text{[round footing]} \\ N_q &= 1.0 \\ N_\gamma &= 0 \\ c &= 1300\ \text{psf} \quad \text{[given]} \\ \gamma &= 115\ \text{lbf/ft}^3 \quad \text{[given]} \\ z &= 65\ \text{ft} \quad \text{[given]} \end{aligned}$$

Use Eq. 10.24.

$$\begin{aligned} W_{end} &= \left[\frac{\pi(0.896)^2}{4}\right][0 + (1300)(6.8) + 0] \\ &= 5574\ \text{lbf} \end{aligned}$$

• Friction capacity:

From Table 10.6, for steel-silt, $\delta \approx 11°$.

Assume

$$C_{a,\min} = 0.5c = (0.5)(1300 \text{ psf}) = 650 \text{ psf}$$

Assume $k = 1$ because $\phi = 0°$ below the water table. Since relative density = 75%, from Eq. 10.27,

$$z_{\text{critical}} = 20B = (20)\left(\frac{10.75 \text{ in}}{12 \frac{\text{in}}{\text{ft}}}\right)$$
$$= 17.9 \text{ ft} \quad [\text{use } 18]$$

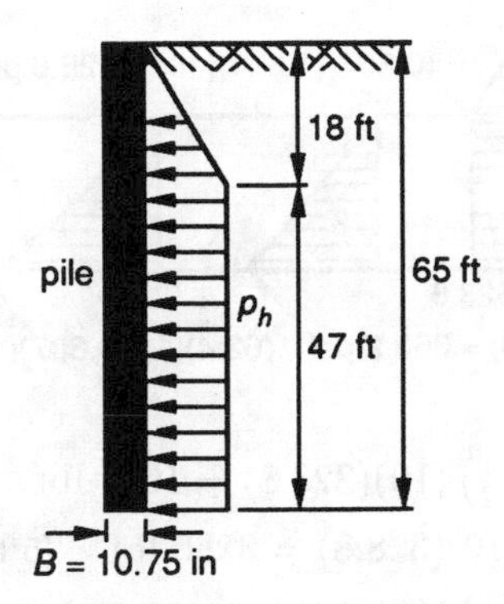

For 0 ft $\le z \le$ 18 ft,

$$z_{\text{ave}} = 9 \text{ ft}$$

From Eqs. 10.26 and 10.28,

$$p_{h,\text{ave}} = k(\gamma z - \mu) = k(\gamma_{\text{soil}} z - \gamma_{\text{water}} z)$$
$$= (1)(115 - 62.4)(9 \text{ ft})$$
$$= 473 \text{ psf}$$

From Eq. 10.25,

$$f_o = \text{minimum}\left\{\begin{array}{l} 1300 + (473)(\tan 0°) = 1300 \text{ psf} \\ 650 + (473)(\tan 11°) = 742 \text{ psf} \end{array}\right\}$$
$$= 742 \text{ psf}$$

From Eq. 10.24,

$$W_{\text{friction},1} = \pi B z f_o$$
$$= \pi \left(\frac{10.75 \text{ in}}{12 \frac{\text{in}}{\text{ft}}}\right)(18 \text{ ft})(742 \text{ psf})$$
$$= 37{,}588 \text{ lbf}$$

For 18 ft $\le z \le$ 65 ft, p_h is constant at the z = 18 ft level.

$$p_h = (1)(115 - 62.4)(18) = 947 \text{ psf}$$

From Eq. 10.25,

$$f_o = \text{minimum}\left\{\begin{array}{l} 1300 + (947)(\tan 0°) = 1300 \text{ psf} \\ 650 + (947)(\tan 11°) = 834 \text{ psf} \end{array}\right\}$$
$$= 834 \text{ psf}$$

$$W_{\text{friction},2} = \pi B z f_o = \pi \left(\frac{10.75}{12}\right)(65 - 18)(834)$$
$$= 110{,}317 \text{ lbf}$$

The total capacity is

$$5574 + 37{,}588 + 110{,}317 = \boxed{153{,}479 \text{ lbf}}$$

Timed

1. From Table 10.3, for $\phi = 25°$,

$$N_c = 25.1$$
$$N_q = 12.7$$
$$N_\gamma = 9.7$$

• Wall footing:

From Eq. 10.3 (per foot),

$$p_{\text{net}} = F p_a = F\left(\frac{\text{load}}{B}\right) = (2.5)\left(\frac{30{,}000}{B}\right) = \frac{75{,}000}{B}$$

Use Eqs. 10.1 and 10.2 with $p_q = 0$.

$$p_{\text{net}} = \tfrac{1}{2}\gamma B N_\gamma + c N_c + \gamma D_f (N_q - 1)$$
$$\frac{75{,}000}{B} = \left(\tfrac{1}{2}\right)(108)(B)(9.7) + (400)(25.1) + (108)(6)(12.7 - 1)$$
$$523.8B^2 + 17{,}622B = 75{,}000$$
$$B^2 + 33.64B = 143.18$$
$$B = \boxed{3.82 \text{ ft}}$$

• Square footing:

From Tables 10.2 and 10.4,

$$N_c = (1.25)(25.1) = 31.4$$
$$N_q = 12.7$$
$$N_\gamma = (0.85)(9.7) = 8.2$$

$$p_{\text{net}} = F p_a = F\left(\frac{\text{load}}{B^2}\right) = (2.5)\left(\frac{215{,}000}{B^2}\right) = \frac{537{,}500}{B^2}$$
$$\frac{537{,}500}{B^2} = \left(\tfrac{1}{2}\right)(108)(B)(82) + (400)(31.4) + (108)(2)(12.7 - 1)$$
$$537{,}500 = 443B^3 + 15{,}087B^2$$

By trial and error,

$$B = \boxed{5.5 \text{ ft}}$$

• Round footing:

From Tables 10.2 and 10.4,

$$N_c = (1.20)(25.1) = 30.1$$
$$N_q = 12.7$$
$$N_\gamma = (0.70)(9.7) = 6.8$$
$$p_{\text{net}} = \frac{(2.5)(300{,}000)}{\pi R^2} = \frac{238{,}732}{R^2}$$
$$\frac{238{,}732}{R^2} = \left(\tfrac{1}{2}\right)(108)(2R)(6.8) + (400)(30.1) + (108)(4)(12.7 - 1)$$
$$238{,}732 = 734R^3 + 17{,}094R^2$$

By trial and error,

$$R = \boxed{3.5 \text{ ft}}$$

2. The effective pressure at the midpoint of the clay layer is

$$p_o = (5)(100) + (10)(105 - 62.4) + \left(\tfrac{1}{2}\right)(102 - 62.4)(8) = 1084.4 \text{ psf}$$

The final effective pressure is

$$p = (10)(110) + (10)(100) + (5)(105 - 62.4) + \left(\tfrac{1}{2}\right)(102 - 62.4)(8) = 2471.4 \text{ psf}$$

From Eq. 10.40,

$$S = \left(\frac{0.38}{1 + 1.60}\right)(8)\log_{10}\left(\frac{2471.4}{1084.4}\right) = \boxed{0.42 \text{ ft}}$$

Note that C_c can also be estimated from e_o, but the data provided in the problem takes priority.

3.

• As designed:

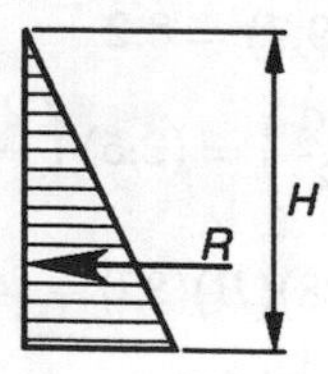

From Eq. 10.54,

$$k_A = \frac{1 - \sin 32^\circ}{1 + \sin 32^\circ} = 0.307$$

From Eq. 9.10,

$$\gamma = (1 + 0.05)(102) = 107.1 \text{ pcf}$$

From Eq. 10.60,

$$R = \left(\tfrac{1}{2}\right)(0.307)(107.1)(20)^2 = \boxed{6576 \text{ lbf/ft}}$$

This acts

$$\left(\tfrac{1}{3}\right)(20) = \boxed{6.67 \text{ ft up from the bottom}}$$

• As it happened:

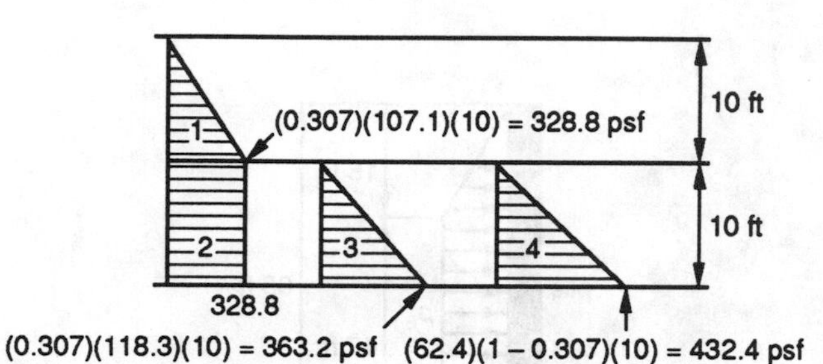

$$R_1 = \left(\tfrac{1}{2}\right)(10)(328.8) = 1644 \text{ lbf} \quad [13.33 \text{ ft up}]$$
$$R_2 = (10)(328.8) = 3288 \text{ lbf} \quad [5 \text{ ft up}]$$
$$R_3 + R_4 = \left(\tfrac{1}{2}\right)(10)(363.2 + 432.4) = 3978 \text{ lbf} \quad [10/3 \text{ ft up}]$$
$$R_{\text{total}} = 1644 + 3288 + 3978 = \boxed{8910 \text{ lbf}}$$

This acts

$$\frac{(1644)(13.33) + (3288)(5) + (3978)\left(\frac{10}{3}\right)}{8910} = \boxed{5.79 \text{ ft up}}$$

4.
$$k_{A,\text{sand}} = \frac{1 - \sin 32^\circ}{1 + \sin 32^\circ} = 0.307$$
$$p_{\text{vertical}} = (110)(25) = 2750 \text{ psf}$$
$$p_{\text{active},25} = (0.307)(2750) = 844 \text{ psf}$$
$$R = \left(\tfrac{1}{2}\right)(25)(844) = 10{,}550 \text{ lbf}$$

R acts $10 + 25/3 = 18.33$ ft up from bedrock.

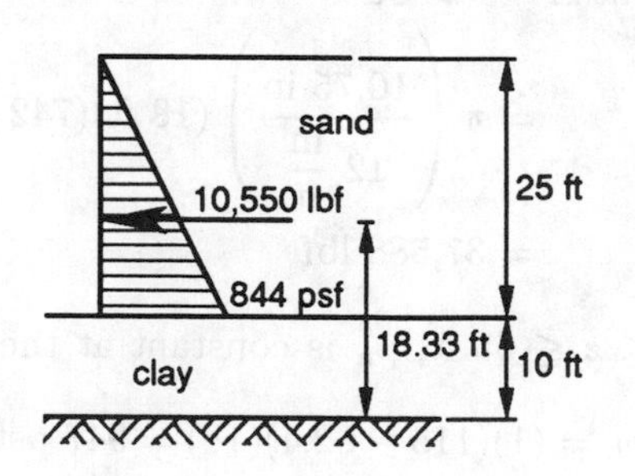

Equation 10.51 is used to determine the horizontal clay pressure.

$$\phi = 0^\circ$$
$$k_A = 1$$
$$p_{\text{active,clay}} = p_{\text{vertical}} - 2c$$

p_{vertical} is the surcharge at the top of the clay layer plus the self-load (γH), which is duplicated on the passive side.

$$p_{\text{active}} = 2750 + (\text{clay depth})(120) - 2c$$

From Eq. 10.55,

$$p_{\text{passive}} = p_{\text{vertical}} + 2c = (\text{clay depth})(120) + 2c$$

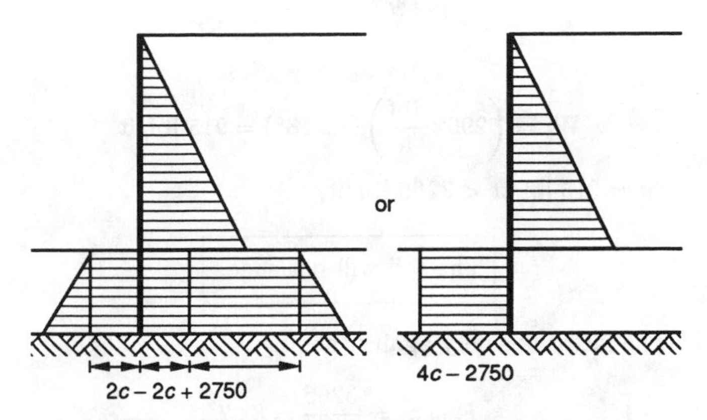

$$4c - 2750 = (4)(750) - 2750 = 250$$
$$R_p = (250)(10) = 2500 \text{ lbf}$$

R_p acts 5 ft above bedrock.

Refer to p. 10-22 to determine the inflection point. This occurs (approximately) at h below the mud line, where

$$844 = 250h$$
$$h = 3.38 \text{ ft} \quad [1.62 \text{ ft above bedrock}]$$

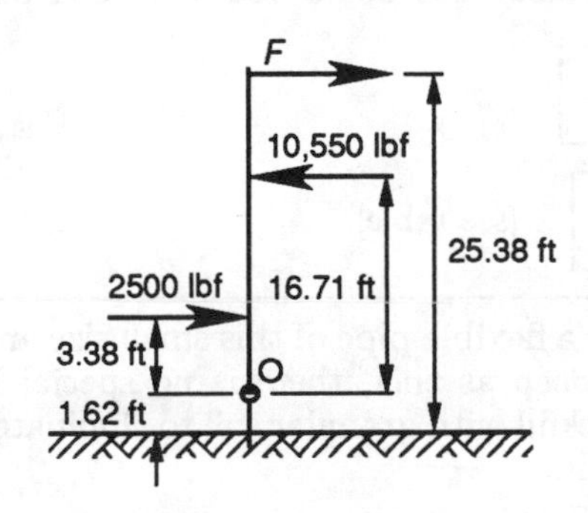

Take clockwise moments as positive.

$$\sum M_O = (2500)(3.38) + 25.38F - (10{,}550)(16.71) = 0$$

$$F = \boxed{6613 \text{ lbf/ft of wall}}$$

(This is fairly insensitive to the choice of inflection point location. If the bedrock is taken as point O, then $F =$ 6700 lbf.)

Note that

$$\sum F_x = 2500 + 6613 - 10{,}550 = -1437 \text{ lbf}$$

Since $\sum F_x \neq 0$, neither the method of calculating the soil forces or the method of determining the inflection point is exact.

5. no. of piles = 36

$$\text{pile group dimension} = (5)(3.5) = 17.5 \text{ ft}$$
$$\text{pile group area} = (17.5)^2 = 306.25 \text{ ft}^2$$
$$p_{\text{max}} = \frac{(500)(36) - 600}{306.25} = 56.8 \text{ tons/ft}^2 \text{ (tsf)}$$
$$p_{\text{min}} = \frac{-[(150)(36) + 600]}{306.25} = -19.6 \text{ tons/ft}^2 \text{ (tsf)}$$

(a)

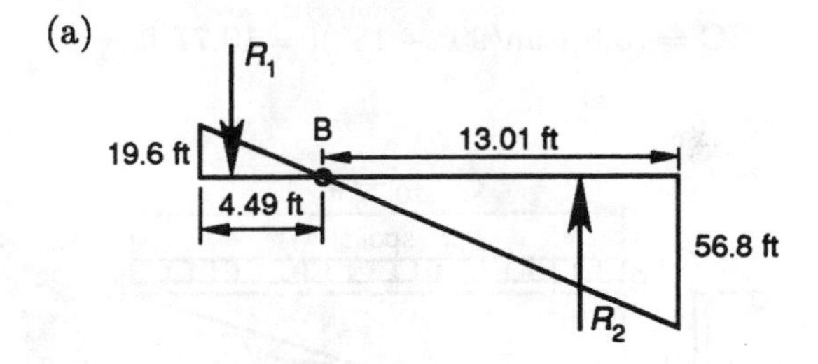

Calculate and locate the resultants.

$$\left(\frac{19.6}{19.6 + 56.8}\right)(17.5) = 4.49 \text{ ft}$$
$$\left(\frac{56.8}{19.6 + 56.8}\right)(17.5) = 13.01 \text{ ft}$$

$$R_1 = \left(\tfrac{1}{2}\right)(19.6)(4.49) = 44 \text{ tons, located } \left(\tfrac{2}{3}\right)(4.49) = 2.99 \text{ ft from point B}$$

$$R_2 = \left(\tfrac{1}{2}\right)(56.8)(13.01) = 369.5 \text{ tons, located } \left(\tfrac{2}{3}\right)(13.01) = 8.67 \text{ ft from point B}$$

Taking clockwise moments as positive,

$$\sum M_B = M_{\text{ext}} - (44)(2.99) - (369.5)(8.67) = 0$$
$$M_{\text{ext}} = 3335 \text{ ft-tons per foot of width}$$

The total moment is

$$M_{\text{total}} = (3335)(17.5) = \boxed{58{,}360 \text{ ft-tons}}$$

(b) **The plate will act as a hinge, and no moment will be transmitted across the plate.**

6. From Eq. 9.10,

$$\gamma = (1 + 0.20)(115) = 138 \text{ pcf}$$

The weight of the soil excavated per foot of trench is

$$\left(\frac{18}{12}\right)(3.5)(138) = 725 \text{ lbf/ft}$$

Half of the spoils will appear on each side of the trench.

$$\frac{725}{2} = 362 \text{ lbf/ft per side}$$

(a) Since $\phi > 5°$, failure will be toe slope (see p. 10-15). Investigate the stability with the trial wedge method. Since the soil is somewhat cohesive ($c > 0$), the failure plane would make an angle of 18° or greater. The horizontal distance BC is

$$\text{BC} = (3.5)[\tan(90° - 18°)] = 10.77 \text{ ft}$$

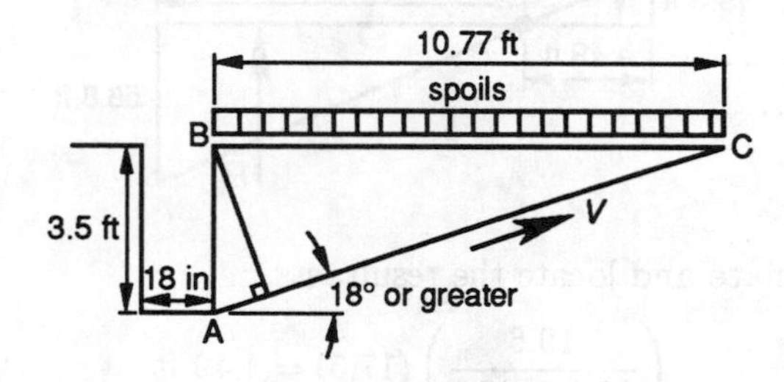

The length of line AC is

$$\sqrt{(3.5)^2 + (10.77)^2} = 11.3 \text{ ft}$$

The soil weight in ABC is

$$\left(\tfrac{1}{2}\right)(3.5)(10.77)(138) = 2600 \text{ lbf}$$

The average pressure occurs at the average depth.

$$p_h = \left(\frac{3.5}{2}\right)(138) = 241.5 \text{ psf}$$

The surcharge from the spoils is

$$p_q = \frac{362}{10.77} = 33.6 \text{ psf}$$

From Eq. 9.35, the shear strength of the soil is

$$\begin{aligned} S_{us} &= c + p \tan\phi \\ &= 200 + (241.5 + 33.6)(\tan 18°) \\ &= 289 \text{ psf} \end{aligned}$$

The ultimate resisting shear force is

$$V = (289)(11.3) = 3266 \text{ lbf/ft}$$

The total vertical weight is

$$W_{\text{wedge}} + W_{\text{spoils}} = 2600 + 362 = 2962 \text{ lbf/ft}$$

Resolve the soil weight into a force parallel to the assumed shear plane.

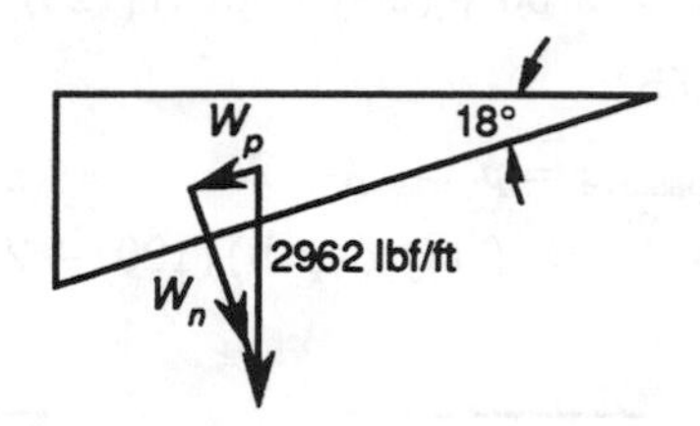

$$W_p = \left(2962 \ \frac{\text{lbf}}{\text{ft}}\right)(\sin 18°) = 915 \text{ lbf/ft}$$

Since 915 lbf/ft < 3266 lbf/ft,

The soil will not slide.

The factor of safety against sliding is

$$F_{\text{sliding}} = \frac{3266}{915} = 3.57$$

This process can be duplicated for other angles.

ϕ	BC	AC	soil mass	p_q	S_{us}	V	W_p	F	failure?
18°	10.77	11.3	2600	33.6	289	3266	915	3.57	no
30°	6.06	7.0	1463	59.7	298	2086	913	2.29	no
40°	4.17	5.44	1007	86.8	307	1670	880	1.90	no
60°	2.02	4.04	488	179.2	337	1361	736	1.84	no
80°	0.62	3.55	150	584.0	468	1661	504	3.29	no

(b) **2 ft**

(c) **yes** [see table]

(d) For a flexible pipe of this small size and buried as deep as this, there is no special bedding. Backfill with granular soil to eliminate settling of fill.

(e)
- place barricades along open trench
- join pipes above ground, not in trench
- have someone outside of trench spotting (less than 5 ft deep does not normally get shoring)
- place exit ladders every 50 ft

7. (a) $A = (90)(30) + (30)(30) = 3600 \text{ ft}^2$

$$p_{\text{applied}} = \frac{(20)(2000)}{3600} = 11.11 \text{ psf}$$

Use Fig. 10.11.

Draw the slab such that 1 in = 30 ft. Approximately 46 squares are covered.

$$p_{30 \text{ ft}} = (46)(0.005)(11.11) = \boxed{2.56 \text{ psf}}$$

(b) Draw the slab such that 1 in = 45 ft. Approximately 63 squares are covered.

$$p_{45 \text{ ft}} = (63)(0.005)(11.11) = \boxed{3.5 \text{ psf}}$$

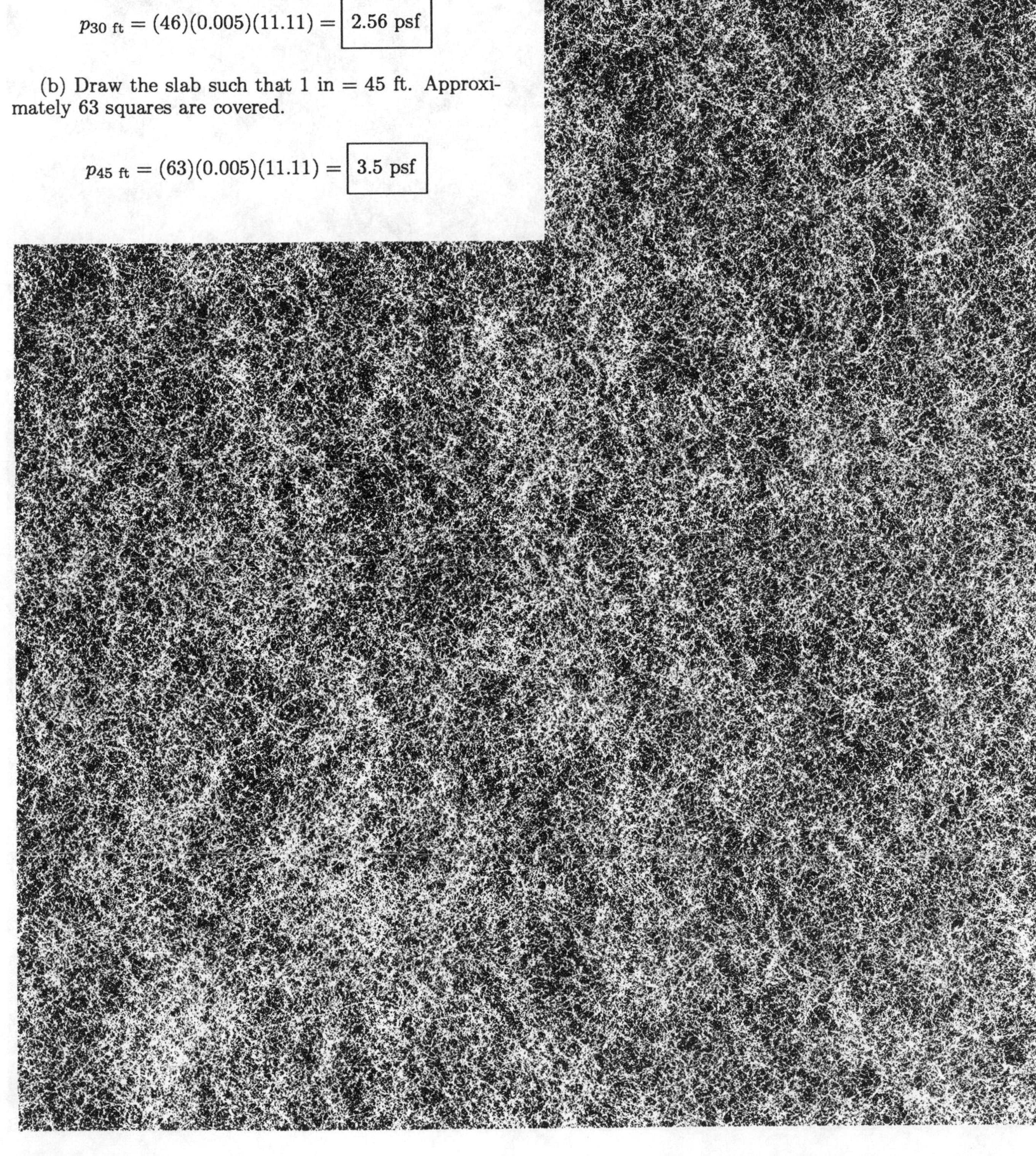

STATICS

Untimed

1.

DE:	3733 lbf (compression)
DC:	1600 lbf (tension)
CE:	2667 lbf (compression)
CB:	2133 lbf (compression)
BE:	0
BA:	2133 lbf (compression)
AE:	2667 lbf (tension)
AF:	0
FE:	8000 lbf (compression)

2. Use the procedure on p. 11-12, first moving the origin to the apex of the tripod so that the coordinates of the tripod bases become

A: (5,–12,0)
B: (0,–8,–8)
C: (–4,–7,6)

By inspection, $F_x = 1200$, $F_y = 0$, and $F_z = 0$.

$$L_A = \sqrt{(5)^2 + (-12)^2 + (0)^2} = 13$$

$$\cos\theta_{xA} = \frac{5}{13} = 0.385$$

$$\cos\theta_{yA} = \frac{-12}{13} = -0.923$$

$$\cos\theta_{zA} = \frac{0}{13} = 0$$

Similarly, $L_B = \sqrt{(0)^2 + (-8)^2 + (-8)^2} = 11.31$.

$$\cos\theta_{xB} = \frac{0}{11.31} = 0$$

$$\cos\theta_{yB} = \frac{-8}{11.31} = -0.707$$

$$\cos\theta_{zB} = \frac{-8}{11.31} = -0.707$$

And, $L_C = \sqrt{(-4)^2 + (-7)^2 + (6)^2} = 10.05$.

$$\cos\theta_{xC} = \frac{-4}{10.05} = -0.398$$

$$\cos\theta_{yC} = \frac{-7}{10.05} = -0.697$$

$$\cos\theta_{zC} = \frac{6}{10.05} = 0.597$$

From Eqs. 11.65, 11.66, and 11.67,

$$\begin{aligned} 0.385F_A \qquad\qquad\quad -0.398F_C &= -1200 \\ -0.923F_A \quad -0.707F_B \quad -0.697F_C &= 0 \\ -0.707F_B \quad +0.597F_C &= 0 \end{aligned}$$

Solving these simulanteously,

$$\boxed{\begin{aligned} F_A &= -1793 \text{ (compression)} \\ F_B &= 1080 \text{ (tension)} \\ F_C &= 1279 \text{ (tension)} \end{aligned}}$$

3. Divide into three areas. (Figure is not drawn to scale.)

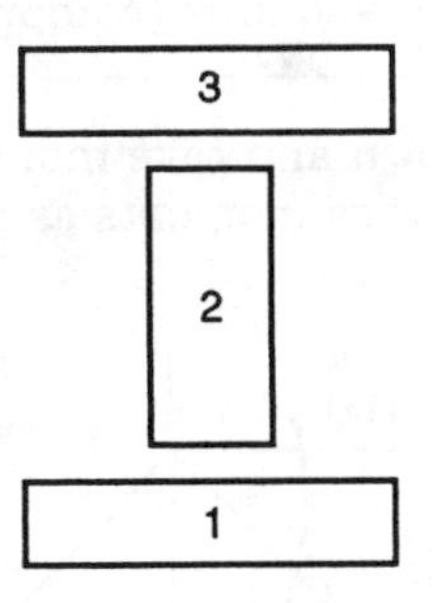

$$A_1 = (4)(1) = 4 \text{ ft}^2$$
$$\bar{y}_1 = \tfrac{1}{2} \text{ ft}$$
$$A_2 = (2)(12) = 24 \text{ ft}^2$$
$$\bar{y}_2 = 1 + 6 = 7 \text{ ft}$$
$$A_3 = (6)\left(\tfrac{1}{2}\right) = 3 \text{ ft}^2$$
$$\bar{y}_3 = 13.25 \text{ ft}$$
$$\bar{y}_c = \frac{(4)(0.5) + (24)(7) + (3)(13.25)}{4 + 24 + 3} = 6.77 \text{ ft}$$
$$I_{c1} = \frac{bh^3}{12} = \frac{(4)(1)^3}{12} = 0.333 \text{ ft}^4$$
$$d_1 = 6.77 - 0.5 = 6.27 \text{ ft}$$
$$I_{c2} = \frac{(2)(12)^3}{12} = 288 \text{ ft}^4$$
$$d_2 = 0.23 \text{ ft}$$
$$I_{c3} = \frac{(6)\left(\frac{1}{2}\right)^3}{12} = 0.0625 \text{ ft}^4$$
$$d_3 = 6.48 \text{ ft}$$

Using the parallel axis theorem three times,

$$\begin{aligned} I_{\text{total}} &= 0.333 + (4)(6.27)^2 + 288 + (24)(0.23)^2 \\ &\quad + 0.0625 + (3)(6.48)^2 \\ &= \boxed{572.89 \text{ ft}^4} \end{aligned}$$

4. By symmetry, $A_y = L_y = 160$ kips.

For DE, cut as shown and sum the vertical forces.

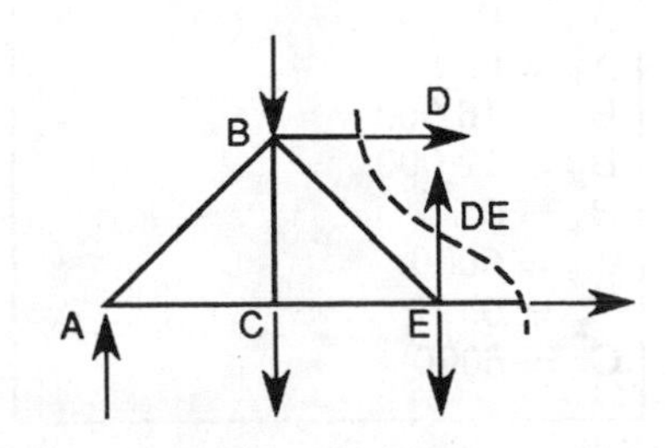

Taking upward forces as positive,

$$\sum F_y = 160 + \text{DE} - 60 - 60 - 4 = 0$$

$$\text{DE} = \boxed{-36 \text{ kips (compression)}}$$

For HJ, cut as shown and take moments about point I. Take counterclockwise moments as positive.

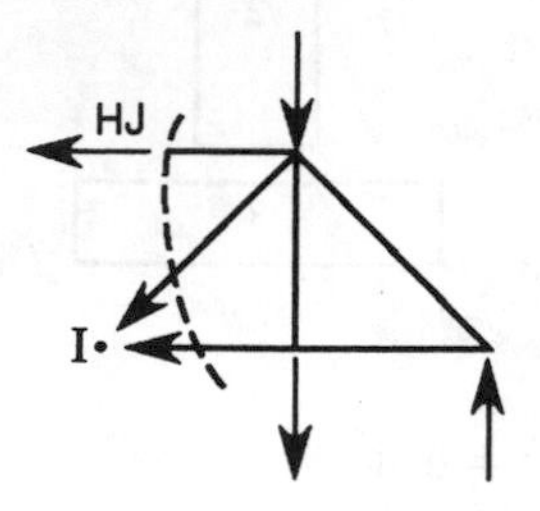

$$\sum M_{\text{I}} = (160)(60) - (60)(30) - (4)(30) + 20\text{HJ} = 0$$

$$\text{HJ} = \boxed{-384 \text{ kips (compression)}}$$

5. In the x-y plane,

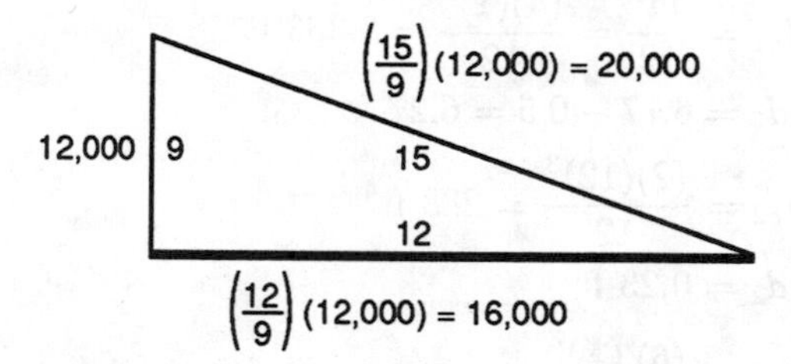

In the x-z plane,

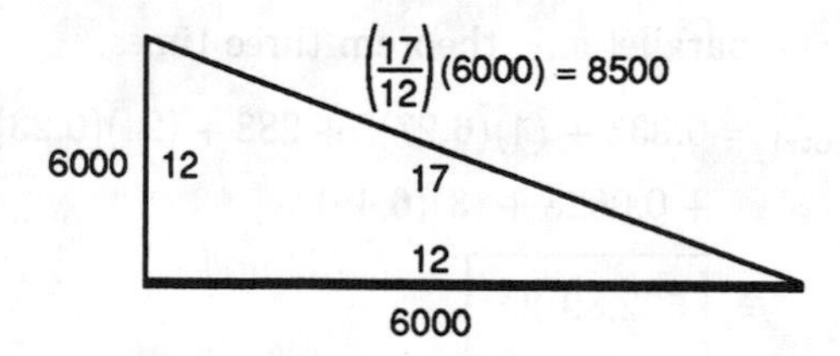

$$\boxed{\begin{aligned}
&\text{A}_x = 6000 + 16{,}000 = 22{,}000\\
&\text{A}_y = 0\\
&\text{A}_z = 0\\
&\text{B}_x = 16{,}000\\
&\text{B}_y = 12{,}000\\
&\text{B}_z = 0\\
&\text{C}_x = 6000\\
&\text{C}_y = 0\\
&\text{C}_z = 6000
\end{aligned}}$$

6. Refer to pp. 11-10 and 11-11. From Fig. 11.12,

$$w = 2 \text{ lbf/ft}$$
$$S = 10 \text{ ft}$$
$$a = 50 \text{ ft}$$

Solving Eq. 11.51 by trial and error gives distance c.

$$S = c\left[\cosh\left(\frac{a}{c}\right) - 1\right]$$
$$c = 126.6 \text{ ft}$$

Minimum tension occurs where there is no vertical component. The midpoint (horizontal) tension is given by Eq. 11.53.

$$H = wc = (2)(126.6) = \boxed{253.2 \text{ lbf}}$$

The endpoint (maximum) tension is given at a cable support, $y = c + S$. From Eq. 11.55,

$$T = wy = w(c + S)$$
$$= (2)(126.6 + 10) = \boxed{273.2 \text{ lbf}}$$

7. From Eq. 11.55,

$$y = \frac{T}{w} = \frac{500}{2} = 250 \text{ ft at right support}$$

From Eq. 11.48,

$$250 = c\left[\cosh\left(\frac{50}{c}\right)\right]$$
$$c = 245 \text{ ft by trial and error}$$

From Eq. 11.51,

$$S = (245)\left[\cosh\left(\frac{50}{245}\right) - 1\right] = \boxed{5.12 \text{ ft}}$$

8. By inspection, $\overline{y} = 0$. To find $\overline{x}$, divide the object into three parts. (Figure is not drawn to scale.)

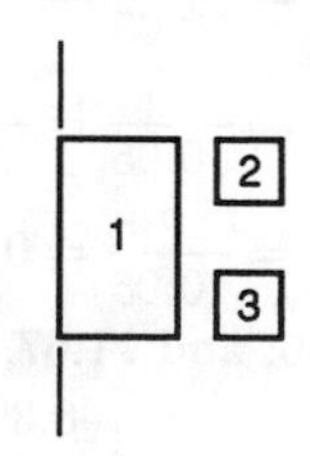

$$A_1 = (8)(4) = 32$$
$$\overline{x}_1 = 2$$
$$A_2 = A_3 = (2)(4) = 8$$
$$\overline{x}_2 = \overline{x}_3 = 6$$
$$\overline{x} = \frac{(32)(2) + (8)(6) + (8)(6)}{32 + 8 + 8} = \boxed{3.333 \text{ ft}}$$

9. For the parabola,

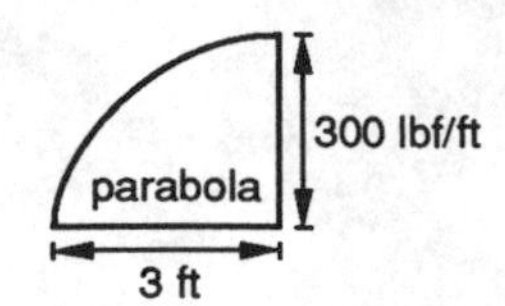

From p. 1-2,

$$A_{\text{parabola}} = \frac{2bh}{3} = \frac{(2)(300)(3)}{3} = \boxed{600 \text{ lbf}}$$

Cutting the parabola in half does not change the y-component of the centroid. From p. 11-20, the centroid is

$$\frac{3h}{5} = \frac{(3)(3)}{5} = \boxed{1.8 \text{ ft from tip}}$$

Divide the remaining area into a triangle and a rectangle.

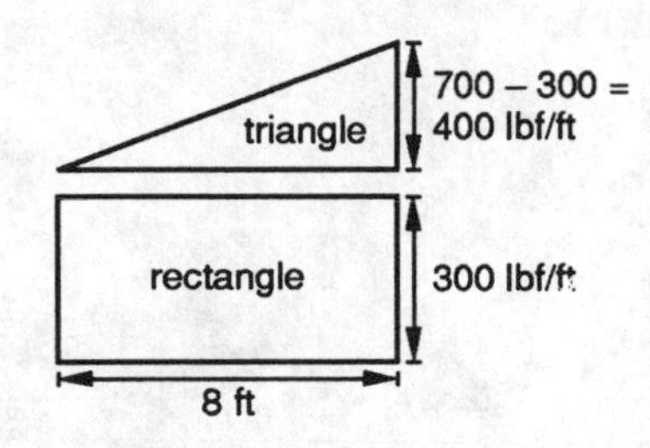

For the rectangle,

$$A_{\text{rectangle}} = (8)(300) = \boxed{2400 \text{ lbf}}$$

The centroid is

$$\boxed{4 \text{ ft from the right end}}$$

For the triangle,

$$A_{\text{triangle}} = \left(\tfrac{1}{2}\right)(8)(700 - 300) = \boxed{1600 \text{ lbf}}$$

The centroid is located

$$\frac{8 \text{ ft}}{3} = \boxed{2.67 \text{ ft from the right end}}$$

10. First, find the reactions. Take clockwise moments about A as positive.

$$\sum M_{\text{A}} = (27)(-\text{D}_y) + (18)(24) + (9)(22.5) = 0$$
$$\text{D}_y = 23.5 \text{ kips}$$
$$\text{A}_y = 18 + 24 + 4.5 - 23.5 = 23 \text{ kips}$$

The general force triangle is

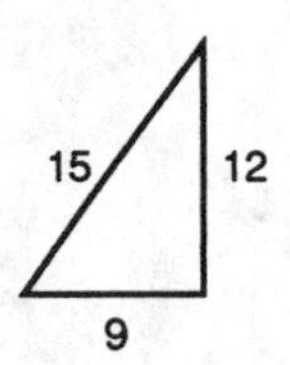

At pin A,

$\text{AE}_y = 23$ kips

$\text{AE}_x = \left(\tfrac{9}{12}\right)(23) = 17.25$ kips

$$\boxed{\begin{array}{l}\text{AE} = \left(\tfrac{15}{12}\right)(23) = 28.75 \text{ kips (compression)} \\ \text{AB} = \text{AE}_x = 17.25 \text{ kips (tension)}\end{array}}$$

At pin B,

$$\boxed{\begin{array}{l}\text{BE} = 4.5 \text{ kips (tension)} \\ \text{BC} = \text{AB} = 17.25 \text{ kips (tension)}\end{array}}$$

At pin D,

$\text{DF}_y = 23.5$ kips

$\text{DF}_x = \left(\tfrac{9}{12}\right)(23.5) = 17.63$ kips

$$\boxed{\begin{array}{l}\text{DF} = \left(\tfrac{15}{12}\right)(23.5) = 29.38 \text{ kips (compression)} \\ \text{DC} = \text{DF}_x = 17.63 \text{ kips (tension)}\end{array}}$$

At pin F,

$$\boxed{\begin{array}{l}\text{FE} = \text{DF}_x = 17.63 \text{ kips (compression)} \\ \text{FC} = 24 - \text{DF}_y = 0.5 \text{ kips (compression)}\end{array}}$$

At pin C,

$\text{CE}_y = \text{FC} = 0.5$ kips

$$\boxed{\text{CE} = \left(\tfrac{15}{12}\right)(0.5) = 0.625 \text{ kips (tension)}}$$

MECHANICS OF MATERIALS

Untimed

1.

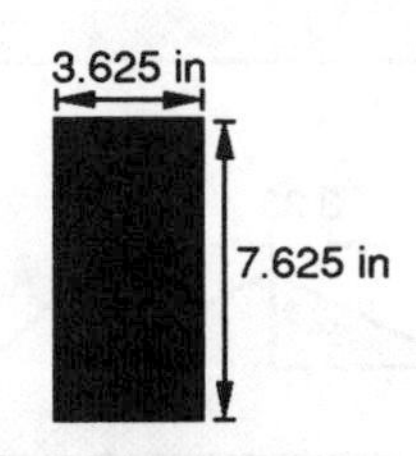

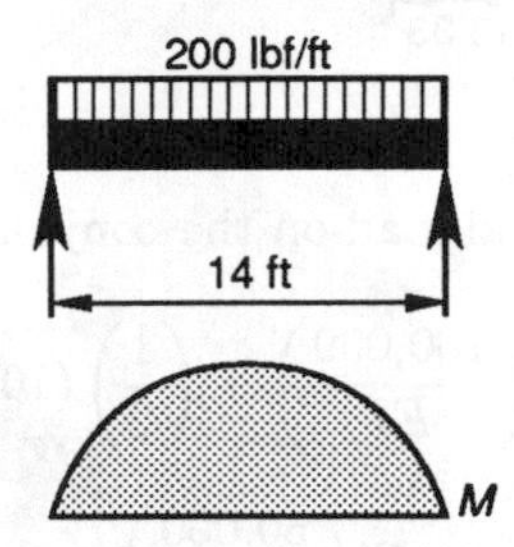

$$I = \frac{(3.625)(7.625)^3}{12} = 133.92 \text{ in}^4$$

$$M_{\text{max}} = (7)\left(\frac{7}{2}\right)(200) = 4900 \text{ ft-lbf (58,800 in-lbf)}$$

$$\sigma_{b,\text{max}} = \frac{(58{,}800)\left(\frac{7.625}{2}\right)}{133.92} = \boxed{1674 \text{ lbf/in}^2 \text{ (psi)}}$$

$$V_{\text{max}} = \frac{(14)(200)}{2} = 1400 \text{ lbf}$$

$$\tau_{\text{max}} = \frac{3V}{2A} = \frac{(3)(1400)}{(2)(3.625)(7.625)} = \boxed{76.0 \text{ lbf/in}^2 \text{ (psi)}}$$

2.

$$n = \frac{2.9 \times 10^7}{2 \times 10^6} = 14.5$$

$$A_s = 1 \text{ in}^2$$

$$nA_s = (14.5)(1) = 14.5 \text{ in}^2$$

The two relevant sections are

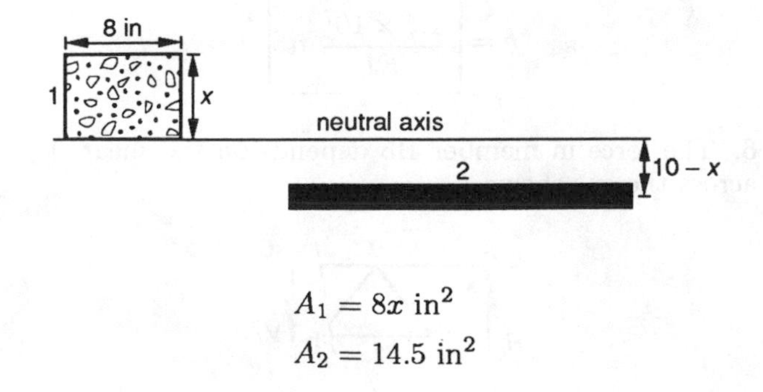

$$A_1 = 8x \text{ in}^2$$

$$A_2 = 14.5 \text{ in}^2$$

For balanced design, the two area moments must be the same. (Concrete and steel will fail simultaneously.)

$$(8x)\left(\tfrac{1}{2}x\right) = (10 - x)(14.5)$$

$$4x^2 + 14.5x - 145 = 0$$

$$x = 4.4752 \text{ in}$$

$$10 - x = 5.5248 \text{ in}$$

h for steel is unknown, so the steel's contribution to I is found from the radius of gyration concept. Disregarding the concrete below the neutral axis, the centroidal moment of inertia is

$$I_c = \frac{(8)(4.4752)^3}{3} + (14.5)(5.5248)^2 = 681.6 \text{ in}^4$$

The maximum concrete bending stress is

$$\sigma_c = \frac{(8125)(12)(4.4752)}{681.6} = \boxed{640.2 \text{ lbf/in}^2 \text{ (psi)}}$$

The maximum steel bending stress is

$$\sigma_s = \frac{(14.5)(8125)(12)(5.5248)}{681.6} = \boxed{11{,}459 \text{ lbf/in}^2 \text{ (psi)}}$$

3. Use the virtual work method. The reactions are

$$R_1 = 7071 \text{ lbf}$$

$$R_2 = 7071 \text{ lbf}$$

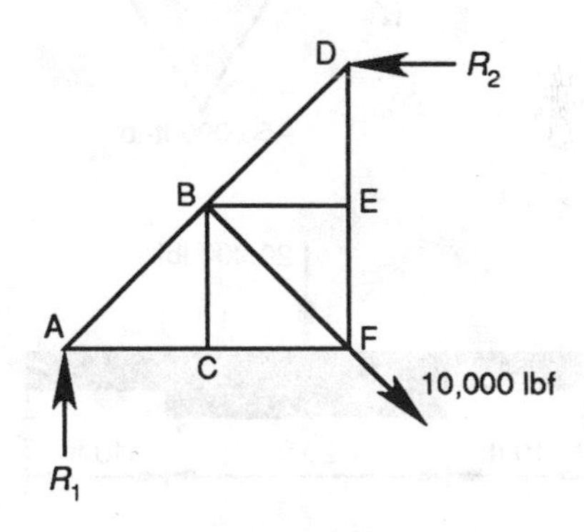

The member forces, S, tabulated below can be found from principles of static equilibrium.

member	force, S	u	Su
AB	−10,000	−1.414	14,140
AC	7071	0	0
CB	0	0	0
CF	7071	0	0
BF	0	0	0
FE	7071	1	7071
EB	0	0	0
ED	7071	1	7071
BD	−10,000	−1.414	14,140
total			42,420

Apply a vertical load of 1 at point F. Calculate the forces u in each member.

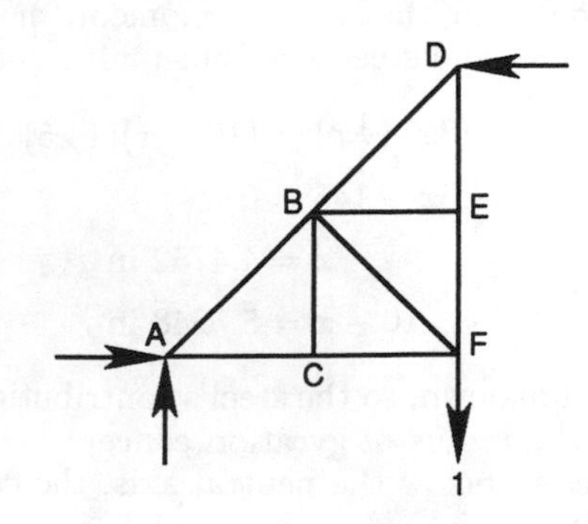

$$\delta = \sum \frac{SuL}{AE} = \frac{(42{,}420)(50)}{2.9 \times 10^7} = \boxed{0.0731 \text{ in}}$$

($\delta = 0.0693$ in if the strain energy method is used. However, that method is less accurate.)

4.

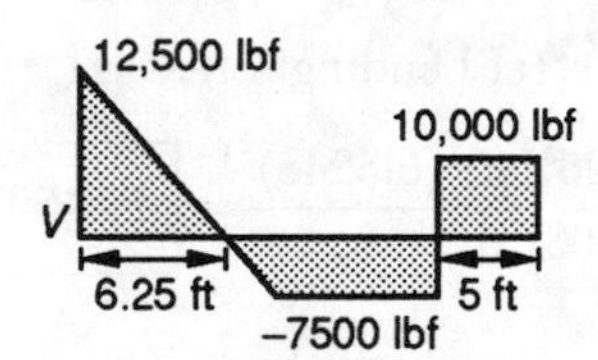

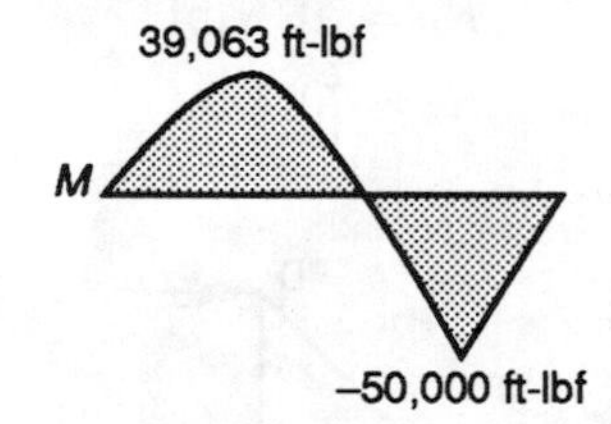

5.

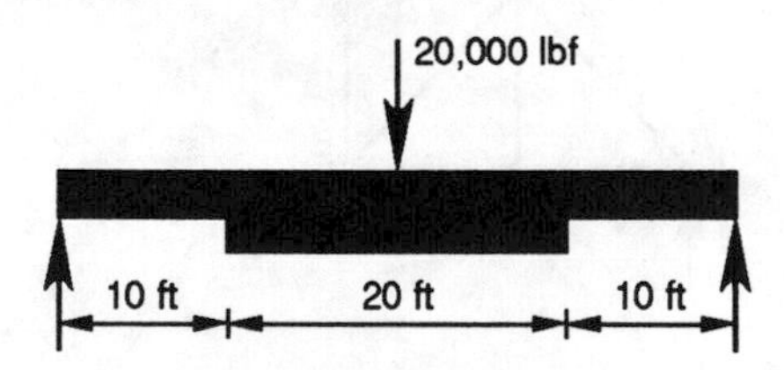

Assume simple supports. Use the conjugate beam method.

step 1:

$$M_{\text{max}} = (20)(10{,}000) = 200{,}000 \text{ ft-lbf}$$

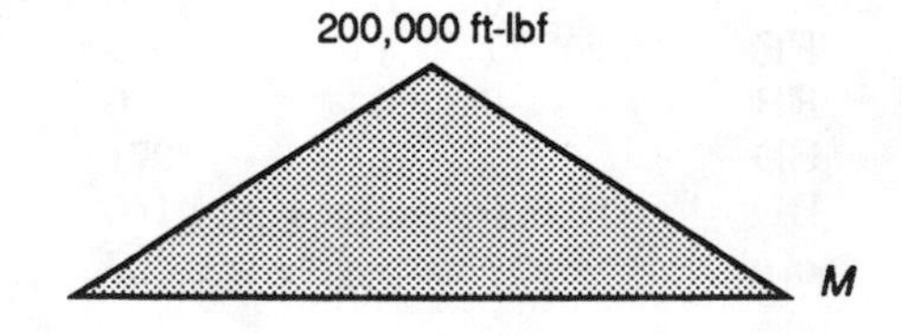

step 2:

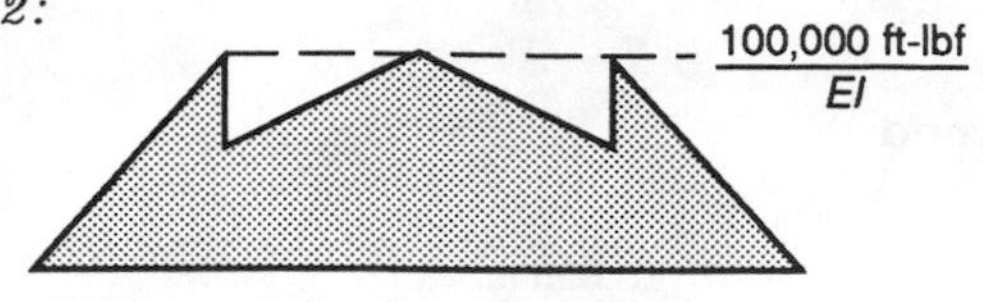

steps 3 and 4:

The total load on the conjugate beam is

$$(2)\left[\left(\frac{1}{2}\right)(10)\left(\frac{100{,}000}{EI}\right) + \left(\frac{1}{2}\right)(10)\left(\frac{50{,}000}{EI}\right) + (10)\left(\frac{50{,}000}{EI}\right)\right]$$

$$= \frac{2.5 \times 10^6 \text{ ft}^2\text{-lbf}}{EI}$$

$$L^* = R^* = \frac{2.5 \times 10^6 \text{ ft}^2\text{-lbf}}{2EI} = \frac{1.25 \times 10^6 \text{ ft}^2\text{-lbf}}{EI}$$

step 5: Taking clockwise moments as positive, the conjugate moment at the midpoint is

$$M_{\text{mid}} = (20)\left(\frac{1.25 \times 10^6}{EI}\right) - \left(\frac{1}{2}\right)(10) \times \left(\frac{100{,}000}{EI}\right)(13.33) - \left(\frac{1}{2}\right)(10)\left(\frac{50{,}000}{EI}\right)(3.33) - (10)\left(\frac{50{,}000}{EI}\right)(5)$$

$$= \frac{1.5 \times 10^7 \text{ ft}^3\text{-lbf}}{EI}$$

If E is in lbf/ft^2 (psf) and I is in ft^4, then

$$\delta = \boxed{\frac{1.5 \times 10^7}{EI} \text{ ft}}$$

6. The force in member Bb depends on the shear, V, across the cut shown.

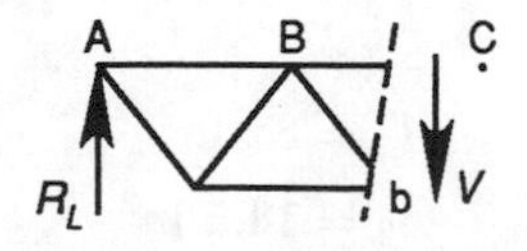

• Influence diagram for shear across panel Bb:

If the unit load is to the right of point C, the reaction R_L is

$$R_L = \frac{x}{120}$$

(x is the distance from right reaction to the unit load.)

$$V = R_L = \frac{x}{120}$$

If the unit load is to the left of point B,

$$R_L = \frac{x}{120}$$
$$V = R_L - 1 = \frac{x}{120} - 1$$

At points B and C,

$$V_B = \frac{90}{120} - 1 = -\frac{1}{4}$$
$$V_C = \frac{60}{120} = 0.5$$

The influence diagram is

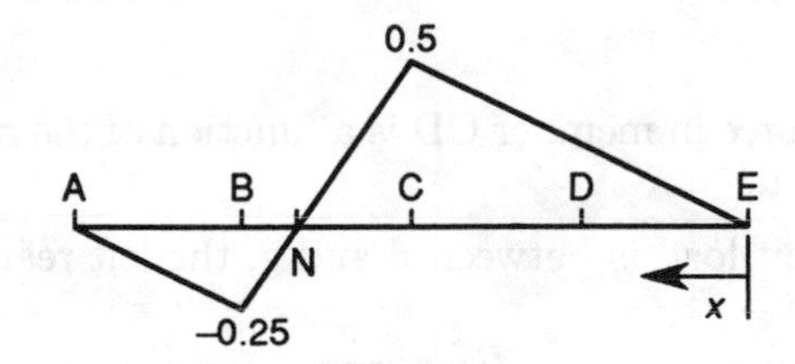

The neutral point, N, is located at

$$x = (2)(30) + \left(\frac{0.5}{0.5 + 0.25}\right)(30) = 80 \text{ ft}$$

• Maximum shear due to moving uniform load:

The moving load, (perhaps) representing a stream of cars, is allowed to be over any part or all of the bridge deck. The shear will be maximum in member Bb if the load is distributed from N to E.

The area under the influence line from N to E is

$$\left(\tfrac{1}{2}\right)[(20)(0.5) + (2)(30)(0.5)] = 20 \text{ ft}$$

The maximum shear, V, is

$$(20 \text{ ft})\left(2 \frac{\text{kips}}{\text{ft}}\right) = 40 \text{ kips}$$

• Maximum shear due to moving concentrated load:

From the influence diagram, maximum shear will occur when the concentrated load is at point C. The shear in panel Bb is

$$(0.5)(15) = 7.5 \text{ kips}$$

• Tension in member Bb:

The force triangle is

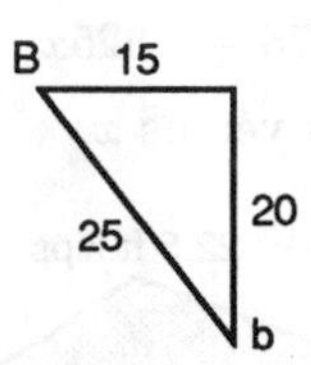

The total maximum shear across panel Bb is

$$40 + 7.5 = 47.5 \text{ kips}$$

The total maximum tension in member Bb is

$$(47.5)\left(\frac{25}{20}\right) = \boxed{59.375 \text{ kips}}$$

• Influence diagram for moment at point b:

Since the horizontal member BC cannot resist vertical shear, the shear influence diagram previously used will not work.

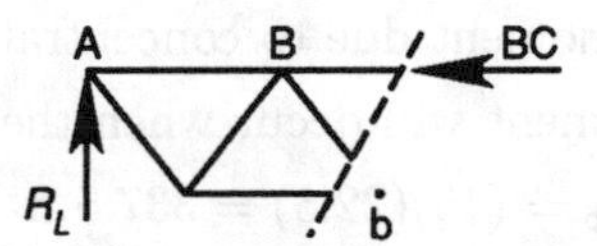

With no loads between A and C, the force in member BC can be found by summing the moments about point b. Taking clockwise moments as positive,

$$\sum M_b = 45R_L - 20(\text{BC}) = 0$$
$$\text{BC} = \frac{45R_L}{20}$$

$45R_L$ is the moment that the moment from force BC opposes. In general,

$$\text{BC} = \frac{M_b}{20}$$

If the load is between C and E,

$$R_L = \frac{x}{120} \quad [x \text{ is measured from E}]$$

The moment caused by R_L is

$$M_b = (45)\left(\frac{x}{120}\right) = 0.375x$$

If the load is between A and B, the reaction is

$$R_L = \frac{x}{120}$$

The moment at b is also affected by the load between A and B.

$$\begin{aligned} M_b &= 0.375x - (1)(x - 75) \\ &= 75 - 0.625x \end{aligned}$$

Plotting these values versus x,

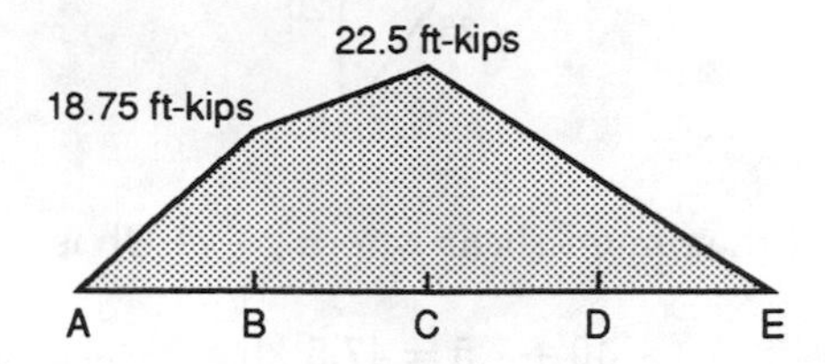

- Maximum moment due to uniform load:

The moment at b is maximum when the entire truss is loaded from A to E. The area under the curve is

$$\begin{aligned} &\left(\tfrac{1}{2}\right)(30)(18.75) + \left(\tfrac{1}{2}\right)(60)(22.5) + (30)(18.75) \\ &\quad + \left(\tfrac{1}{2}\right)(30)(22.5 - 18.75) \\ &= 1575 \text{ ft}^2 \end{aligned}$$

The maximum moment is

$$M_b = \left(2 \ \frac{\text{kips}}{\text{ft}}\right)(1575 \text{ ft}^2) = 3150 \text{ ft-kips}$$

- Maximum moment due to concentrated load:

Maximum moment will occur when the load is at C.

$$M_b = (15)(22.5) = 337.5 \text{ ft-kips}$$

- Total maximum moment:

$$M_b = 337.5 + 3150 = 3487.5 \text{ ft-kips}$$

- Compression in BC:

$$\text{BC} = \frac{M_b}{20} = \frac{3487.5}{20} = \boxed{174.4 \text{ kips}}$$

7. $$I = \frac{(3.5)^4}{12} = 12.51 \text{ in}^4 \quad \text{[finished lumber size]}$$

$$k = \sqrt{\frac{I}{A}} = \sqrt{\frac{12.51}{(3.5)^2}} = 1.01 \text{ in}$$

From Table 12.3, $C = 2$. The slenderness ratio is

$$\frac{L}{k} = \frac{(2)(9)(12)}{1.01} = 213.9$$

Since L/k is well above 100, Eq. 12.58 can be used if the stress is not excessive. (Note that most timber codes limit L/k to 50, so this would not be a permitted application.)

$$F_e = \frac{(\pi)^2(1.5 \times 10^6)(12.51)}{[(2)(9)(12)]^2} = \boxed{3970 \text{ lbf}}$$

$$\sigma = \frac{F_e}{A} = \frac{3970}{(3.5)^2} = 324.1 \text{ lbf/in}^2 \text{ (psi)}$$

This stress is less than $F_{c\|}$ for oak, and well below its elastic limit.

8. Use superposition.

- Uniform load:

Use case 5 on p. 12-31.

$$y_{\text{center}} = y_{\text{max}} = \frac{(5)\left(\frac{500}{12}\right)[(17)(12)]^4}{(384)(3 \times 10^7)(200)} = 0.1566 \text{ in}$$

- Concentrated load:

Use case 11 on p. 12-32.

$$\begin{aligned} L &= (17)(12) = 204 \text{ in} \\ b &= (5)(12) = 60 \text{ in} \\ x_a &= \left(\frac{17}{2}\right)(12) = 102 \text{ in} \end{aligned}$$

$$\begin{aligned} y_{\text{center}} = y_a &= \left[\frac{(2000)(60)(102)}{(6)(3 \times 10^7)(200)(204)}\right] \\ &\quad \times [(204)^2 - (60)^2 - (102)^2] \\ &= 0.0460 \text{ in} \end{aligned}$$

$$y_{\text{total}} = 0.1566 + 0.0460 = \boxed{0.2026 \text{ in}}$$

9. The force in member CD is a function of the moment at point d.

If the unit load is between d and g, the left reaction is

$$R_L = \frac{x}{162}$$

The moment at point d is

$$M_d = (3)(27)\left(\frac{x}{162}\right) = 0.5x \quad \text{[d to g]}$$

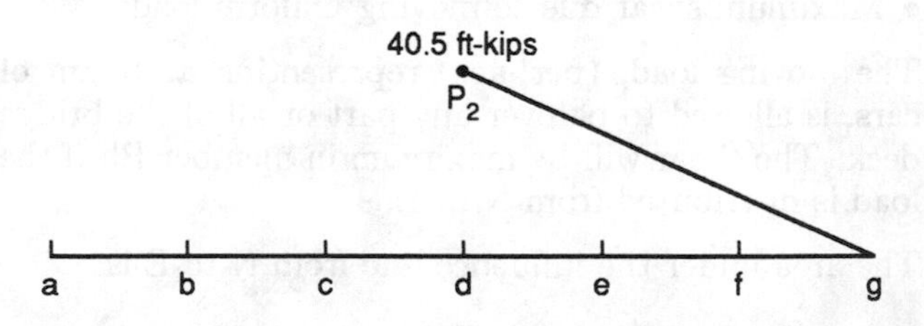

If the unit load is between a and c, the moment at point d is

$$\begin{aligned} M_d &= 0.5x - (1)[x - (3)(27)] \\ &= 81 - 0.5x \quad \text{[a to c]} \end{aligned}$$

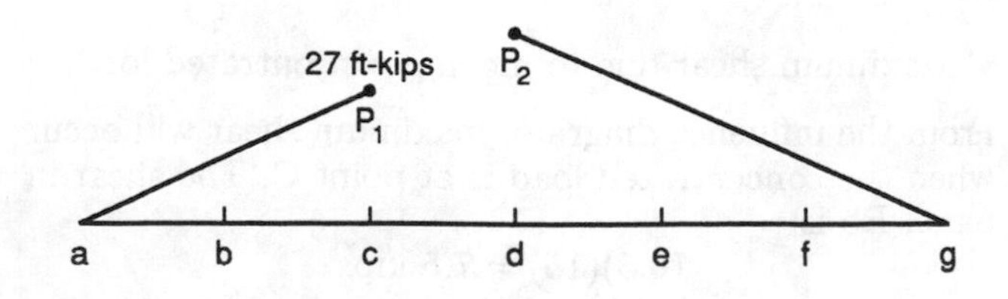

Completing the influence diagram by joining points P_1 and P_2, observe that the slope of a-P_1 is the same as that of P_1-P_2. This is because point d is at the center of the truss.

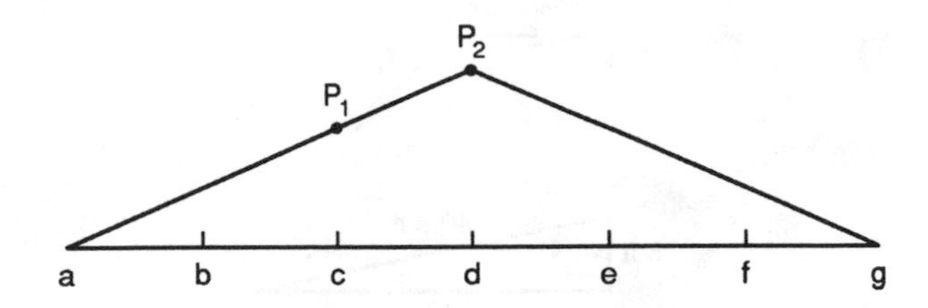

The resultant of the load group and its location are

$$8+40+7+30 = 85 \text{ kips}$$

$$\frac{(40)(16)+(7)(16+28)+(30)(16+28+16)}{85} = 32.33 \text{ ft}$$

The resultant is located 32.33 ft to the right of the 8-kip load.

Assume the load group moves from right to left.

- Case 1:

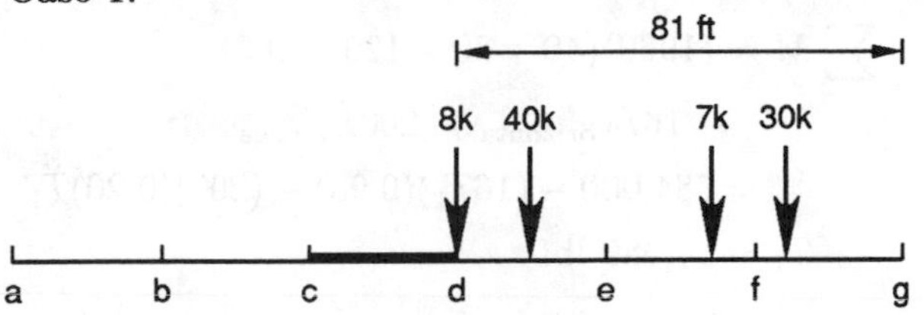

Taking counterclockwise moments as positive,

$$\sum M_g = (85)(81-32.33) - 162R_L = 0$$

$$R_L = 25.54 \text{ kips}$$

The shear does not change the sign under panel cd.

- Alternate test:

$$\frac{40+7+30}{(3)(27)} = 0.95$$

$$\frac{8}{(3)(27)} = 0.10$$

$0.95 > 0.10$, so greater moment is possible.

- Case 2:

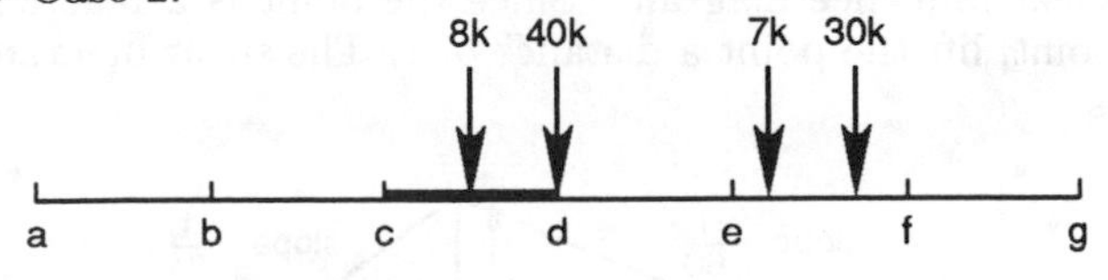

$$R_L = 33.9 \text{ kips}$$

$$33.9 - 8 - 40 = -14.1 \text{ kips}$$

Since the shear changes sign (goes through zero), the moment is maximum when the 40-kip load is at point d.

- Alternate test:

$$\frac{7+30}{(3)(27)} = 0.46$$

$$\frac{8+40}{(3)(27)} = 0.59$$

$0.46 < 0.59$, so no greater moment is possible.

load	x	influence diagram height	moment
8 kips	97	32.5	(8)(32.5) = 260 ft-kips
40 kips	81	40.5	(40)(40.5) = 1620 ft-kips
7 kips	53	26.5	(7)(26.5) = 185.5 ft-kips
30 kips	37	18.5	(30)(18.5) = 555 ft-kips
total			2620.5 ft-kips

The compression in CD is

$$CD = \frac{2620.5}{32} = \boxed{81.9 \text{ kips}}$$

Since $7+30 < 8+40$, if the load moves from left to right, it must reach the same position as case 2 for the moment to be maximum. Therefore, the left-to-right analysis is not needed.

- Alternate solution:

Having determined that the 40-kip load should be at point d, find the left reaction.

$$\sum M_{R_R} = 162R_L - (8)(97) - (40)(81) - (7)(53) - (30)(37) = 0$$

$$R_L = 33.93 \text{ kips}$$

Sum the moments about point d and use the method of sections.

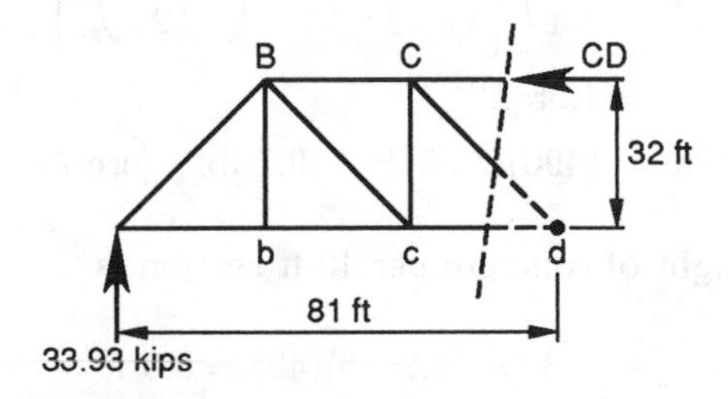

Taking clockwise moments as positive,

$$\sum M_d = (33.93)(81) - 32(CD) - (8)(16) = 0$$

$$CD = \boxed{81.9 \text{ kips}}$$

10. From Eq. 12.19,

$$z = \frac{M}{\sigma} = \frac{(1.5\times10^5)(12)}{20{,}000} = \boxed{90 \text{ in}^3}$$

Timed

1. Assumptions:

   - 6 × 19 standard hoisting rope (Table 12.5)
     weight = 0.4 lbf/ft
     breaking strength = 11 tons
   - vertical wire ropes remain vertical
   - pump and header support their ends
   - concrete density = 150 pcf (wet and dry)
   - infinite horizontal restraint (disregard pumping pulsations)
   - schedule-40 steel pipe (weight may not be significant)
   - ignore weight of pipe section connections
   - steel density = 490 pcf (Table 12.1)
   - vertical hangers are $\frac{1}{2}$-in steel rope also
   - connection stress risers are ignored (see comment at end of solution)
   - ignore cable self-weight
   - cable sections act as tension links

- Pipe:

$$D_o = 6.625 \text{ in}$$
$$D_i = 6.065 \text{ in}$$
$$A_i = 0.2006 \text{ ft}^2$$

The volume and weight of pipe per 40-ft section is

$$V = \left(\frac{\pi}{4}\right)\left[\left(\frac{6.625}{12}\right)^2 - \left(\frac{6.065}{12}\right)^2\right](40)$$
$$= 1.55 \text{ ft}^3$$
$$\text{weight} = (490)(1.55) = 759.7 \text{ lbf} \quad [\text{use } 760 \text{ lbf}]$$

The weight of concrete per 40-ft section is

$$V = (0.2006)(40) = 8.0 \text{ ft}^3$$
$$\text{weight} = (150)(8.0) = 1200 \text{ lbf}$$

The load per vertical hanger is

$$760 + 1200 = 1960 \text{ lbf}$$

Note: Since the horizontal force component is constant, the cable tension will be greatest where the cable is steepest. This appears to be the right-hand support, but insufficient information is given to fully evaluate the right-hand support.

At the left cable support,

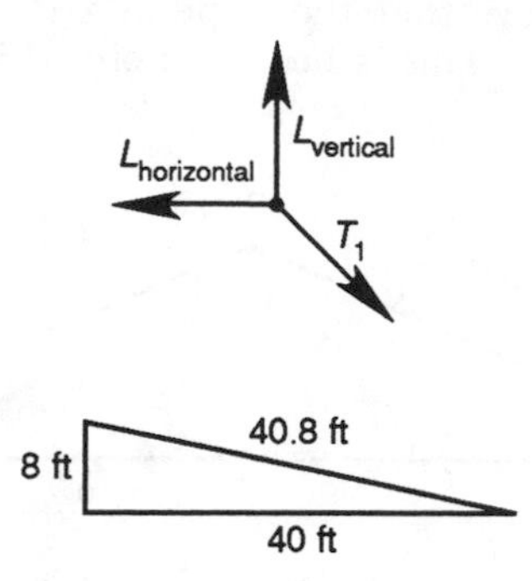

$$L_{\text{vertical}} = \left(\frac{8}{40.8}\right)T_1 = 0.196T_1 \quad [\text{use } 0.20T_1]$$
$$L_{\text{horizontal}} = \left(\frac{40}{40.8}\right)T_1 = 0.98T_1$$

Sum the moments about the right cable support to find T_1. Take counterclockwise moments to be positive.

$$\sum M = (1960)(40 + 80 + 120 + 160) - 16L_{\text{horizontal}} - 200L_{\text{vertical}}$$
$$= 784{,}000 - (16T_1)(0.98) - (200)(0.20)T_1$$
$$T_1 = 14{,}080 \text{ lbf}$$

Since 14,080 lbf < (11)(2000) = 22,000 lbf,

no breakage

Other factors:

- factor of safety for steel rope = 5
  (14,080)(5) = 70,400 lbf [not acceptable]
- A 50% loss of strength at connections (stress risers of 2.0) should also be considered.

2.

- Shear:

Use the method of virtual displacement to draw the shear influence diagram. Since the point is a reaction point, lift the point a distance of 1. The shear diagram is

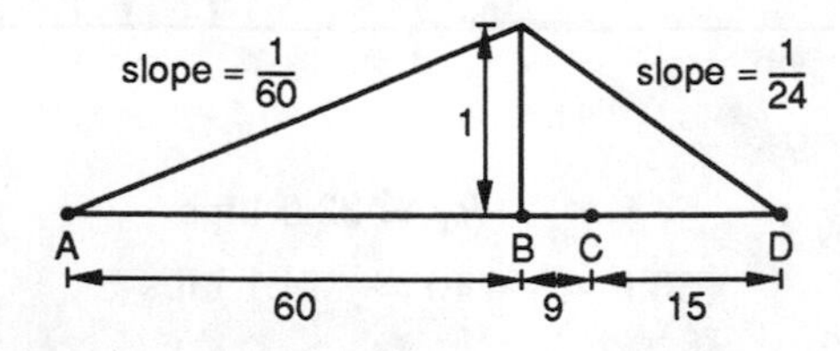

For maximum shear, one load or the other must be at point B.

By inspection, the effect of having both loads to the left of B is greater than having them to the right.

The maximum shear is

$$V_{max} = (30)\left(1 + \frac{54}{60}\right) = \boxed{57 \text{ kips}}$$

- Moment:

Put a hinge at point B and rotate.

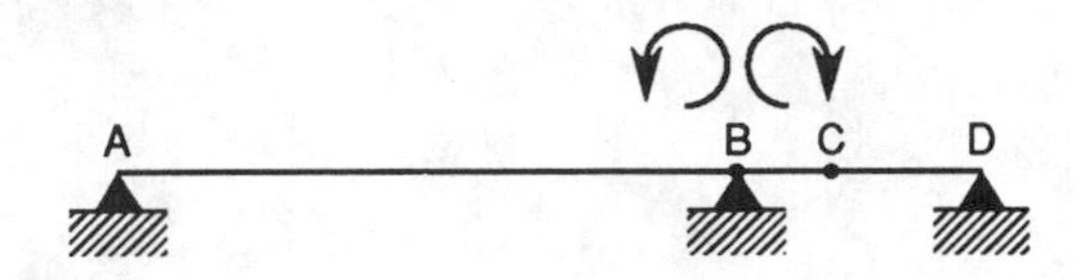

The moment influence diagram is

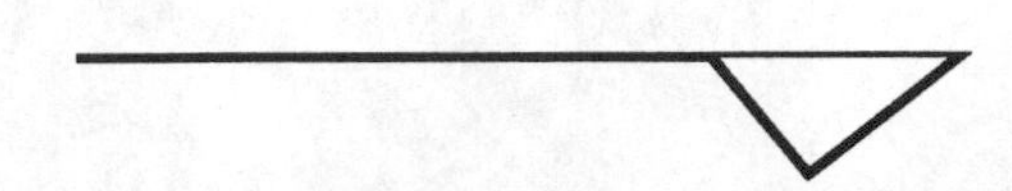

One of the loads should be at point C. Since the slope of the influence line is less between C and D, the ordinate 6 ft to the right of point C will be larger than the ordinate 6 ft to the left point of point C. The reaction at the hinge due to the 30-kip load 6 ft to the right of point C is

$$R = (30 \text{ kips})\left(\frac{15-6}{15}\right) = 18 \text{ kips}$$

The moment at point B is

$$\begin{aligned} M_B &= (9 \text{ ft})\left[30 \text{ kips} + \left(\frac{9}{15}\right)(30 \text{ kips})\right] \\ &= \boxed{432 \text{ ft-kips}} \end{aligned}$$

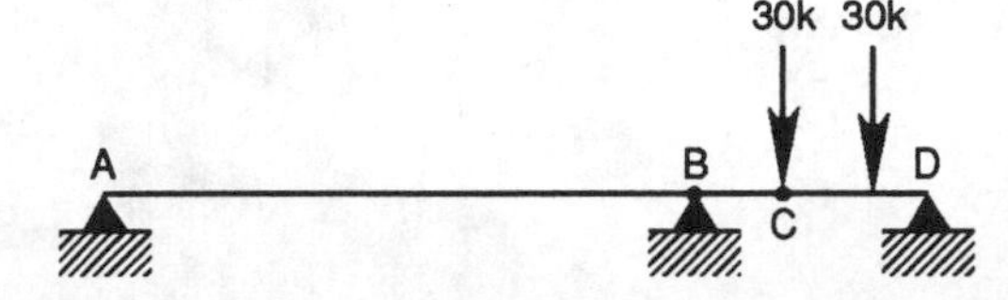

INDETERMINATE STRUCTURES

Untimed

1.

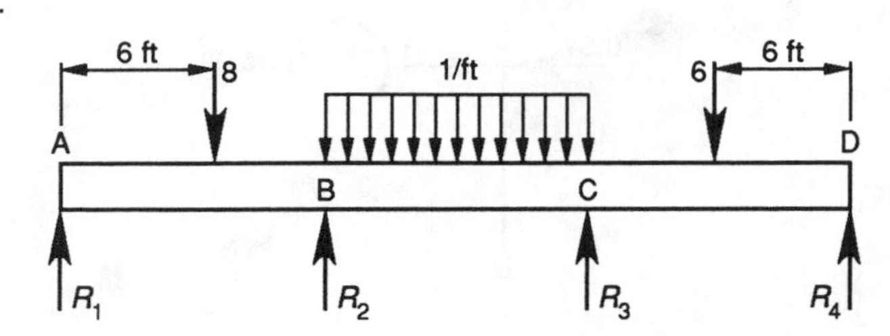

	A	AB	BC	BC	CD	CD
L	12		24		16	
EI	1		1		1	
R	$\frac{1}{12}$		$\frac{1}{24}$		$\frac{1}{16}$	
F	0	1	1	1	1	0
K		$\frac{1}{4}$	$\frac{1}{6}$	$\frac{1}{6}$	$\frac{3}{16}$	
D	1	0.6	0.4	0.471	0.529	1
C	$\frac{1}{2}$	0	$\frac{1}{2}$	$\frac{1}{2}$	0	$\frac{1}{2}$

FEM	−12	12	−48	48	−8.44	−14.1

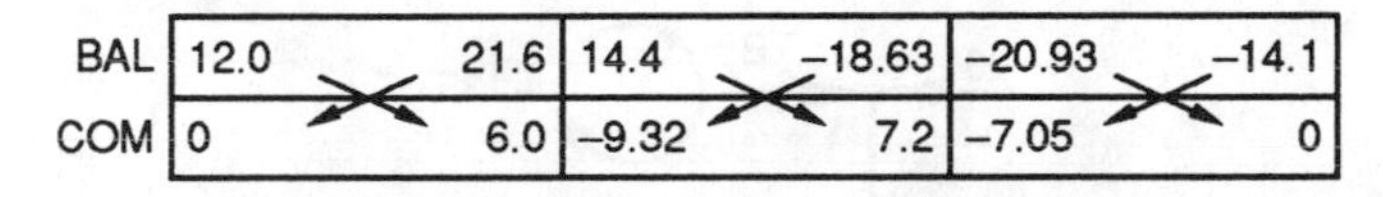

BAL	12.0	21.6	14.4	−18.63	−20.93	−14.1
COM	0	6.0	−9.32	7.2	−7.05	0

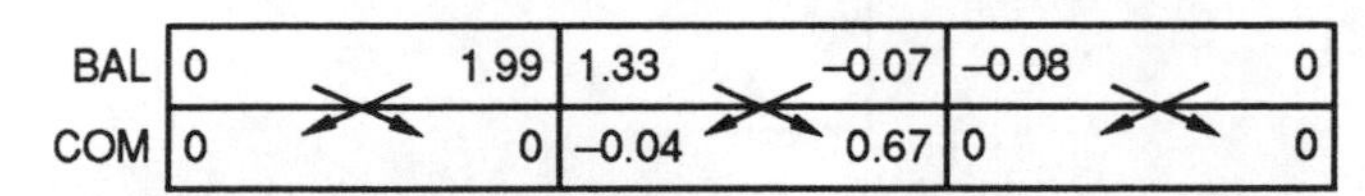

BAL	0	1.99	1.33	−0.07	−0.08	0
COM	0	0	−0.04	0.67	0	0

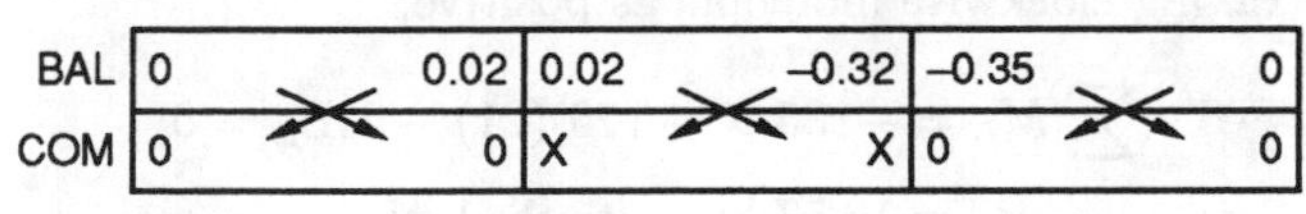

BAL	0	0.02	0.02	−0.32	−0.35	0
COM	0	0	X	X	0	0

total	0	41.61	−41.61	36.85	−36.85	0

Find reactions.

- Member AB:

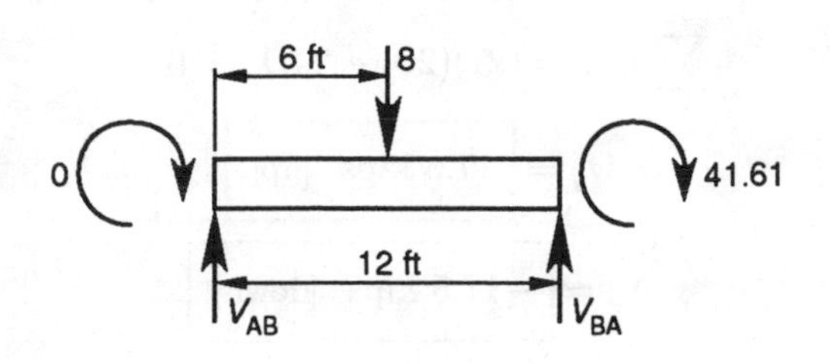

Taking clockwise moments and upward forces as positive,

$$\sum M_A = 0 + 41.61 + (6)(8) - 12V_{BA} = 0$$
$$V_{BA} = 7.47$$
$$\sum F_y = V_{AB} + 7.47 - 8 = 0$$
$$V_{AB} = 0.53$$

- Member BC:

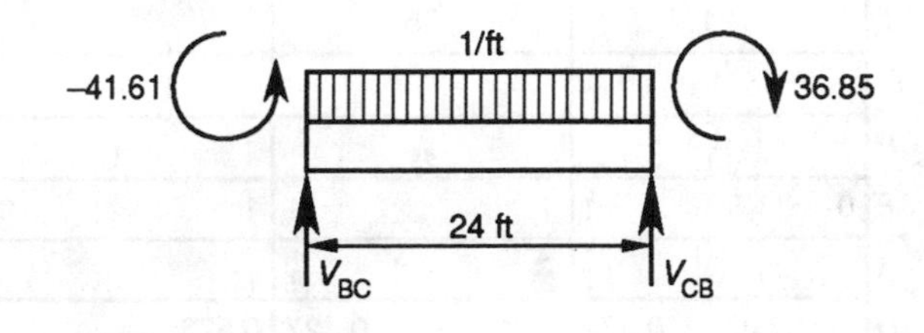

$$\sum M_B = -41.61 + 36.85 + (\tfrac{1}{2})(24)^2(1) - 24V_{CB} = 0$$
$$V_{CB} = 11.8$$
$$\sum F_y = V_{BC} + 11.8 - (24)(1) = 0$$
$$V_{BC} = 12.2$$

- Member CD:

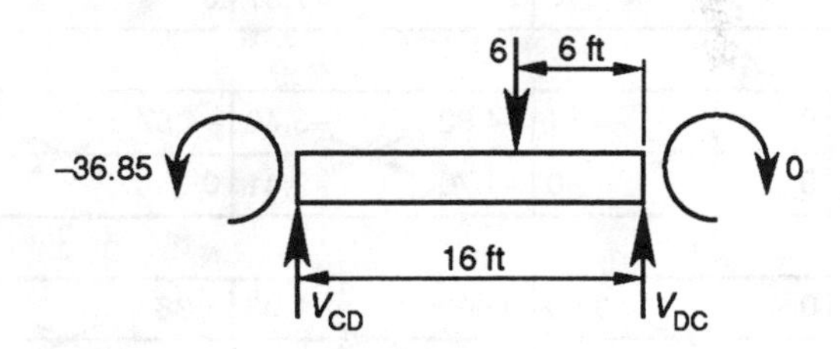

$$\sum M_C = -36.85 + (10)(6) - 16V_{DC} = 0$$
$$V_{DC} = 1.45$$
$$\sum F_y = V_{CD} + 1.45 - 6 = 0$$
$$V_{CD} = 4.55$$

The reactions are

$$R_1 = V_{AB} = 0.53$$
$$R_2 = V_{BA} + V_{BC} = 7.47 + 12.2 = 19.67$$
$$R_3 = V_{CB} + V_{CD} = 11.80 + 4.55 = 16.35$$
$$R_4 = V_{DC} = 1.45$$

2. Use the moment distribution worksheet to calculate the end moments produced when a total moment of 100 is distributed to the two vertical members. Draw free-bodies.

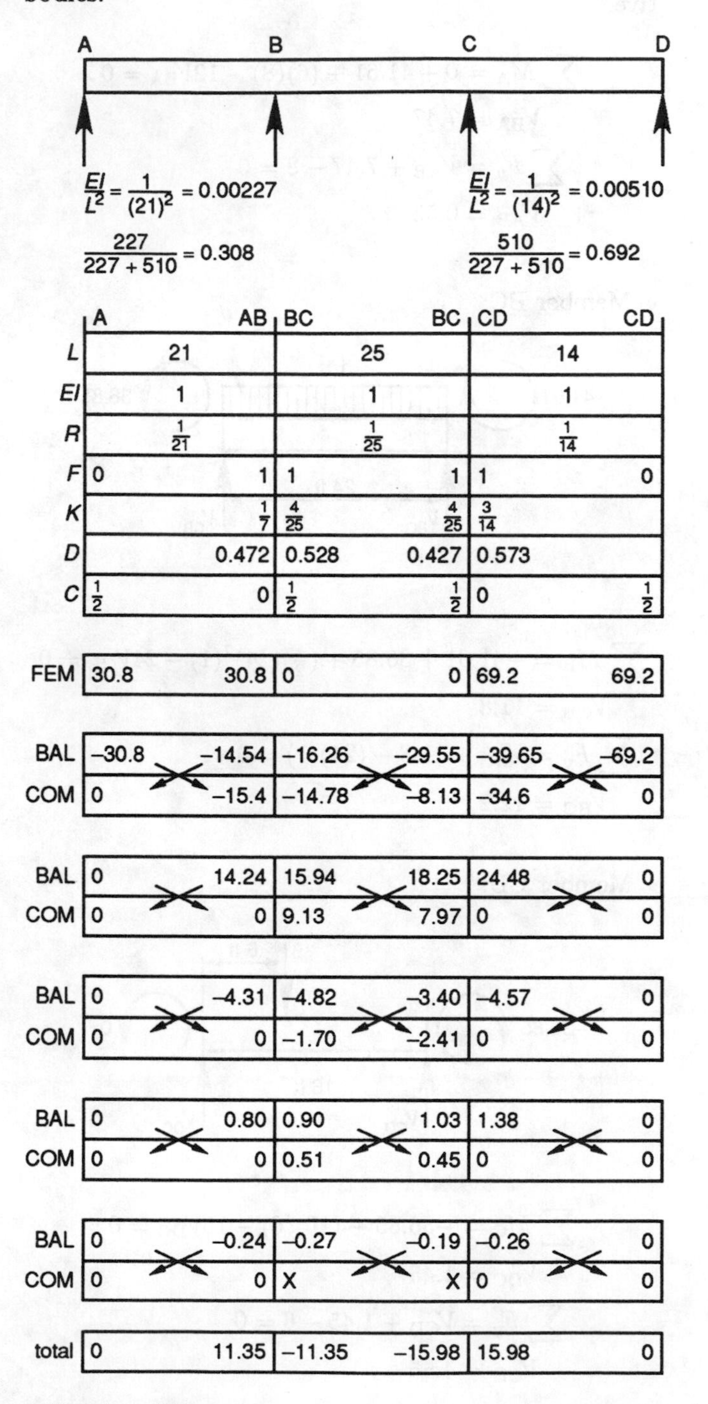

	A	AB	BC	BC	CD	CD
L	21		25		14	
EI	1		1		1	
R	$\frac{1}{21}$		$\frac{1}{25}$		$\frac{1}{14}$	
F	0	1	1	1	1	0
K		$\frac{1}{7}$	$\frac{4}{25}$	$\frac{4}{25}$	$\frac{3}{14}$	
D		0.472	0.528	0.427	0.573	
C	$\frac{1}{2}$	0	$\frac{1}{2}$	$\frac{1}{2}$	0	$\frac{1}{2}$
FEM	30.8	30.8	0	0	69.2	69.2
BAL	−30.8	−14.54	−16.26	−29.55	−39.65	−69.2
COM	0	−15.4	−14.78	−8.13	−34.6	0
BAL	0	14.24	15.94	18.25	24.48	0
COM	0	0	9.13	7.97	0	0
BAL	0	−4.31	−4.82	−3.40	−4.57	0
COM	0	0	−1.70	−2.41	0	0
BAL	0	0.80	0.90	1.03	1.38	0
COM	0	0	0.51	0.45	0	0
BAL	0	−0.24	−0.27	−0.19	−0.26	0
COM	0	0	X	X	0	0
total	0	11.35	−11.35	−15.98	15.98	0

- Member AB:

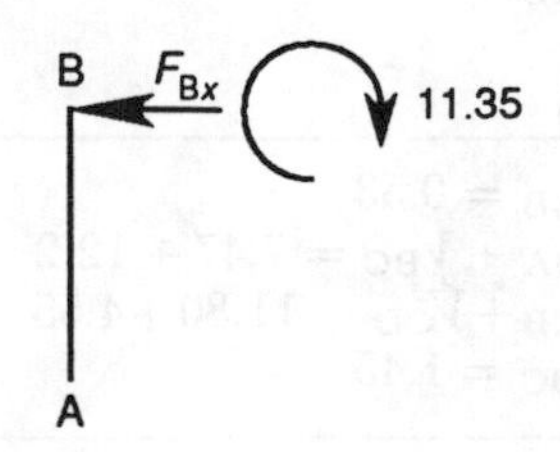

Take clockwise moments as positive.

$$\sum M_A = 11.35 - 21F_{Bx} = 0$$
$$F_{Bx} = 0.54 \text{ kips}$$

- Member DC:

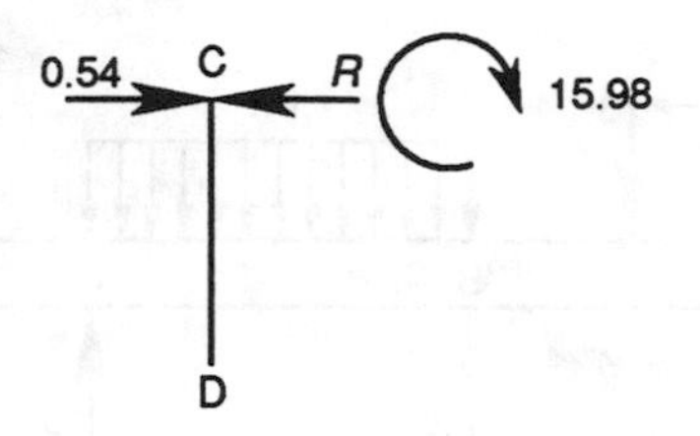

$$\sum M_D = 15.98 + (0.54)(14) - 14R = 0$$
$$R = \boxed{1.68 \text{ kips [to the left]}}$$

P_I was 20 to the left to keep the original frame from swaying.

$$\frac{20}{1.68} = 11.90$$
$$M_{AB} = -(11.90)(11.35) = -135.07 \text{ ft-kips}$$
$$M_{CD} = -(11.90)(15.98) = -190.16 \text{ ft-kips}$$

The actual conditions are developed as follows.

- Member AB:

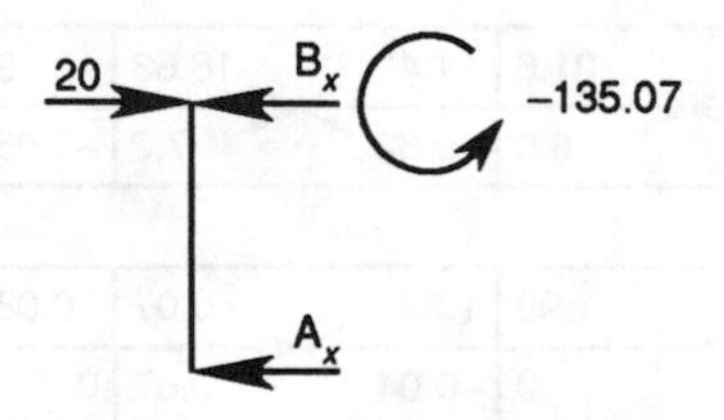

Taking clockwise moments as positive,

$$\sum M_A = -135.07 + (20)(21) - 21B_x = 0$$
$$B_x = 13.57 \text{ kips [to the left]}$$
$$\sum M_B = 21A_x - 135.07 = 0$$
$$A_x = 6.43 \text{ kips [to the left]}$$

Taking the frame as a whole unit with forces to the right and clockwise moments as positive,

$$\sum F_x = 20 - 6.43 - D_x = 0$$
$$D_x = 13.57 \text{ kips [to the left]}$$
$$\sum M_A = (21)(20) - 24D_y = 0$$
$$D_y = \boxed{17.5 \text{ kips [up]}}$$
$$A_y = \boxed{-17.5 \text{ kips [down]}}$$

3. *step 1:* Complete the moment distribution worksheet to find the end moments (assuming sidesway is prevented).

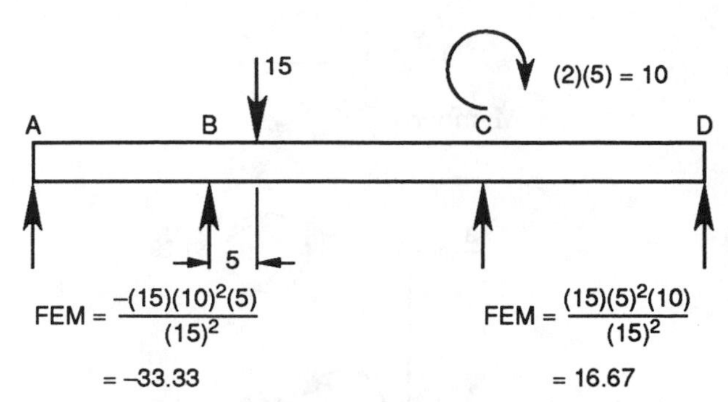

Note: The applied joint moment is clockwise, hence it is positive. The resisting FEM is negative, since the distribution factors are 0.5 for spans BD–DC; –5 goes to each.

	A	AB	BC	BC	CD	CD
L	10		15		15	
EI	240		300		300	
R	24		20		20	
F	0	1	1	1	1	1
K		72	80	80	80	
D		0.474	0.526	0.5	0.5	
C	$\frac{1}{2}$	0	$\frac{1}{2}$	$\frac{1}{2}$	$\frac{1}{2}$	$\frac{1}{2}$
FEM	0	0	–33.33	16.67		0
				–5	–5 (do not add to total)	
BAL	0	15.80	17.53	–3.34	–3.34	0
COM	0	0	–1.67	8.77	0	–1.67
BAL	0	0.79	0.88	–4.39	–4.39	0
COM	0	0	–2.19	0.44	0	–2.19
BAL	0	1.04	1.15	–0.22	–0.22	0
COM	0	0	–0.11	0.58	0	–0.11
BAL	0	0.05	0.06	–0.29	–0.29	0
COM	0	0	X	X	0	X
total	0	17.68	–17.68	18.22	–8.24	–3.97

step 2: Now find the reaction needed to prevent sidesway.

- Member AB:

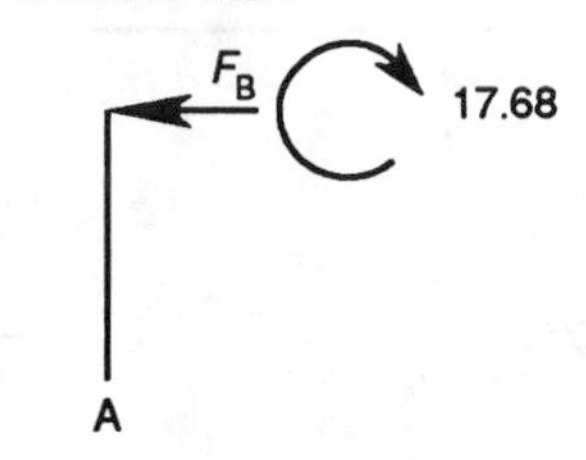

Taking clockwise moments as positive,

$$\sum M_A = 17.68 - 10F_B = 0$$

$$F_B = 1.768 \text{ kips} \quad \text{[to the left]}$$

- Member DC:

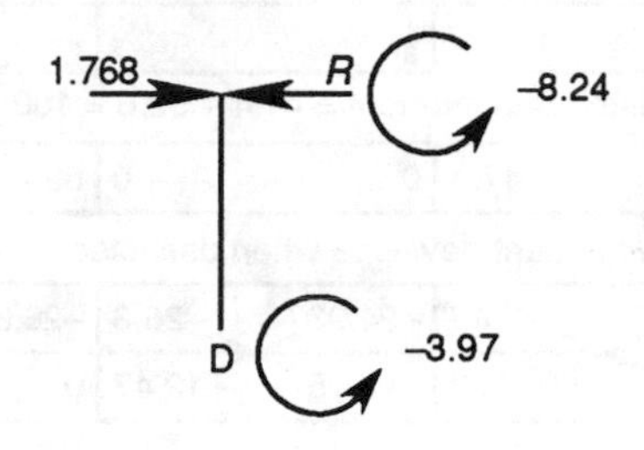

$$\sum M_D = -3.97 - 8.24 + (1.768)(15) - 15R = 0$$

$$R = 0.95 \text{ kips} \quad \text{[to the left]}$$

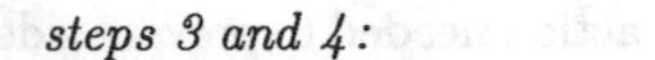

steps 3 and 4:

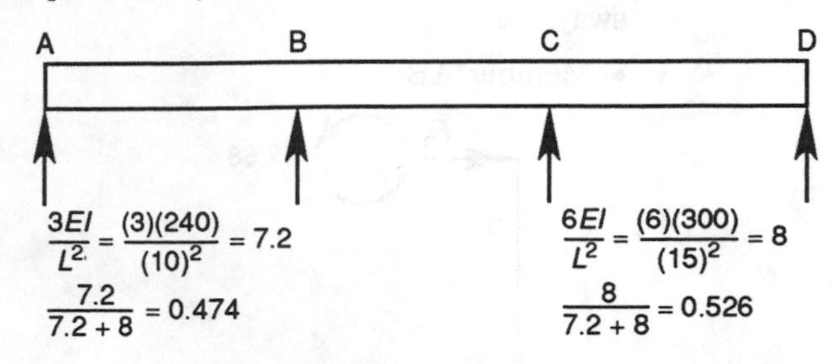

	A	AB	BC	BC	CD	CD
L	10		15		15	
EI	240		300		300	
R	24		20		20	
F	0	1	1	1	1	1
K		72	80	80	80	
D		0.474	0.526	0.5	0.5	
C	$\frac{1}{2}$	0	$\frac{1}{8}$	$\frac{1}{2}$	$\frac{1}{2}$	$\frac{1}{2}$
total distributed moment = 47.4 + 52.6 = 100						
FEM	0	47.4	0	0	52.6	52.6
BAL	0	−22.47	−24.93	−26.3	−26.3	0
COM	0	0	−13.15	−12.47	0	−13.15
BAL	0	6.23	6.92	6.24	6.24	0
COM	0	0	3.12	3.46	0	3.12
BAL	0	−1.48	−1.64	−1.73	−1.73	0
COM	0	0	−0.87	−0.82	0	−0.87
BAL	0	0.41	0.46	0.41	0.41	0
COM	0	0	0.20	0.23	0	0.20
BAL	0	−0.09	−0.11	−0.11	−0.11	0
COM	X	0	X	X	0	X
total	0	30.0	−30.0	−31.09	31.11	41.9

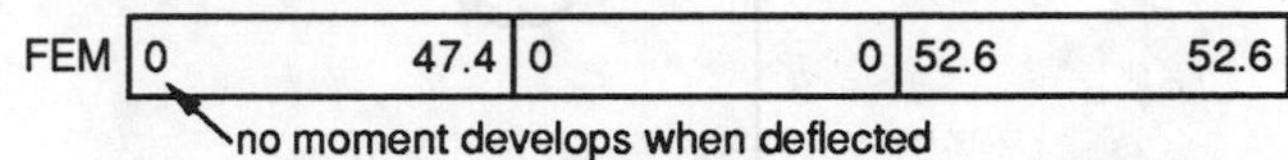

step 5: Calculate the force required to prevent sidesway.

- Member AB:

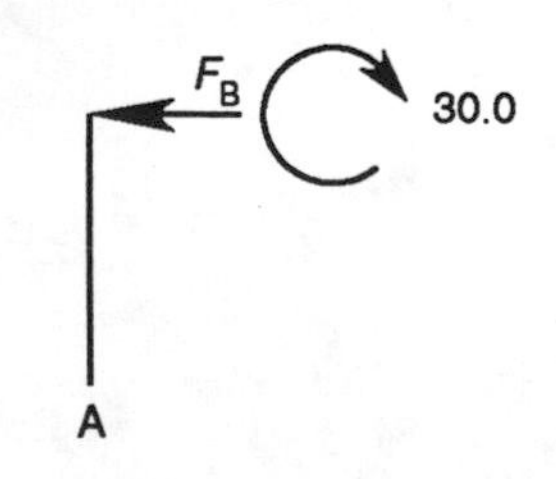

$$\sum M_A = 30.0 - 10F_B = 0$$

$$F_B = \boxed{3.0 \text{ kips}}$$

- Member CD:

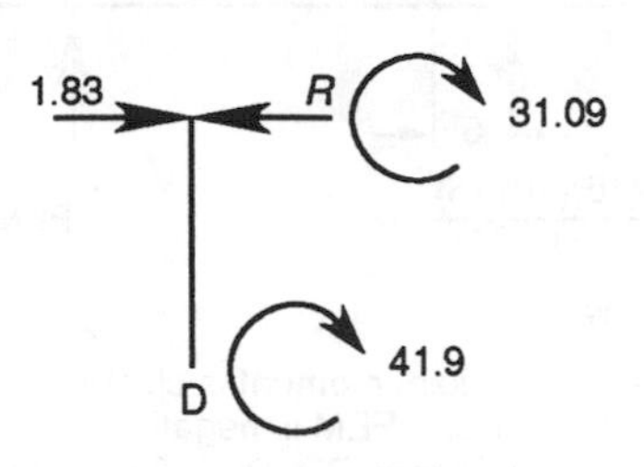

$$\sum M_D = 31.09 + 41.9 + (3)(15) - 15R = 0$$

$$R' = 7.87 \text{ kips} \quad \text{[to the left]}$$

steps 6 and 7:

$$\text{correction ratio: } \frac{0.95}{7.87} = 0.12$$

moment at A: $0 + (0.12)(-1)(0) = 0$

moment at B: $17.68 + (0.12)(-1)(30) = 14.08$ ft-kips

moment at C: $18.22 + (0.12)(-1)(-31.09)$
$= 21.95$ ft-kips

moment at D: $-3.97 + (0.12)(-1)(41.9) = -9.0$ ft-kips

Since R and R' are in the same directions, the derived moments must be reversed in sign.

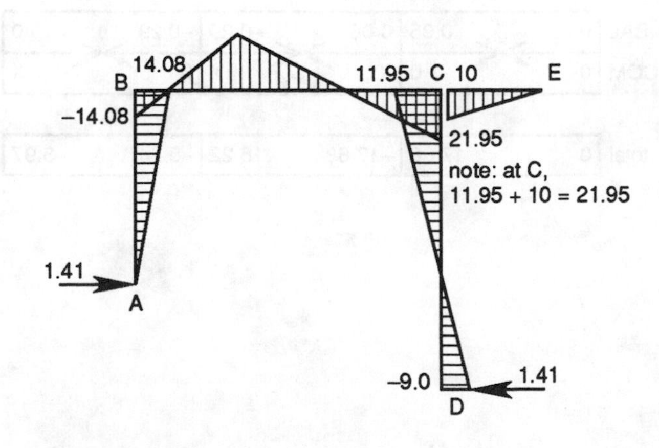

4.

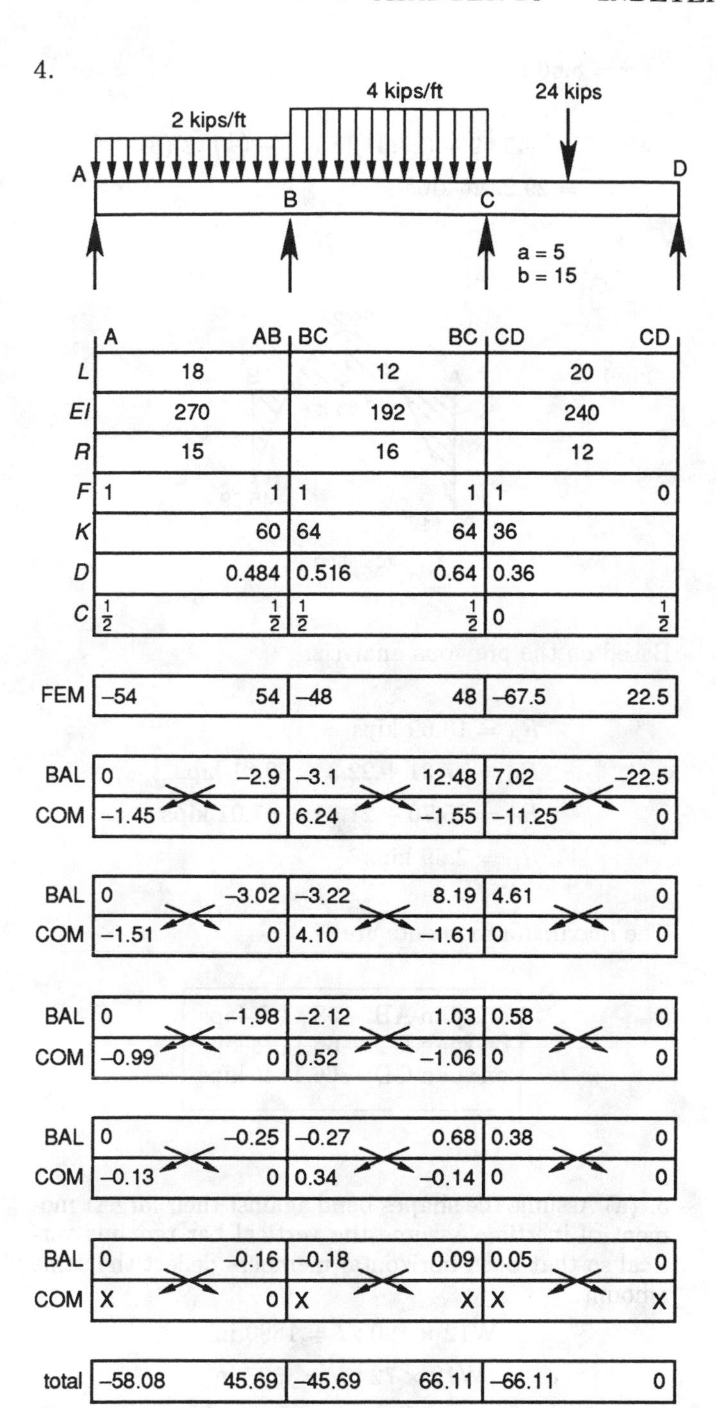

	A	AB	BC	BC	CD	CD
L	18		12		20	
EI	270		192		240	
R	15		16		12	
F	1	1	1	1	1	0
K		60	64	64	36	
D		0.484	0.516	0.64	0.36	
C	$\frac{1}{2}$	$\frac{1}{2}$	$\frac{1}{2}$	$\frac{1}{2}$	0	$\frac{1}{2}$
FEM	−54	54	−48	48	−67.5	22.5
BAL	0	−2.9	−3.1	12.48	7.02	−22.5
COM	−1.45	0	6.24	−1.55	−11.25	0
BAL	0	−3.02	−3.22	8.19	4.61	0
COM	−1.51	0	4.10	−1.61	0	0
BAL	0	−1.98	−2.12	1.03	0.58	0
COM	−0.99	0	0.52	−1.06	0	0
BAL	0	−0.25	−0.27	0.68	0.38	0
COM	−0.13	0	0.34	−0.14	0	0
BAL	0	−0.16	−0.18	0.09	0.05	0
COM	X	0	X	X	X	X
total	−58.08	45.69	−45.69	66.11	−66.11	0

Start at span CD and work to the left.

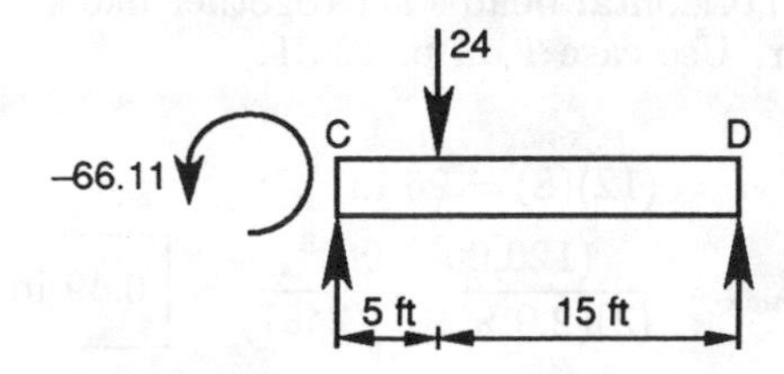

Taking clockwise moments as positive,

$$\text{applied} \sum M_\text{C} = -20R_\text{D} + (5)(24) - 66.11 = 0$$

$$R_\text{D} = 2.69 \text{ kips}$$

$$\text{applied} \sum M_\text{D} = 20R_\text{C} - (15)(24) - 66.11 = 0$$

$$R_\text{C} = 21.31 \text{ kips}$$

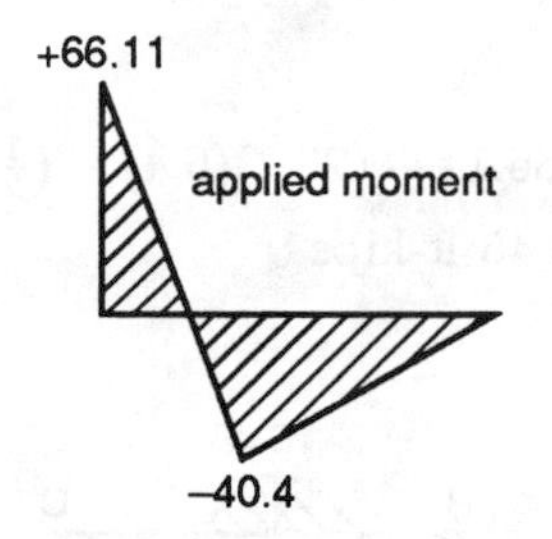

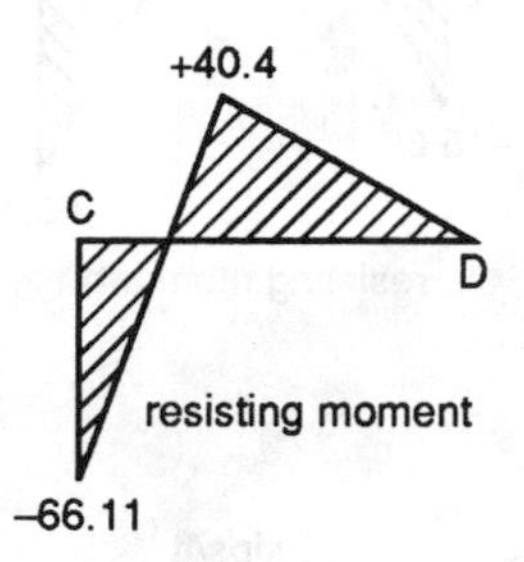

For span BC,

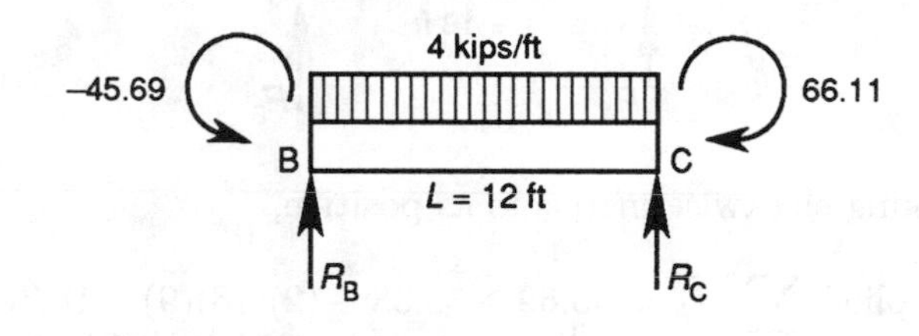

$$\text{applied} \sum M_\text{B} = 66.11 - 45.69 + (4)(12)(6) - 12R_\text{C} = 0$$

$$R_\text{C} = 25.7 \text{ kips}$$

$$\text{applied} \sum M_\text{C} = 66.11 - 45.69 - (4)(12)(6) + 12R_\text{B} = 0$$

$$R_\text{B} = 22.3 \text{ kips}$$

Taking counterclockwise moments as positive,

$$\text{resisting} \sum M = -66.11 + 25.7x - \left(\tfrac{1}{2}\right)(4)x^2$$

At $x = 12$ ft,

$$M = -66.11 + (25.7)(12) - \left(\tfrac{1}{2}\right)(4)(12)^2 = -45.71 \text{ ft-kips}$$

The first derivative is

$$\frac{dM}{dx} = 25.7 - 4x$$

M is maximum at

$$x = \frac{25.7}{4} = 6.4 \text{ ft}$$

At $x = 6.4$ ft,

$$\begin{aligned} M_B &= -66.11 + (25.7)(6.4) - \left(\tfrac{1}{2}\right)(4)(6.4)^2 \\ &= 16.45 \text{ ft-kips} \end{aligned}$$

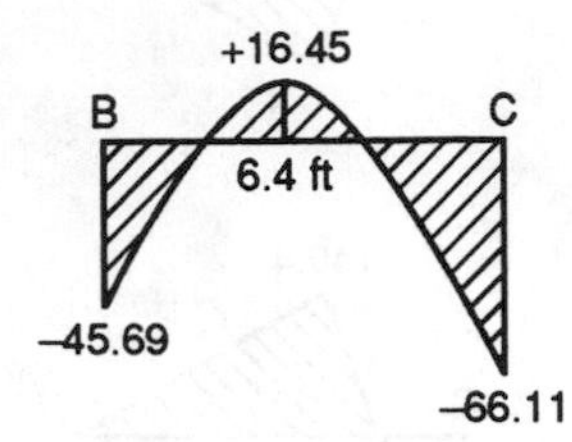

For span AB,

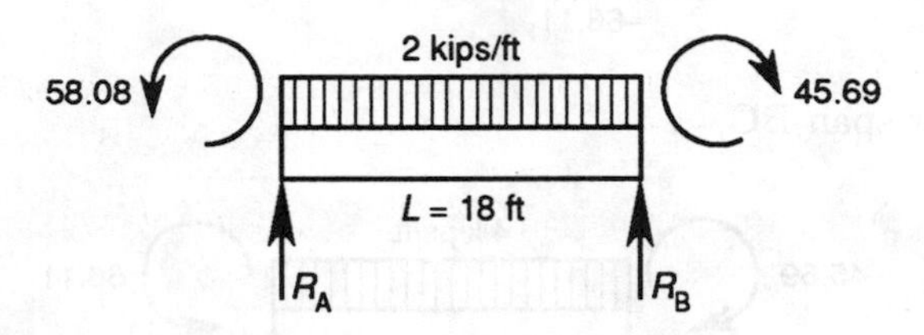

Taking clockwise moments as positive,

$$\begin{aligned} \text{applied} \sum M_A &= 45.69 - 58.08 + (2)(18)(9) - 18R_B \\ &= 0 \\ R_B &= 17.31 \text{ kips} \end{aligned}$$

$$\begin{aligned} \text{applied} \sum M_B &= 45.69 - 58.08 - (2)(18)(9) + 18R_A \\ &= 0 \\ R_A &= 18.69 \text{ kips} \end{aligned}$$

Taking counterclockwise moments as positive,

$$\text{resisting} \sum M = -45.69 + 17.31x - \left(\tfrac{1}{2}\right)(2)x^2$$

At $x = 18$ ft,

$$\begin{aligned} M &= -45.69 + (17.31)(18) - \left(\tfrac{1}{2}\right)(2)(18)^2 \\ &= 58.11 \text{ ft-kips} \end{aligned}$$

The first derivative is

$$\frac{dM}{dx} = 17.31 - 2x$$

M is maximum at

$$x = \frac{17.31}{2} = 8.66 \text{ ft}$$

At $x = 8.66$ ft,

$$\begin{aligned} M &= -45.69 + (17.31)(8.66) - \left(\tfrac{1}{2}\right)(2)(8.66)^2 \\ &= 29.22 \text{ ft-kips} \end{aligned}$$

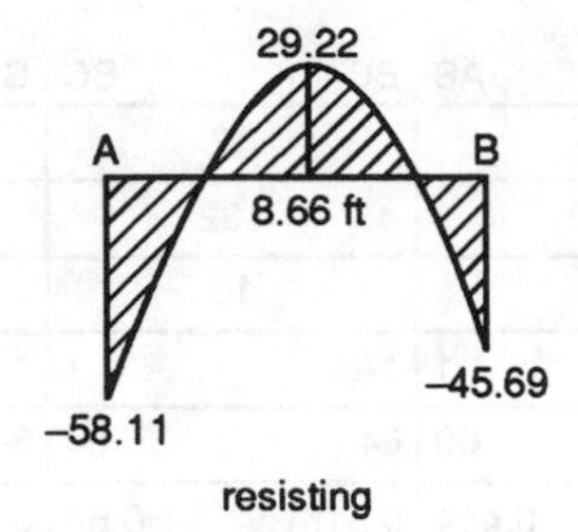

Based on the previous analysis,

$$\begin{aligned} R_A &= 18.69 \text{ kips} \\ R_B &= 17.31 + 22.3 = 39.61 \text{ kips} \\ R_C &= 25.70 + 21.31 = 47.01 \text{ kips} \\ R_D &= 2.69 \text{ kips} \end{aligned}$$

The maximum moments are

on span AB: –58.11 ft-kips on span BC: –66.11 ft-kips on span CD: –66.11 ft-kips

5. (a) Assume the shapes bend against their largest moment of inertia. Assume the vertical bar remains vertical so that both horizontal members deflect the same amount.

$$\begin{aligned} \text{W12} \times 190 : I &= 1890 \text{ in}^4 \\ \text{W12} \times 72 : I &= \underline{\ 597 \text{ in}^4} \\ \text{total} &\quad 2487 \text{ in}^4 \end{aligned}$$

The two horizontal beams act together like a combined cantilever. Use case 1 on p. 12-31.

$$L = (12)(8) = 96 \text{ in}$$

$$y_{max} = \frac{(120{,}000)(96)^3}{(3)(2.9 \times 10^7)(2487)} = \boxed{0.49 \text{ in}}$$

(b) Proceed as shown on p. 13-15.

step 1: Since there are no fixed-end moments, all total moments are zero.

step 2: In order to prevent all of the sidesway, all of the applied lateral load must be canceled. Therefore, $P_I = 120$ to the left.

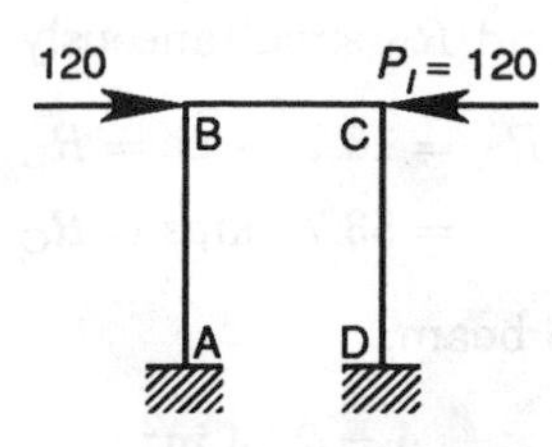

step 3: Apply a total FEM of 100.

For AB,

$$\frac{I}{L^2} = \frac{1890}{(8)^2} = 29.53$$

For DC,

$$\frac{I}{L^2} = \frac{597}{(8)^2} = 9.33$$

$$\text{give } \frac{29.53}{29.53 + 9.33} = 0.76 \text{ (76\%) to AB}$$

give $1 - 0.76 = 0.24$ (24%) to DC

step 4: The moment distribution worksheet for the application of these FEMs is

	A	AB	BC	BC	CD	CD
L	8		4.5		8	
EI	1890		310		597	
R	236.25		68.89		74.625	
F	1	1	1	1	1	1
K	945	945	275.56	275.56	298.5	298.5
D		0.774	0.226	0.48	0.52	
C	$\frac{1}{2}$	$\frac{1}{2}$	$\frac{1}{2}$	$\frac{1}{2}$	$\frac{1}{2}$	$\frac{1}{2}$
FEM	−76	−76	0	0	−24	−24
BAL	0	58.82	17.18	11.52	12.48	0
COM	29.41	0	5.76	8.59	0	6.24
BAL	0	−4.46	−1.30	−4.12	−4.47	0
COM	−2.23	0	−2.06	−0.65	0	−2.23
BAL	0	1.59	0.47	0.31	0.34	0
COM	0.79	0	0.15	0.24	0	0.17
BAL	0	−0.12	−0.03	−0.11	−0.13	0
COM	X	0	X	X	0	X
total	−48.03	−20.17	20.17	15.78	−15.78	−19.95

step 5: Take member AB as a free-body.

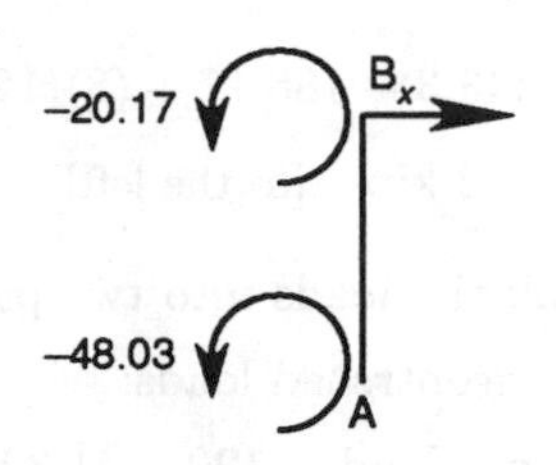

Taking clockwise moments as positive,

$$\sum M_A = 8B_x - 20.17 - 48.03 = 0$$

$$B_x = 8.53 \text{ kips [to the right]}$$

Take member DC as a free-body.

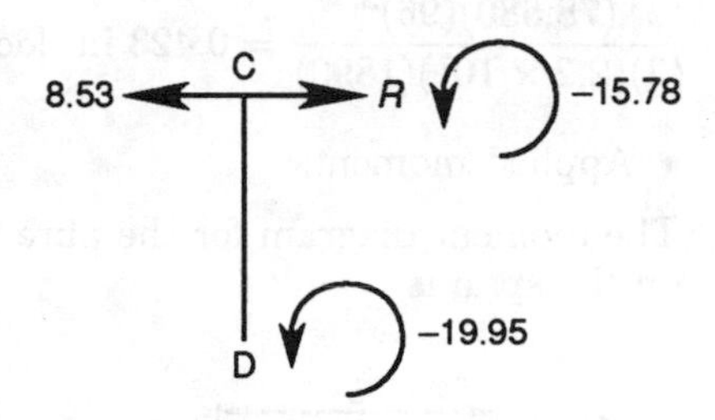

$$\sum M_D = 8R - (8)(8.53) - 15.78 - 19.95 = 0$$

$$R = 13.0 \text{ kips [to the right]}$$

step 6: From Eq. 13.67,

$$-\left(\frac{-120}{13.0}\right) = 9.23$$

$$M_A = (9.23)(-48.03) = -443.3 \text{ ft-kips}$$

$$M_B = (9.23)(\pm 20.17) = \pm 186.17 \text{ ft-kips}$$

$$M_C = (9.23)(\pm 15.78) = \pm 145.6 \text{ ft-kips}$$

$$M_D = (9.23)(-19.95) = -184.1 \text{ ft-kips}$$

step 7: Draw the free-body of member AB.

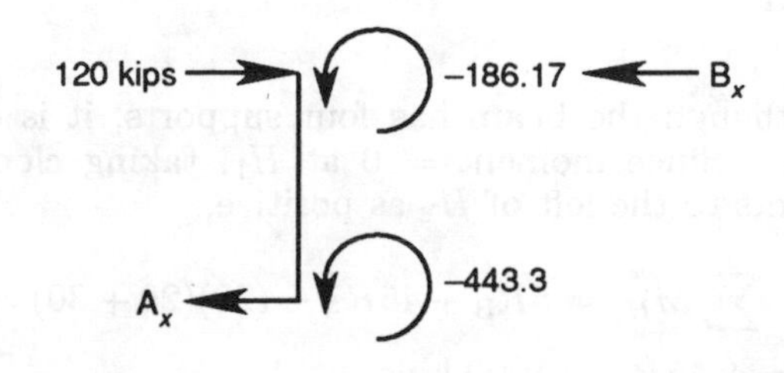

B_x can be found by summing moments about A.

$$\sum M_A = -443.3 - 186.17 + (8)(120) - 8B_x = 0$$

$$B_x = 41.32 \text{ kips} \quad \text{[to the left]}$$

Break the loads into two parts.

- Concentrated loads:

The net load is $120 - 41.32 = 78.68$ kips.

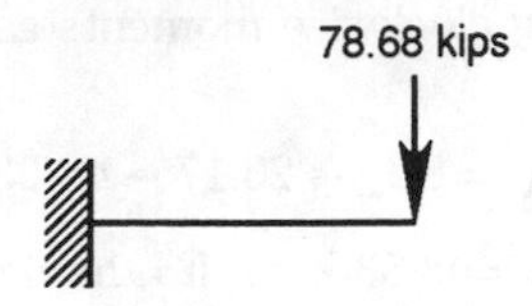

From p. 12-31, case 1,

$$y_1 = \frac{(78{,}680)(96)^3}{(3)(2.9 \times 10^7)(1890)} = 0.423 \text{ in} \quad \text{[down]}$$

- Applied moments:

The moment diagram for the pure moments on the span is

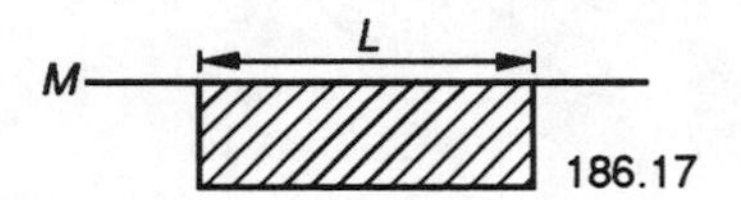

From the area-moment method,

$$EIy = ML\left(\frac{L}{2}\right)$$

$$y_2 = \frac{ML^2}{2EI}$$

The moment at the built-in end does not cause a deflection, so $M = 186.17$ ft-kips.

$$y_2 = \frac{(-186{,}170)(12)(96)^2}{(2)(2.9 \times 10^7)(1890)} = -0.188 \text{ in} \quad \text{[up]}$$

$$y_{\text{total}} = 0.423 - 0.188 = \boxed{0.235 \text{ in} \quad \text{[down]}}$$

Timed

1. Although the beam has four supports, it is determinate. Since moment $= 0$ at H_1, taking clockwise moments to the left of H_1 as positive,

$$\sum M_{H_1} = 5R_B + 45R_A - (20)(20 + 30) = 0$$

$$5R_B + 45R_A = 1000 \text{ kips}$$

Since the beam and loading are symmetrical,

$$R_A + R_B = \left(\tfrac{1}{2}\right)(7)(20) = 70 \text{ kips}$$

Solving for R_A and R_B simultaneously,

$$R_A = 16.25 \text{ kips} = R_D$$

$$R_B = 53.75 \text{ kips} = R_C$$

For a W24 × 76 beam,

$$A = 22.4 \text{ in}^2$$
$$d = 23.92 \text{ in}^2$$
$$S = 176 \text{ in}^3$$
$$t_w = 0.440 \text{ in}$$
$$I = 2100 \text{ in}^4$$

The shear diagram is

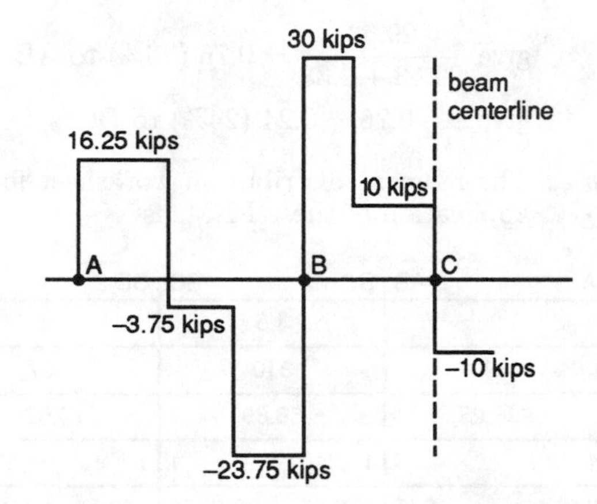

The web takes all the shear. The maximum shear stress is

$$\tau_{\max} = \frac{30 \times 10^3}{(23.92)(0.440)} = \boxed{2850 \text{ lbf/in}^2 \text{ (psi)}}$$

The moment is maximum where $V = 0$. Under the first inboard load,

$$M = (15)(16.25) = 243.75 \text{ ft-kips}$$

At R_B,

$$M = (40)(16.25) - (15 + 25)(20) = -150 \text{ ft-kips}$$

At the center,

$$M = (15)\left(\frac{20}{2}\right) = 150 \text{ ft-kips}$$

$$M_{\max} = 243.75 \text{ ft-kips}$$

$$\sigma_{\max} = \frac{M}{S} = \frac{(243.75 \times 10^3)\left(12 \ \frac{\text{in}}{\text{ft}}\right)}{176}$$

$$= \boxed{16{,}619 \text{ lbf/in}^2 \text{ (psi)}}$$

From case 4 (p. 12-31) between hinges,

$$y_{max} = \frac{(20 \times 10^3)(30 \times 12)^3}{(48)(2.9 \times 10^7)(2100)} = 0.319 \text{ in}$$

The deflection at the midpoint is comprised of four terms.

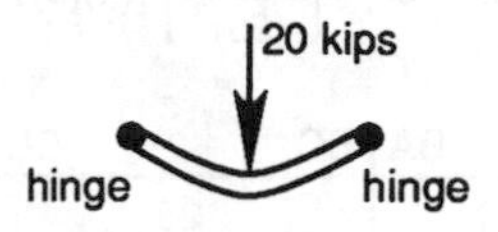

This is case 4, p. 12-31.

$$y_1 = \frac{FL^3}{48EI} = \frac{-(20 \times 10^3)(30 \times 12)^3}{(48)(2.9 \times 10^7)(2100)} = -0.319 \text{ in}$$

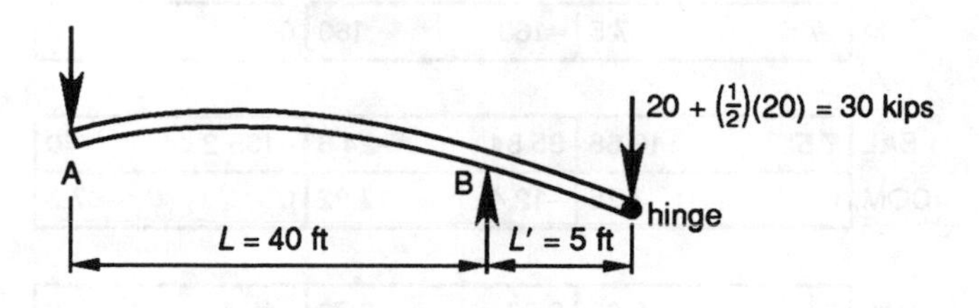

$$\begin{aligned} y_2 &= \left(\frac{FL^2}{3EI}\right)(L + L') \\ &= \frac{-(30 \times 10^3)(5 \times 12)^2(45 \times 12)}{(3)(2.9 \times 10^7)(2100)} \\ &= -0.319 \text{ in} \end{aligned}$$

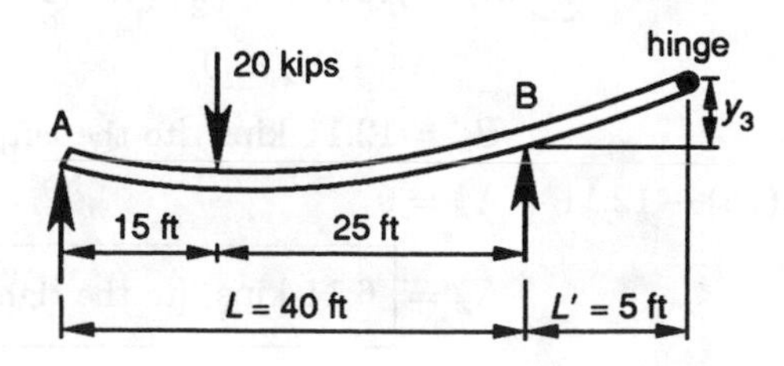

The slope at B (from Merritt's *Standard Handbook for Civil Engineers*) is

$$m_B = \left(\frac{FL^2k}{6EI}\right)(1 - k^2)$$

$$k = \frac{15}{40} = 0.375$$

$$\begin{aligned} m_B &= \frac{(20 \times 10^3)(40 \times 12)^2(0.375)\left[1 - (0.375)^2\right]}{(6)(2.9 \times 10^7)(2100)} \\ &= 0.00406 \end{aligned}$$

$$y_3 = m_B L' = (0.00406)(5)(12) = 0.244 \text{ in}$$

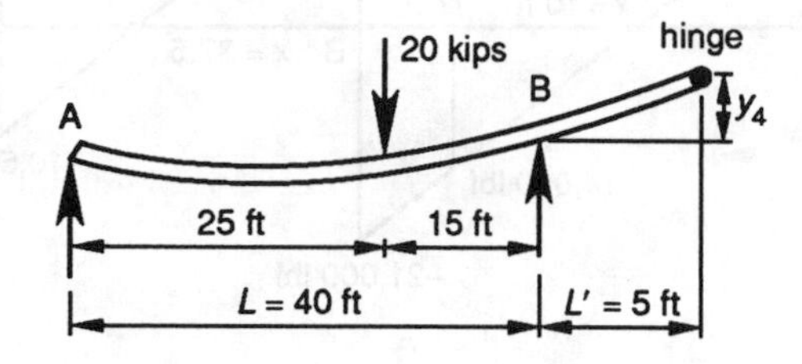

$$k = \frac{25}{40} = 0.625$$

$$\begin{aligned} m_B &= \frac{FL^2k(1 - k^2)}{6EI} \\ &= \frac{(20 \times 10^3)(40 \times 12)^2(0.625)\left[1 - (0.625)^2\right]}{(6)(2.9 \times 10^7)(2100)} \\ &= 0.00480 \end{aligned}$$

$$y_4 = M_B L' = (0.00480)(5)(12) = 0.288 \text{ in}$$

The total midpoint deflection is

$$\begin{aligned} y_t &= y_1 + y_2 + y_3 + y_4 \\ &= -0.319 - 0.319 + 0.244 + 0.288 \\ &= \boxed{-0.106 \text{ in}} \end{aligned}$$

2. This is a determinate beam. Taking clockwise moments from the hinge to the left as positive,

(a) $\sum M_H = 20R_A - \left(\tfrac{1}{2}\right)(1400)(20)^2 = 0$

$$R_A = \boxed{14{,}000 \text{ lbf}}$$

$$\begin{aligned} V_H &= (14{,}000) - (20)(1400) \\ &= -14{,}000 \text{ lbf} \end{aligned}$$

$$\begin{aligned} R_B + R_C &= (5 + 20 + 5)(1400) + 14{,}000 \\ &= 56{,}000 \text{ lbf} \end{aligned}$$

$$\begin{aligned} \sum M_C &= 20R_B - (25)(14{,}000) \\ &\quad - \left(\tfrac{1}{2}\right)(1400)(25)^2 + \left(\tfrac{1}{2}\right)(1400)(5)^2 \\ &= 0 \end{aligned}$$

$$R_B = \boxed{38{,}500 \text{ lbf}}$$

$$R_C = 56{,}000 - 38{,}500 = \boxed{17{,}500 \text{ lbf}}$$

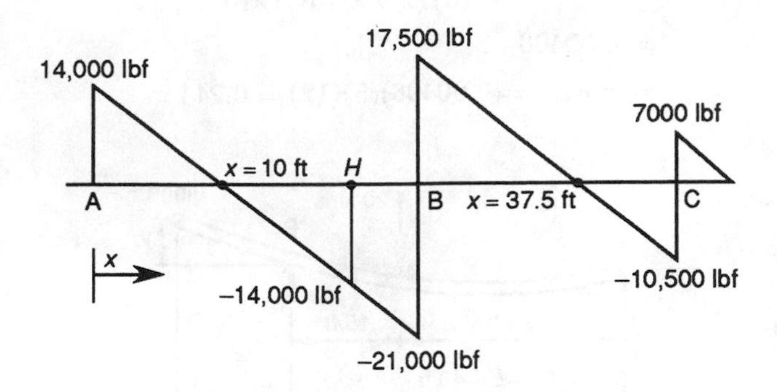

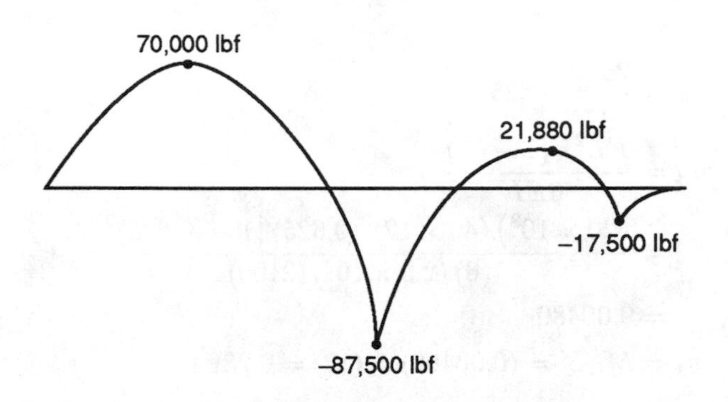

(c) $M_{\text{max}} = 87{,}500 \text{ ft-lbf} = 87.5 \text{ ft-kips}$

Since there is lateral bracing only at points of support,

$$L_b = 25$$

From the allowable moment beam chart (assuming A-36 steel),

W14 × 53

(d) $L_u =$ **25.9 ft [from table]**

3. The moment distribution worksheet is

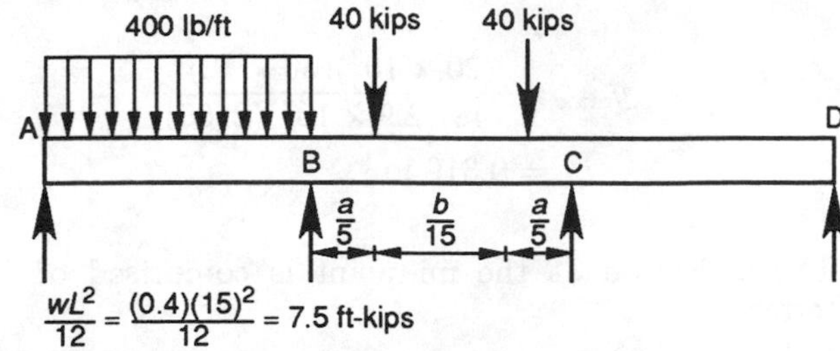

$$\frac{wL^2}{12} = \frac{(0.4)(15)^2}{12} = 7.5 \text{ ft-kips}$$

$$\left(\frac{Pa}{L^2}\right)(2a^2 + 3ab + b^2) = \left[\frac{(40)(5)}{(25)^2}\right]\left[(2)(5)^2 + (3)(5)(15) + (15)^2\right] = 160 \text{ ft-kips}$$

	AB	BA	BC	CB	CD	DC
L	15		25		12	
EI	890		341		890	
R	59.33		13.64		74.17	
F	0	1	1	1	1	1
K	237.3	178	54.6	54.6	296.7	296.7
D	1	0.765	0.235	0.155	0.845	1
C	$\frac{1}{2}$	0	$\frac{1}{2}$	$\frac{1}{2}$	$\frac{1}{2}$	$\frac{1}{2}$
FEM	–7.5	7.5	–160	160	0	0
BAL	7.5	116.66	35.84	–24.8	–135.2	0
COM	0	3.75	–12.4	17.92	0	–67.6
BAL	0	6.62	2.03	–2.78	–15.14	0
COM	0	0	–1.39	1.01	0	–7.57
BAL	0	1.06	0.33	–0.16	–0.85	0
COM	X	0	X	X	0	X
total	0	136.71	–136.71	151.19	–151.19	–75.17

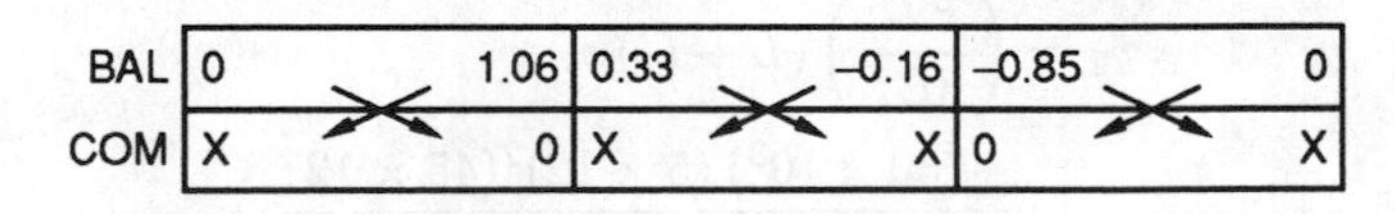

Taking clockwise moments as positive,

$$\sum M_A = 136.71 + \left(\tfrac{1}{2}\right)(0.4)(15)^2 - 15B_x = 0$$

$$B_x = 12.11 \text{ kips [to the left]}$$

$$(0.4)(15) - 12.11 + A_x = 0$$

$$A_x = \boxed{6.11 \text{ kips [to the right]}}$$

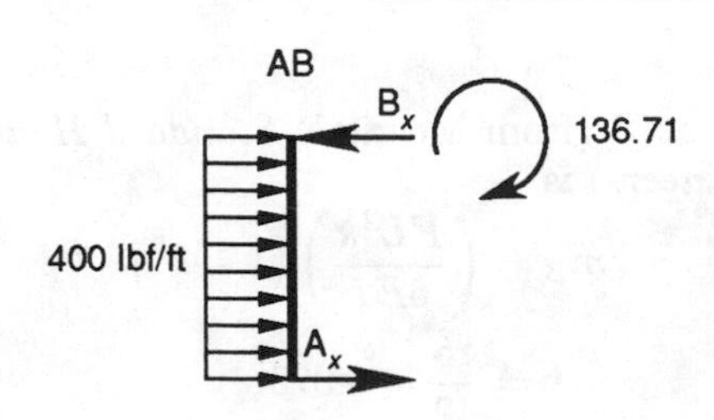

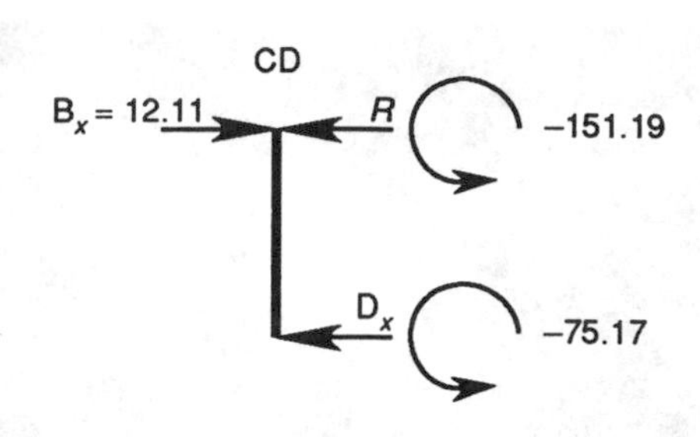

(b)

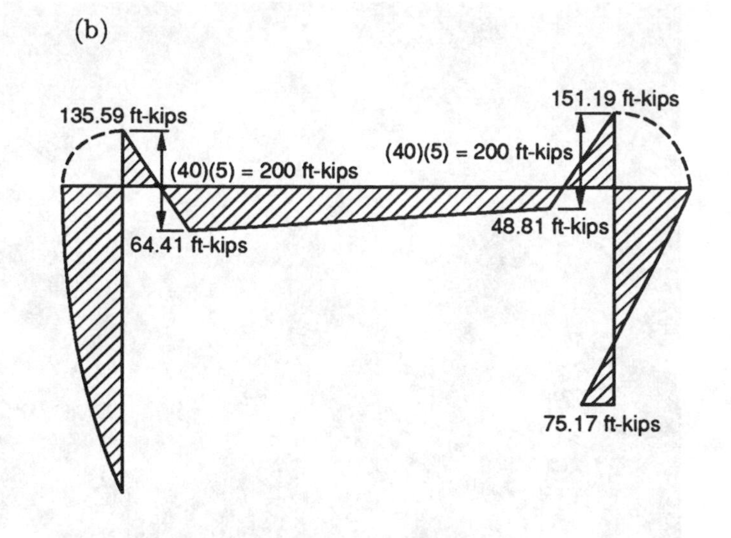

$$\sum M_{\text{D}} = (12)(12.11) - 12R - 151.19 - 75.17 = 0$$

$$R = \boxed{-6.75 \text{ kips} \quad [\text{to the right}]}$$

$$12.11 + 6.75 - \text{D}_x = 0$$

$$\text{D}_x = \boxed{18.86 \text{ kips} \quad [\text{to the left}]}$$

REINFORCED CONCRETE DESIGN

Untimed

1. On a per cubic yard basis,

$$\text{volume of cement used} = \frac{\left(6.5\ \frac{\text{bags}}{\text{yd}^3}\right)\left(94\ \frac{\text{lbf}}{\text{bag}}\right)}{\left(62.4\ \frac{\text{lbf}}{\text{ft}^3}\right)(3.10)}$$

$$= 3.16\ \text{ft}^3/\text{yd}^3$$

$$\text{volume of water used} = \frac{\left(6.5\ \frac{\text{bags}}{\text{yd}^3}\right)\left(5.75\ \frac{\text{gal}}{\text{bag}}\right)}{7.48\ \frac{\text{gal}}{\text{ft}^3}}$$

$$= 5.00\ \text{ft}^3/\text{yd}^3$$

$$\text{volume of air} = (0.05)\left(27\ \frac{\text{ft}^3}{\text{yd}^3}\right) = 1.35\ \text{ft}^3/\text{yd}^3$$

$$\text{total: } 3.16 + 5.00 + 1.35 = 9.51\ \text{ft}^3/\text{yd}^3$$

The remainder of the cubic yard ($27 - 9.51 = 17.49$ ft^3/yd^3) must be aggregate. The volumes of fine and coarse aggregates are

$$\begin{aligned}\text{fine:}\quad & (17.49)(0.30) = 5.25\ \text{ft}^3\\ \text{coarse:}\quad & (17.49)(0.70) = 12.24\ \text{ft}^3\end{aligned}$$

The weights are

$$\begin{aligned}\text{cement:}\quad & (3.16)(3.10)(62.4) = 611.3\ \text{lbf/yd}^3\\ \text{water:}\quad & (5.0)(62.4) = 312.0\ \text{lbf/yd}^3\\ \text{fine aggregate:}\quad & (5.25)(2.65)(62.4) = 868.1\ \text{lbf/yd}^3\\ \text{coarse aggregate:}\quad & (12.24)(2.00)(62.4) = 1527.6\ \text{lbf/yd}^3\end{aligned}$$

The mix ratio is (1:1.42:25) by weight. Use the procedure outlined in Ex. 14.2.

constituent	ratio	weight per sack cement	weight density	absolute volume
cement	1	94	193.4	0.486
fine	1.42	133.5	165.4	0.807
coarse	2.5	235.0	124.8	1.883
water			$\left(\frac{5.75}{7.48}\right)$ =	0.769
				3.945 ft^3

The solid yield is 3.945 ft^3/sack. The yield with 5% air is

$$\frac{3.945}{1 - 0.05} = 4.153\ \text{ft}^3/\text{sack}$$

The number of 1-sack batches is

$$\frac{(1.5)(27)}{4.153} = 9.752$$

The required cement weight is

$$(9.752)(94) = \boxed{916.7\ \text{lbf}}$$

The required fine aggregate weight is

$$(9.752)(1.015)(94)(1.42)$$

$$= \boxed{1321.2\ \text{lbf as delivered [1301.7 lbf SSD]}}$$

Note: Dividing by 1.015 converts to SSD conditions.

The required coarse aggregate is

$$(9.752)(0.97)(94)(2.5)$$

$$= \boxed{2223.0\ \text{lbf as delivered [2291.7 lbf SSD]}}$$

The weight of the excess water in the fine aggregate is

$$\left(\frac{1321.2}{1.015}\right)(0.015) = 19.53\ \text{lbf}$$

The water needed to bring the coarse aggregate to SSD condition is

$$(2223.0)\left(\frac{0.03}{0.97}\right) = 68.75\ \text{lbf}$$

The total weight of water required is

$$\left[\frac{(5.75)(9.752)}{7.48}\right](62.4) + 68.75 - 19.53 = \boxed{517.0\ \text{lbf}}$$

2. *steps 1 and 2:* Assume a footing thickness, d_f, of 20 in. The unfactored (net) allowable soil pressure (Eq. 14.144) is

$$p_{a,\text{net}} = 4000 - \left(\frac{20}{12}\right)(150)$$

$$= 3750\ \text{psf}$$

The required footing area is

$$\frac{240{,}000}{3750} = 64.0\ \text{ft}^2$$

Use $(8.0\ \text{ft})(8.0\ \text{ft}) = 64.0\ \text{ft}^2$.

$$p'_{\text{actual}} = \frac{240{,}000}{64.0} = 3750\ \text{psf}$$

Assume a depth to reinforcement of d = 16 in. From Fig. 14.19, the critical area is

$$(8.0)^2 - \left(\frac{16+16}{12}\right)^2 = 56.89 \text{ ft}^2$$

The critical perimeter is

$$\frac{(4)(16+16)}{12} = 10.67 \text{ ft}$$

step 3: Assuming all of the load is live, the required ultimate shear is

$$V_u = (1.7)(3750)(56.89) = 362{,}674 \text{ lbf}$$

step 4: From Eqs. 14.130 and 14.131, the allowable shear is

$$\beta_c = \frac{16}{16} = 1 \quad \text{[must be a minumum of 2]}$$

$$V_c = \left[\left(2+\frac{4}{2}\right)\sqrt{3000}\right](144)\left[(10.67)\left(\frac{16}{12}\right)\right] = 448{,}834 \text{ lbf}$$

step 5:

$$\phi V_c = (0.85)(448{,}834) = 381{,}509 \text{ lbf}$$

$$V_u < \phi V_c \quad \text{[acceptable]}$$

Square footings should also be checked for 1-way shear.

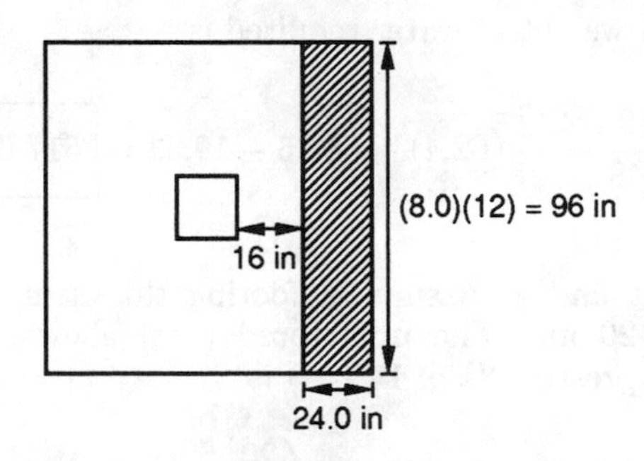

$$\text{critical area} = \frac{(25)(96)}{144} = 16.67 \text{ ft}^2$$

From Eq. 14.128,

$$V_u = (1.7)(3750)(16.67) = 106{,}271 \text{ lbf}$$

The allowable shear is

$$V_c = \left(2\sqrt{3000}\right)(144)\left[(8.0)\left(\frac{16}{12}\right)\right] = 168{,}260 \text{ lbf}$$

$$\phi V_c = (0.85)(168{,}260) = 143{,}021 \text{ lbf}$$

$$V_u < \phi V_c \quad \text{[acceptable]}$$

step 6: From Fig. 14.21,

$$b_m = \frac{8.0 - \frac{16}{12}}{2} = 3.33 \text{ ft} \quad [40.0 \text{ in}]$$

From Eq. 14.140,

$$M_u = \frac{[(1.7)(3750)](8.0)(3.33)^2}{2} = 282{,}767 \text{ ft-lbf}$$

step 7: From Eq. 14.141,

$$R_u = \frac{282{,}767}{(0.90)(8.0)\left(\frac{16}{12}\right)^2} = 22{,}091 \text{ psf} \quad [153.4 \text{ psi}]$$

step 8: From Eq. 14.142,

$$\rho = \left[\frac{(0.85)(3000)}{40{,}000}\right]\left[1 - \sqrt{1 - \frac{(2)(153.4)}{(0.85)(3000)}}\right] = 0.00396 \quad [\text{assume } 200/f_y \text{ does not apply}]$$

step 9: $A_{st} = (0.00396)(16)(8.0)(12) = 6.08 \text{ in}^2$

Use 11 #7 bars (6.6 in^2).

step 10: Minimum spacing is not a problem.

step 11: From Eq. 14.2,

$$l_d = \text{maximum}\begin{cases} \dfrac{(0.04)(0.60)(40{,}000)}{\sqrt{3000}} = 17.5 \text{ in} \\ \dfrac{(0.03)(0.875)(40{,}000)}{\sqrt{3000}} = 19.2 \text{ in} \\ 12 \text{ in} \end{cases}$$

$$= 19.2 \text{ in}$$

$$19.2 \text{ in} < b_m = 40.0 \text{ in} \quad \text{[acceptable]}$$

3. Refer to the procedure on p. 14-11.

step 1: Calculate ρ_{max} from Eq. 14.9.

$$\rho_{max} = 0.75\rho_{balanced} = (0.75)\left[\frac{(0.85)(0.85)(3000)}{50{,}000}\right]\left(\frac{87{,}000}{87{,}000+50{,}000}\right) = 0.0206$$

From Eq. 14.28, try

$$\rho = \frac{(0.18)(3000)}{50{,}000} = 0.0108 \quad [< 0.0206, \text{ so acceptable}]$$

step 2: From Eqs. 14.29 and 14.30,

$$m = \frac{50{,}000}{(0.85)(3000)} = 19.608$$

$$R_u = (0.0108)(50{,}000)\,[1 - (0.5)(0.0108)(19.608)] = 482.8 \text{ psi}$$

step 3:

$$M_u = (1.4)(50{,}000) + (1.7)(200{,}000) = 410{,}000 \text{ ft-lbf}$$

From Eq. 14.31,

$$bd^2 = \frac{(410{,}000)(12)}{(0.90)(482.8)} = 11{,}323 \text{ in}^3$$

step 4: Choose $d = 1.75b$.

$$(1.75)^2 b^3 = 11{,}323$$

$$b = 15.46 \text{ in} \quad [\text{round to } 15.5 \text{ in}]$$

$$d = (1.75)(15.46) = 27.06 \text{ in} \quad [27 \text{ in}]$$

step 5:

$$\text{actual } bd^2 = (15.5)(27)^2 = 11{,}300 \text{ in}^3 \quad [\text{close}]$$

No change in ρ is needed.

step 6:

$$A_{st} = (0.0108)(15.5)(27) = 4.52 \text{ in}^2$$

step 7: Use six #8 bars (4.74 in^2). These can all fit in a single layer.

step 8: Assume #4 stirrups (see p. 14-37).

$$d_c = 1.5 + \tfrac{1}{2} + 0.5 = 2.5 \text{ in}$$

4. Refer to App. B.

$$k = \frac{1}{1 + \left[\dfrac{20{,}000}{(10)(1350)}\right]} = 0.403$$

$$j = 1 - \frac{0.403}{3} = 0.866$$

Converting M_c into units of in-lbf,

$$bd^2 = \frac{(2)(50{,}400)\left(12\ \frac{\text{in}}{\text{ft}}\right)}{(1350)(0.403)(0.866)} = 2567 \text{ in}^3$$

Select $d = 2b$. Then, $4b^3 = 2567$ in^3.

$$d = 17.251 \text{ in} \quad [\text{round to } 17\tfrac{1}{4} \text{ in}]$$

$$b = 8.626 \text{ in} \quad [\text{round to } 8\tfrac{3}{4} \text{ in}]$$

Allow 2-in cover on the steel. The beam is

$$\boxed{8\tfrac{3}{4} \text{ in} \times 19\tfrac{1}{4} \text{ in}}$$

The required steel area is

$$A_{st} = \frac{(50{,}400)\left(12\ \frac{\text{in}}{\text{ft}}\right)}{(20{,}000)(0.866)(17.25)} = 2.02 \text{ in}^2$$

Use two #9 bars, giving 2 in^2 of steel. From Table 14.5, this is acceptable.

5. Assume zero eccentricity and proceed as in Ex. 14.7. From Eq. 14.108,

$$P_{\text{factored}} = (1.4)(175{,}000) + (1.7)(300{,}000) = 755{,}000 \text{ lbf}$$

Assume $\rho_g = 0.02$. From Eq. 14.113,

$$755{,}000 = A_g(0.85)(0.75) \times [(0.85)(1 - 0.02)(3000) + (40{,}000)(0.02)]$$

$$A_g = 359 \text{ in}^2$$

$$A_g = \frac{\pi}{4} D_g^2$$

$$D_g = \sqrt{\frac{(4)(359)}{\pi}} = 21.38 \text{ in} \quad [\text{round to } 21.5 \text{ in}]$$

Use $1\frac{1}{2}$-in cover. Assuming #8 bars and #3 spiral wire, the steel diameter is

$$21.5 - (2)\left(1\tfrac{1}{2}\right) - (2)\left(\tfrac{3}{8}\right) - 1 = 16.75 \text{ in}$$

The core diameter is

$$21.5 - (2)\left(1\tfrac{1}{2}\right) = 18.5 \text{ in}$$

The required steel area is

$$(0.02)(359) = 7.18 \text{ in}^2$$

Some possibilities are

23 #5 bars
16 #6 bars
12 #7 bars
9 #8 bars
8 #9 bars
6 #10 bars

Try the 12 #7 bars. The steel circumference is

$$\pi(16.75) = 52.62 \text{ in}$$

The clear spacing between bars is

$$\frac{52.62 - (12)(0.875)}{12} = 3.51 \text{ in}$$

Since 3.51 in > (1.5)(0.875) = 1.31 in, this is acceptable.

From Eq. 14.117, using the core diameter, as opposed to the steel diameter,

$$\rho_s = (0.45)\left[\left(\frac{21.5}{18.5}\right)^2 - 1\right]\left(\frac{3000}{40{,}000}\right) = 0.0118$$

From Eq. 14.118, with a spiral pitch of 2 in,

$$\rho_s = \frac{(4)(0.11)}{(2)(18.5)} = 0.0119 \text{ in} > 0.0118 \quad \text{[acceptable]}$$

The clear spacing between spirals is

$$2 - \tfrac{3}{8} = 1.625 \text{ in}$$

Since 1.0 in < 1.625 in < 3 in, this is acceptable.

6. From Eq. 14.108,

$$\begin{aligned} P_{\text{factored}} &= (1.4)(100{,}000) + (1.7)(125{,}000) \\ &= 352{,}500 \text{ lbf} \end{aligned}$$

Using Eqs. 14.113 and 14.115, with $\phi = 0.70$ and $\rho_g = 0.02$,

$$\begin{aligned} 352{,}500 &= (0.80)(0.70)A_g \\ &\quad \times [(0.85)(1 - 0.02)(3500) + (40{,}000)(0.02)] \\ A_g &= 169.4 \text{ in}^2 \end{aligned}$$

The column width is

$$\sqrt{169.4} \approx 13 \text{ in}$$

Actual $A_g = (13)^2 = 169 \text{ in}^2$.

Assume #6 bars and #3 ties. With $1\frac{1}{2}$-in cover, the steel core size is

$$13 - (2)\left(1\tfrac{1}{2}\right) - (2)\left(\tfrac{3}{8}\right) - (2)\left(\frac{0.75}{2}\right) = 8.5 \text{ in}$$

The required steel area is

$$(0.02)(169.4) = 3.39 \text{ in}^2$$

To distribute bars evenly around the column, the number of bars must be in multiples of fours. Try 12 bars.

$$A_{\text{bar}} = \frac{3.39}{12} = 0.28 \text{ in}^2$$

0.28 in² is too small since #5 bars or larger need to be used. Try 8 bars.

$$A_{\text{bar}} = \frac{3.39}{8} = 0.42 \text{ in}^2$$

Use #6 (0.44 in² each) bars.

The clear spacing between long bars is

$$\frac{(4)(8.5) - (8)(0.750)}{8} = 3.5 \text{ in} \quad \text{[acceptable]}$$

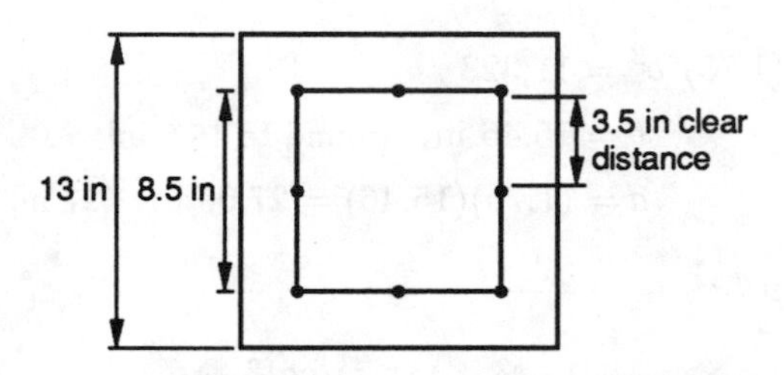

Use #3 ties. The tie spacing should be less than the minimum of

$$\left\{\begin{array}{l} (48)\left(\tfrac{3}{8}\right) = 18 \text{ in} \\ (16)(0.75) = 12 \text{ in} \\ \text{gross width} = 13 \text{ in} \end{array}\right\} = 12 \text{ in}$$

With only 8 bars, every alternate bar is supported. The spacing between bars is less than 6 in, so no additional cross ties are needed.

7. This is similar to problem 2 except for the information about the column steel.

steps 1 and 2: Assume a 25-in-thick footing and 150 pcf concrete. The approximate footing size is

$$A = \frac{200{,}000 + 145{,}000}{4000 - \left(\frac{25}{12}\right)(150)} = 93.6 \text{ ft}^2$$

Try a 10 ft × 10 ft footing (100 ft²). (9.75 ft × 9.75 ft could also be used.)

$$\begin{aligned} p_{\text{net}} &= \frac{(1.4)(200{,}000) + (1.7)(145{,}000)}{100} \\ &= 5265 \text{ psf} \end{aligned}$$

Assume a depth to reinforcement of 21 in. From Fig. 14.19, the critical area is

$$(10)^2 - \left(\frac{18+21}{12}\right)^2 = 89.44 \text{ ft}^2$$

The critical perimeter is

$$(4)\left(\frac{18+21}{12}\right) = 13 \text{ ft}$$

step 3: The ultimate shear is

$$\begin{aligned} V_u \approx P_u &= (1.4)(200{,}000) + (1.7)(145{,}000) \\ &= 526{,}500 \text{ lbf} \end{aligned}$$

step 4: The allowable two-way shear is

$$\beta_2 = 2 \quad \text{[minimum value]}$$

$$\begin{aligned} \phi V_c &= \left[\left(2+\frac{4}{2}\right)(0.85)\sqrt{3000}\right](144)\left[(13)\left(\frac{21}{12}\right)\right] \\ &= 610{,}075 \text{ lbf} \end{aligned}$$

step 5: Since $V_u < \phi V_c$, depth and size are acceptable. (The one-way shear should also be checked.)

step 6: From Fig. 14.21,

$$b_m = \frac{10 - \frac{18}{12}}{2} = 4.25 \text{ ft} \quad [51.0 \text{ in}]$$

From Eq. 14.140,

$$M_u = \frac{(5265)(10)(4.25)^2}{2} = 475{,}495 \text{ ft-lbf}$$

step 7: From Eq. 14.141,

$$\begin{aligned} R_u &= \frac{475{,}495}{(0.90)(10)\left(\frac{21}{12}\right)^2} \\ &= 17{,}252 \text{ psf} \quad [119.8 \text{ psi}] \end{aligned}$$

step 8: From Eq. 14.142,

$$\begin{aligned} \rho &= \left[\frac{(0.85)(3000)}{40{,}000}\right]\left[1 - \sqrt{1 - \frac{(2)(119.8)}{(0.85)(3000)}}\right] \\ &= 0.00307 \quad [\text{assume } 200/f_y \text{ does not apply}] \end{aligned}$$

step 9:

$$\begin{aligned} A_{st} &= (0.00307)(21)(10)(12) \\ &= 7.74 \text{ in}^2 \end{aligned}$$

Use 10 #8 bars (7.9 in^2).

step 10: Minimum spacing is not a problem.

step 11: From Eq. 14.2,

$$l_d = \text{maximum}\left\{\begin{array}{l} \frac{(0.04)(0.79)(40{,}000)}{\sqrt{3000}} = 23.08 \text{ in} \\ \frac{(0.03)(1)(40{,}000)}{\sqrt{3000}} = 21.91 \text{ in} \\ 12 \text{ in} \end{array}\right\}$$

Since 23 in $< b_m = 51.0$ in, it is acceptable.

Finally, since column steel was specified, check the need for dowel bars.

The column area is

$$A_g = (18)^2 = 324 \text{ in}^2$$

From Eq. 14.133, the bearing strength is

$$f_b = (0.85)(3000) = 2550 \text{ psi}$$

The bearing capacity of concrete is

$$\phi f_b A_g = (0.70)(2550)(324) = 578{,}340 \text{ lbf}$$

Since 578,340 > 526,500 ($\phi P_n > P_u$), only the minimum dowel steel is required.

From Eq. 14.137,

$$A_{st,min} = (0.005)(18)(18) = 1.62 \text{ in}$$

Use four #6 bars (1.76 in^2).

Check the development length of the dowels in compression. From Eq. 14.5,

$$l_d = \text{maximum}\left\{\begin{array}{l} \frac{(0.02)(0.75)(40{,}000)}{\sqrt{3000}} = 10.95 \text{ in} \\ (0.0003)(0.75)(40{,}000) = 9.0 \text{ in} \\ 8 \text{ in} \end{array}\right\}$$

$l_d = 10.95$ in controls

Since 10.95 in $< d = 21$ in, it is acceptable.

8. Use the procedure on p. 14-11.

step 1: From Eq. 14.28,

$$\rho \approx \frac{(0.18)(4000)}{40{,}000} = 0.018$$

step 2: From Eqs. 14.29 and 14.30,

$$m = \frac{40{,}000}{(0.85)(4000)} = 11.76$$

$$\begin{aligned} R_u &= (0.018)(40{,}000)\left[1 - (0.5)(0.018)(11.76)\right] \\ &= 643.8 \text{ psi} \end{aligned}$$

step 3: From Eq. 14.31,

$$bd^2 = \frac{(400{,}000)(12)}{(0.90)(643.8)} = 8284 \text{ in}^3$$

step 4: Choose $d = 1.75b$.

$$8284 = b(1.75b)^2$$

$$b = \boxed{13.93 \text{ in} \quad [\text{round to } 14 \text{ in}]}$$

$$d = \boxed{(1.75)(14) = 24.5 \text{ in}}$$

step 5: R_u and ρ are not substantially different.

step 6: $A_{st} = (0.018)(14)(24.5) = 6.17 \text{ in}^2$

step 7: From Tables 14.5 and 14.6, use four #11 bars (6.24 in²) in one layer. Use #3 stirrups.

The overall beam depth will be

$$24.5 + \frac{1.410}{2} + 0.375 + 1.5$$

$$= \boxed{27.08 \text{ in} \quad [\text{round to 27.5-in deep}]}$$

9. The effective beam width is limited by Eq. 14.65.

$$b_e = \text{minimum}\left\{\begin{array}{l} \frac{1}{4}l_{\text{beam}} \\ 8 + (16)(3) = 56 \text{ in} \\ 8 + 22 = 30 \text{ in} \end{array}\right\}$$

$$= 30 \text{ in}$$

Work with the following beam.

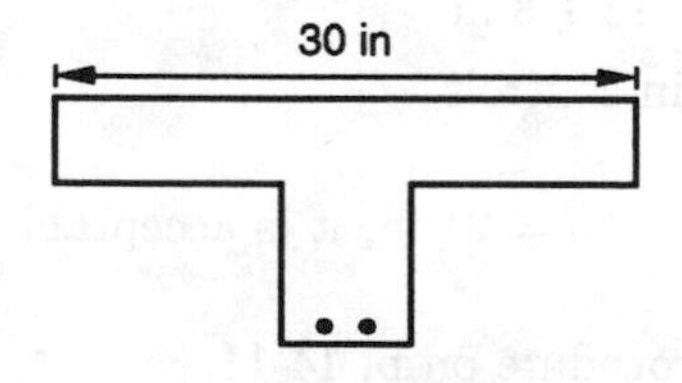

Use the procedure on p. 12-8.

step 1: $E_{\text{steel}} = 2.9 \times 10^7$ psi

From Eq. 14.1, using $w = 145 \text{ lbf/ft}^3$,

$$E_{\text{concrete}} = (145)^{1.5}(33)\sqrt{3000} = 3.16 \times 10^6 \text{ psi}$$

Note: Specific weight of concrete varies from 145–150 lbf/ft³. Use the conservative value. In this case, using 145 lbf/ft³ will give the lowest E.

step 2: $n = \dfrac{2.9 \times 10^7}{3.16 \times 10^6} = 9.18$

Note: The ACI code allows the use of the rounded value of n, but does not require it.

step 3: The expanded steel area is

$$(9.18)(2)(1) = 18.36 \text{ in}^2$$

step 4: Assume the neutral axis is below the flange.

step 5: (skip)

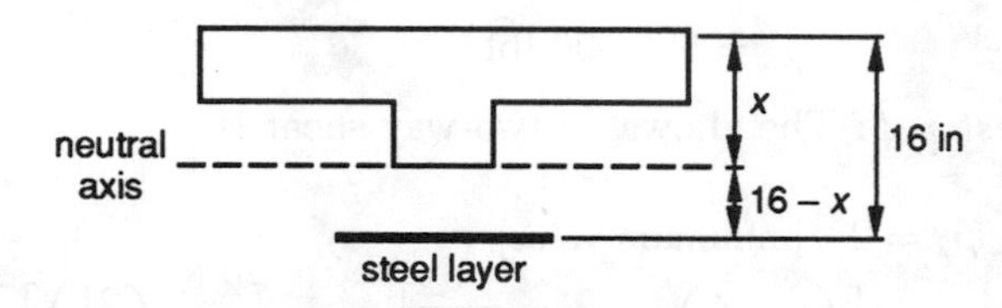

step 6: The area of concrete is

- Stem to flange bottom:

$$A = (8)(x-3) = 8x - 24 \text{ in}^2$$

- All of flange:

$$A = (30)(3) = 90 \text{ in}^2$$

Balance the moments of the transformed areas.

$$(8x - 24)\left(\frac{x-3}{2}\right) + (90)(x - 1.5) = (16 - x)(18.36)$$

$$4x^2 + 84.36x - 392.76$$

$$x = 3.92 \text{ in}$$

$$16 - x = 16 - 3.92 = 12.08 \text{ in}$$

step 7: The moment of inertia of the beam is

$$I_{\text{steel}} = (18.36)(12.08)^2 = 2679.2 \text{ in}^4$$

$$I_{\text{concrete}} = \frac{(8)(0.92)^3}{3} + \frac{(30)(3)^3}{12} + (3)(30)(2.42)^2 = 594.8 \text{ in}^4$$

$$I_{\text{total}} = 2679.2 + 594.8 = 3274.0 \text{ in}^2$$

step 8:

$$M = \left(20{,}000 \ \frac{\text{ft-lbf}}{\text{ft}}\right)\left(\frac{30}{12} \text{ ft}\right) = 50{,}000 \text{ ft-lbf}$$

step 9:

$$\sigma_{\text{concrete}} = \frac{(50{,}000)(12)(3.92)}{3274.0} = \boxed{718.4 \text{ psi}}$$

$$\sigma_{\text{steel}} = \frac{(9.18)(50{,}000)(12)(12.08)}{3274.0} = \boxed{20{,}323 \text{ psi}}$$

10.

sieve	cumulative % retained	
4	0 + 4 =	4
8	4 + 11 =	15
16	15 + 21 =	36
30	36 + 22 =	58
50	58 + 24 =	82
100	82 + 17 =	99
		294

$$\text{fineness modulus} = \frac{294}{100} = \boxed{2.94}$$

Timed

1. All loads are dead loads.

Total distributed load = (30)(2) = 60 kips.

Taking clockwise moments as positive,

$$\sum M_L = -30R + (9)(20) + (24)(60) + (60)\left(\frac{30}{2}\right) = 0$$

$$R = 84 \text{ kips}$$

$$\sum M_R = 30L - (60)(6) - (20)(21) - (60)\left(\frac{30}{2}\right) = 0$$

$$L = 56 \text{ kips}$$

Check: 84 + 56 = 20 + 60 + 60 = 140 kips [correct]

Taking clockwise moments as positive, the moment between the two center loads is

$$M_L = 56x - x^2 - (20)(x - 9)$$

$$\frac{dM}{dx} = 56 - 2x - 20 = 36 - 2x = 0$$

$$x = 18 \text{ ft} \quad \text{[at maximum moment]}$$

$$M_{18} = (56)(18) - (18)^2 - (20)(9) = 504 \text{ ft-kips} = M_w$$

$$M_u = 1.4\,(M_w) = (1.4)(504) = 705.6 \text{ ft-kips}$$

Follow the procedure on p. 14.11.

step 1: Find ρ_{balanced} from Eq. 14.10.

$$\beta_1 = 0.85, \text{ since } f'_c = 4000 \text{ psi}$$

$$\rho_{\text{balanced}} = \left[\frac{(0.85)(0.85)(4000)}{60{,}000}\right]\left(\frac{87{,}000}{87{,}000 + 60{,}000}\right) = 0.0285$$

choose $\rho = 0.375\rho_{\text{balanced}} = (0.375)(0.0285) = 0.0107$

From Eq. 14.9, check ρ_{min}.

$$\frac{200}{60{,}000} = 0.00333 \quad \text{[acceptable]}$$

step 2: From Eq. 14.29, calculate the coefficient of resistance.

$$R_u = (0.0107)(60{,}000)\left[1 - \frac{(0.0107)(60{,}000)}{(2)(0.85)(4000)}\right] = 581.4 \text{ psi}$$

step 3: From Eq. 14.31,

$$bd^2 = \frac{(705.6)(1000)(12)}{(0.90)(581.4)} = 16{,}182 \text{ in}^3$$

step 4: Assume $\frac{d}{b} = 2.0$.

$$bd^2 = b \times 4b^2 = 16{,}182 \text{ in}^3$$

$$b = 15.93 \text{ in} \quad \text{[round to 16 in]}$$

$$d = (2)(16) = 32 \text{ in}$$

step 5:

$$R_{u,\text{revised}} = \frac{(705{,}600)(12)}{(0.90)(16)(32)^2} = 574.2 \text{ psi}$$

$$\rho_{\text{revised}} = (0.0107)\left(\frac{574.2}{581.4}\right) = 0.0106$$

Check the limits.

$$0.00333 < 0.0106 < (0.75)(0.0285) = 0.0214 \quad \text{[acceptable]}$$

step 6: $A_{st} = (0.0106)(16)(32) = 5.43 \text{ in}^2$

step 7: Use four #11 bars (6.24 in^2).

From Table 14.5,

$$\text{beam width: } 16 \text{ in} > 13.7 \text{ in} \quad \text{[acceptable]}$$

$$\text{actual } \rho = \frac{6.24}{(16)(32)} = 0.0122 < \rho_{\text{max}} \quad \text{[acceptable]}$$

step 8: Use 3-in cover (given), which includes room for stirrups.

$$h = 32 + 3 = 35 \text{ in}$$

step 9: (skip)

step 10: (skip)

step 11: Refer to Fig. 14.6.

$$d_c = 3 \text{ in}$$
$$d_{st} = 3 \text{ in} \quad [\text{same as } d_c]$$

From Eq. 14.37,

$$f_{st} = (0.60)(60{,}000) = 36{,}000 \text{ psi}$$

From Eq. 14.35,

$$z = 36{,}000\sqrt[3]{\frac{(3)(2)(3)(16)}{4}} = 149{,}766 \text{ lbf/in}$$

Since $z > 145$ kips/in, recalculate f_{st} more accurately.

Locate the centroid. From Eq. 14.17,

$$c = \left[\frac{(8)(6.24)}{16}\right]\left[\sqrt{1 + \frac{(2)(16)(32)}{(8)(6.24)}} - 1\right]$$
$$= 11.35 \text{ in}$$

From Eq. 14.39,

$$\text{moment arm} = 32 - \frac{11.35}{3} = 28.22 \text{ in}$$

From Eq. 14.38,

$$f_{st} = \frac{(504{,}000)(12)}{(6.24)(28.22)} = 34{,}346 \text{ psi}$$

Using Eq. 14.35 again,

$$z = 34{,}346\sqrt[3]{\frac{(3)(2)(3)(16)}{4}}$$
$$= 142{,}885 \text{ lbf/in} < 145 \text{ kips/in} \quad [\text{acceptable}]$$

step 12: (skip)

2. Follow the procedure on p. 14-11.

step 1: See Timed problem 1.

$$\rho_{\text{balanced}} = 0.0285$$
$$\rho_{\text{max}} = (0.75)(0.0285) = 0.0214$$
$$\rho_{\text{min}} = 0.00333$$

Use $\rho = 0.0214$ as directed, although this is not a good idea.

step 2: From Eq. 14.29,

$$R_u = (0.0214)(60{,}000)\left[1 - \frac{(0.0214)(60{,}000)}{(2)(0.85)(4000)}\right]$$
$$= 1041.6 \text{ psi}$$

step 3:

$$\text{factored load} = (1.4)(1) + (1.7)(2) = 4.8 \text{ kips/ft}$$
$$M_u = \frac{wL^2}{8} = \frac{(4.8)(27)^2}{8} = 437.4 \text{ ft-kips}$$

From Eq. 14.31,

$$bd^2 = \frac{(437{,}400)(12)}{(0.90)(1041.6)} = 5599.1 \text{ in}^3$$

step 4:

$$b = 14 \text{ in} \quad [\text{given}]$$
$$d = \sqrt{\frac{5599.1}{14}} = 20.0 \text{ in}$$

step 5: (skip)

step 6: $A_{st,max} = (0.0214)(14)(20) = 5.99 \text{ in}^2$ [maximum]

This is a maximum value, and steel chosen should be slightly less than this.

step 7: Try four #11 bars.

$$A_{st} = (4)\left(1.56 \text{ in}^2\right) = 6.24 \text{ in}^2$$

Since 6.24 in^2 > $A_{st,max}$, increase beam depth to 21 in.

$$\rho = \frac{A_{st}}{bd} = \frac{6.24}{(14)(21)} = 0.02122 < 0.0214 \quad [\text{acceptable}]$$

step 8:

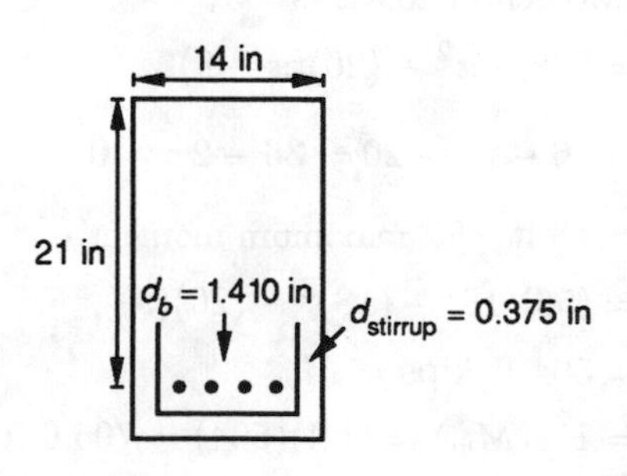

$$\text{beam depth} = 21 + \frac{1.410}{2} + 0.375 + 1.5$$
$$= 23.58 \text{ in} \quad [\text{round to 24 in}]$$

step 9: (skip)

step 10: (skip)

step 11: From Eq. 14.37,

$$f_{st} = (0.60)(60{,}000) = 36{,}000 \text{ psi}$$
$$d_c = 24 - 21 = 3 \text{ in}$$
$$d_{st} = 24 - 21 = 3 \text{ in}$$

From Eq. 14.35,

$$z = 36{,}000\sqrt[3]{\frac{(3)(2)(3)(14)}{4}}$$
$$= 143{,}246 \text{ lbf/in} < 175 \text{ kips/in}$$

acceptable for interior use

step 12: From Eq. 14.43,

$$I_g = \frac{(14)(24)^3}{12} = 16{,}128 \text{ in}^4$$
$$M_{max} = \frac{wL^2}{8} = \frac{(2+1)(27)^2}{8}$$
$$= 273.4 \text{ ft-kips} \quad \text{[unfactored]}$$
$$= (273.4)(12)(1000) = 3.28 \times 10^6 \text{ in-lbf}$$

From Eq. 14.42,

$$f_r = 7.5\sqrt{4000} = 474.3 \text{ psi}$$

From Eq. 14.44,

$$y_t = \frac{24}{2} = 12 \text{ in}$$

From Eq. 14.41,

$$M_{cr} = \frac{(474.3)(16{,}128)}{12} = 6.37 \times 10^5 \text{ in-lbf}$$
$$\left(\frac{M_{cr}}{M_{max}}\right)^3 = \left(\frac{0.637}{3.28}\right)^3 = 0.0073$$

To calculate I_{cr}, find the neutral axis.

$$n = \frac{2.9 \times 10^7}{3.6 \times 10^6} = 8.06$$

From Eq. 14.17,

$$c = \left[\frac{(8.06)(6.24)}{14}\right]\left[\sqrt{1 + \frac{(2)(14)(21)}{(8.06)(6.24)}} - 1\right]$$
$$= 9.21 \text{ in}$$

$$I_{cr} = \frac{(14)(9.21)^3}{3} + (8.06)(6.24)(21 - 9.21)^2$$
$$= 10{,}637 \text{ in}^3$$

From Eq. 14.40,

$$I_e = (0.0073)(16{,}128) + (1 - 0.0073)(10{,}637)$$
$$= 10{,}677 \text{ in}^4$$

Find deflection for a uniform load.

$$\delta_{inst} = \frac{5wL^4}{384EI} = \frac{(5)(3)(1000)\,[(27)(12)]^4}{(384)(3.6 \times 10^6)(10{,}677)(12)}$$
$$= 0.933 \text{ in} \quad \text{[with 100\% live load]}$$

For a long-term loading, the percentage of the load to be sustained is

$$\frac{1 + (0.30)(2)}{1 + 2} = 0.53 \; (53\%)$$

The factored long-term deflection is

$$\delta = (0.53)(0.933) = 0.494 \text{ in}$$

From Eq. 14.47, $\rho' = 0$ (no compression steel). $\lambda = 2$.

$$\delta_{\text{long term}} = (2)(0.494) = 0.988 \text{ in}$$
$$\delta_{\text{total}} = 0.933 + 0.988$$
$$= 1.92 \text{ in} \quad \text{[with 30\% of live load sustained]}$$

3. Follow the procedure on p. 14-23.

step 1: $w_{DL} = 150$ lbf/ft [given]

step 2: $M_{DL} = \frac{wL^2}{8} = \frac{(150)(100)^2}{8} = 187{,}500$ ft-lbf

step 3: Use the top as the reference. Neglect the tension holes.

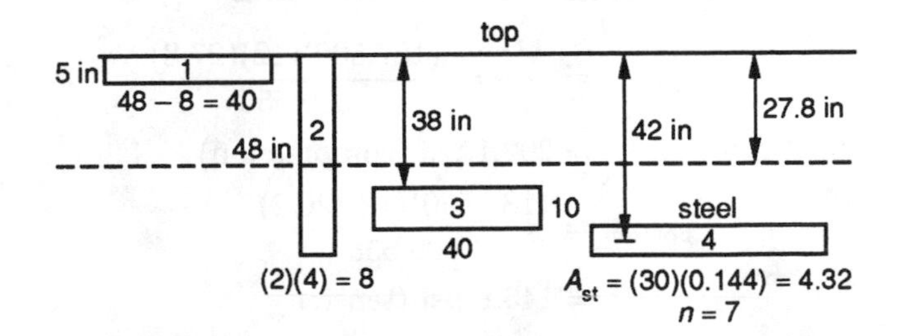

$$\overline{x} = \frac{\sum A_i\overline{x}_i}{\sum A_i}$$
$$= \frac{(5)(40)(2.5) + (48)(8)(24) + (10)(40)(38+5)}{(5)(40) + (48)(8) + (10)(40) + (4.32)(7)}$$
$$+ \frac{(4.32)(7)(42)}{(5)(40) + (48)(8) + (10)(40) + (4.32)(7)}$$
$$= \frac{28{,}186}{1014.2} = 27.8 \text{ in} \quad \text{[from top]}$$

step 4: Assume the entire beam is in compression. Set up a table to calculate I.

section	$I_c = \frac{bh^3}{12}$	A	d	Ad^2
1	416.7	200	25.3	128,018
2	73,728	384	3.8	5545
3	3333.3	400	15.2	92,416
4	0	30.24	14.2	6098
totals	77,478 in^4			232,077 in^4

$$I = 77{,}478 + 232{,}077 = 309{,}555 \text{ in}^4$$

step 5: $e = 42 - 27.8 = 14.2$ in

step 6: Disregarding time-dependent losses immediately after transfer (the conservative assumption), the final prestress is

$$\begin{aligned} f_s &= \text{initial prestress} - \text{losses} \\ &= (0.8)(250{,}000) - 0 = 200{,}000 \text{ psi} \end{aligned}$$

$$\begin{aligned} p_s &= \left(200{,}000 \ \frac{\text{lbf}}{\text{in}^2}\right)\left(0.144 \ \frac{\text{in}^2}{\text{strand}}\right)(30 \text{ strands}) \\ &= 864{,}000 \text{ lbf} \end{aligned}$$

$$A_c = 200 + 384 + 400 = 984 \text{ in}^2$$

$$c_{\text{top}} = 27.8 \text{ in} \quad \text{[for top fibers]}$$

$$c_{\text{bottom}} = 48 - 27.8 = 20.2 \text{ in} \quad \text{[for bottom fibers]}$$

From Eq. 14.103,

$$\begin{aligned} f_{\text{top}} &= -\frac{864{,}000}{984} + \frac{(864{,}000)(14.2)(27.8)}{309{,}555} \\ &= 223.8 \text{ psi (tension)} \end{aligned}$$

$$\begin{aligned} f_{\text{bottom}} &= -\frac{864{,}000}{984} - \frac{(864{,}000)(14.2)(20.2)}{309{,}555} \\ &= -1678.6 \text{ psi (compression)} \end{aligned}$$

step 7: The stress due to dead load is

$$\begin{aligned} f_{b,\text{top}} &= \frac{Mc}{I} = \frac{(187{,}500)(12)(27.8)}{309{,}555} \\ &= 202.1 \text{ psi (compression)} \end{aligned}$$

$$\begin{aligned} f_{b,\text{bottom}} &= \frac{(187{,}500)(12)(20.2)}{309{,}555} \\ &= 146.8 \text{ psi (tension)} \end{aligned}$$

step 8:

$$f_{\text{top}} = 223.8 - 202.1 = -21.7 \text{ psi (compression)}$$

$$f_{\text{bottom}} = -1678.6 + 146.8 = -1531.8 \text{ (compression)}$$

step 9: No tension is present.

step 10: From Eq. 14.98,

$$f_c = (0.60)(3800) = 2280 \text{ psi}$$

Both 21.8 psi and 1531.8 psi are less than 2280 psi, so the stresses are

acceptable

(b) *step 11:* The prestress from step 6 is reduced.

$$\frac{(0.6)(250{,}000)}{(0.8)(250{,}000)} = 0.75 \text{ of step 6 prestress}$$

Recalculate the stresses with the lower prestress.

$$f_{\text{top}} = (0.75)(223.8) = 167.9 \text{ psi (tension)}$$

$$f_{\text{bottom}} = (0.75)(-1678.6) = -1259.0 \text{ psi (compression)}$$

$$M_{LL} = \frac{wL^2}{8} = \frac{(540)(100)^2}{8} = 675{,}000 \text{ ft-lbf}$$

$$f_{\text{top}} = \frac{(675{,}000)(12)(27.8)}{309{,}555} = -727.4 \text{ psi (compression)}$$

$$f_{\text{bottom}} = \frac{(675{,}000)(12)(20.2)}{309{,}555} = 528.6 \text{ psi (tension)}$$

From Eq. 14.107,

$$\begin{aligned} f_{\text{top,total}} &= 167.9 - 202.1 - 727.4 \\ &= -761.6 \text{ psi (compression)} \end{aligned}$$

$$\begin{aligned} f_{\text{bottom,total}} &= -1259.0 + 146.8 + 528.6 \\ &= -583.6 \text{ psi (compression)} \end{aligned}$$

From Eq. 14.100,

$$f_c = (0.45)(5000) = 2250 \text{ psi}$$

Since 761.5 psi < 2250 psi, compressive stress is

acceptable [There is no tensile stress.]

Note: Prestress losses would usually be included in step 11. In this problem, the effective prestress is given.

4. (a) The factored load is

$$(1.4)(1300) + (1.7)(1900) = 5050 \text{ lbf/ft}$$

Total factored load = (28)(5050) = 141,400 lbf. Each reaction is 70,700 lbf (factored). Load is continuous, so the shear diagram is

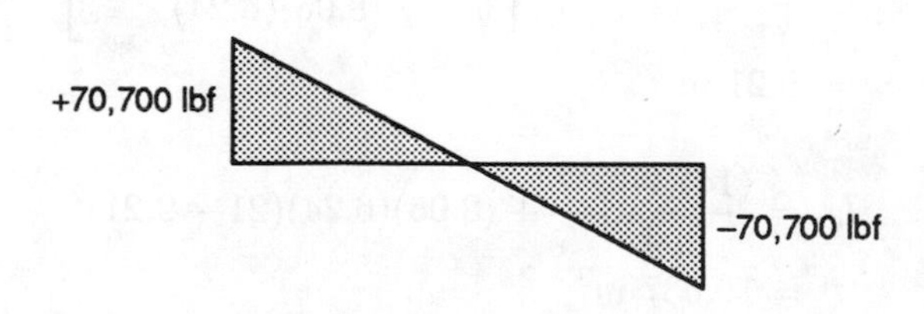

ACI-318 says shear over the distance, d, from the support face can be considered constant. At that point, shear is

$$70{,}700 - \left(\frac{21}{12}\right)(5050) = 61{,}860 \text{ lbf}$$

$$V_u = 61{,}860 \text{ lbf}$$

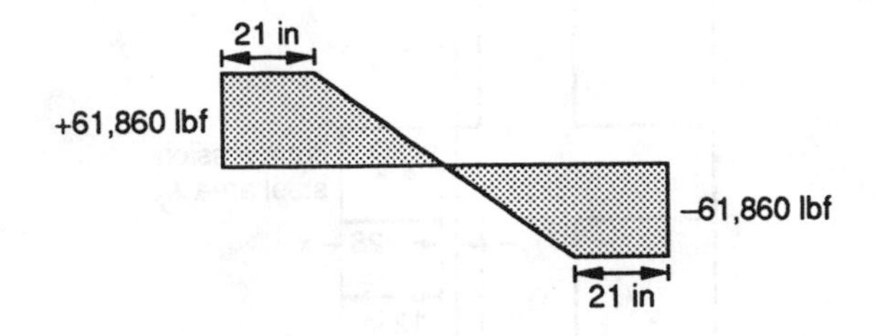

From Eq. 14.48, the nominal concrete shear strength is

$$V_c = 2\sqrt{3000}(12)(21) = 27{,}605 \text{ lbf}$$

Since $V_u > \phi V_c$, the beam needs extra shear reinforcement until V_u drops to ϕV_c. From Eq. 14.50, the steel shear contribution is

$$V_{st} = \frac{61{,}860}{0.85} - 27{,}605 = 45{,}171 \text{ lbf}$$

Use Eq. 14.51 to see if the beam should be redesigned.

$$V_{st,max} = 8\sqrt{3000}(12)(21) = 110{,}421 \text{ lbf} \quad \text{[acceptable]}$$

(b) From Eq. 14.52, no reinforcement is needed after V_u drops below $\frac{1}{2}\phi V_c$.

$$\tfrac{1}{2}\phi V_c = (0.5)(0.85)(27{,}605) = 11{,}732 \text{ lbf}$$

$$70{,}700 - x_{feet}(5050) = 11{,}732 \text{ lbf}$$

$$x = 11.68 \text{ ft from face of each suppport.}$$

The center $28 - (2)(11.68) = 4.64$ ft requires no shear reinforcement; the adjacent 2.32 ft on either side of this zone requires only code minimum reinforcement.

(c) From Eq. 14.56,

$$4\sqrt{3000}(12)(21) = 55{,}210 \text{ lbf}$$

Since $V_{st} < 55{,}210$, Eq. 14.55 holds.

$$\text{maximum spacing} = \frac{21}{2} = 10.5 \text{ in}$$

However, even closer spacing may be needed. For #3 bars,

$$A_{stirrup} = (2)(0.11) = 0.22 \text{ in}^2$$

Eq. 14.57 can be solved for spacing.

$$s = \frac{(2)(0.11)(60{,}000)(21)}{45{,}171} = \boxed{6.14 \text{ in}}$$

Use 6-in spacing starting 10.5 in from the face of support.

5. (a) Find the centroid of the load group. (Ignore dead load.)

$$\overline{x} = \frac{(3)(60) + (13)(80)}{60 + 80} = 8.71 \text{ ft} \quad \text{[from right end]}$$

Calculate the eccentricities that would produce the same moments.

$$\epsilon_1 = \frac{M}{P} = \frac{120}{80} = 1.5 \text{ ft}$$

$$\epsilon_2 = \frac{160}{60} = 2.67 \text{ ft}$$

Move the loads.

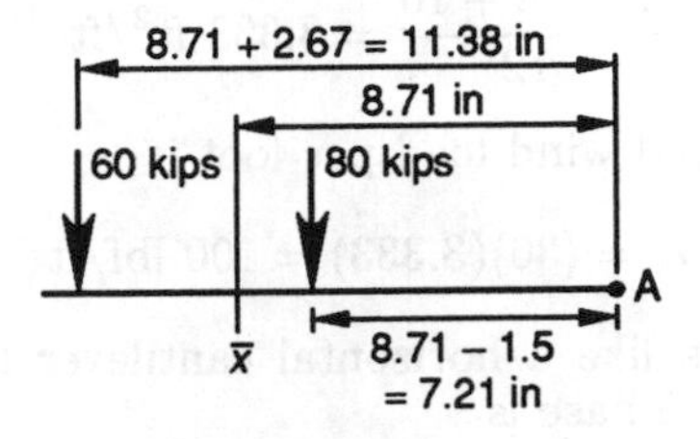

The total net vertical reaction must be $60 + 80 = 140$ kips. This can be assumed to be at $\frac{1}{2}L$ if it is uniform.

$$\sum M_A = (60)(11.38) + (80)(7.21) - (140)\left(\frac{L}{2}\right) = 0$$

$$L = \boxed{17.99 \text{ ft} \quad \text{[round to 18 ft]}}$$

- Alternate solution:

Sum moments about either free end.

$$(80)(L+3) + 120 + (60)(L-3) - 160 - (140)\left(\frac{L}{2}\right) = 0$$

$$L = \boxed{18 \text{ ft}}$$

(b) $w = \text{ soil pressure } = \dfrac{140}{18.0} = 7.78 \text{ kips/ft}$

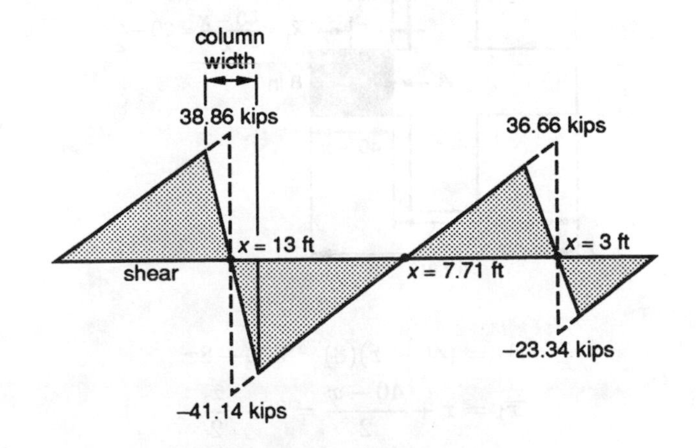

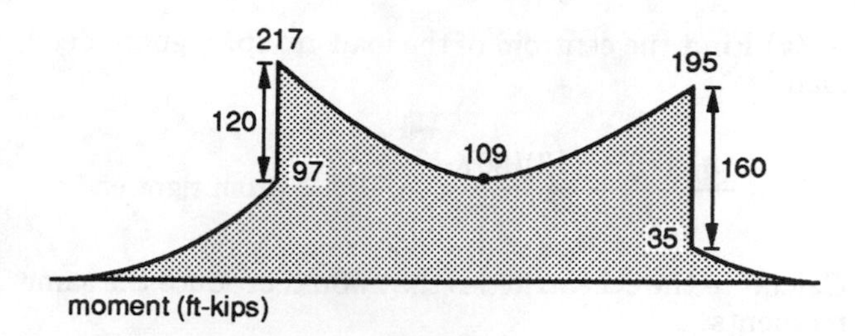

(c) No top steel is needed because the entire footing is in single-curvature bending.

6. The area projected to the wind is

$$\frac{16+8+16}{12} = 3.333 \text{ ft}^2/\text{ft}$$

The distributed wind load per foot is

$$w = (30)(3.333) = 100 \text{ lbf/ft}$$

The pier acts like a horizontal cantilever beam. The moment at the base is

$$M = \left(\tfrac{1}{2}\right)(100)(36)^2 = 64{,}800 \text{ ft-lbf}$$

Masonry problems are solved using working stress concepts.

$$\begin{aligned} E_m &= \text{modulus of elasticity} \\ &= 1000f'_m = 1.5 \times 10^6 \text{ psi} \end{aligned}$$

$$n = \frac{E_s}{E_m} = \frac{3 \times 10^7}{1.5 \times 10^6} = 20$$

The west half of the column will be in tension, and masonry cannot support tension. Assume the neutral axis will be in the east arm, placing the north-south steel in tension.

Disregard the effect of the horizontal #4 bars, which contribute to crack control. Use the most westerly end as the reference point to find the neutral axis.

The masonry in compression is A_1.

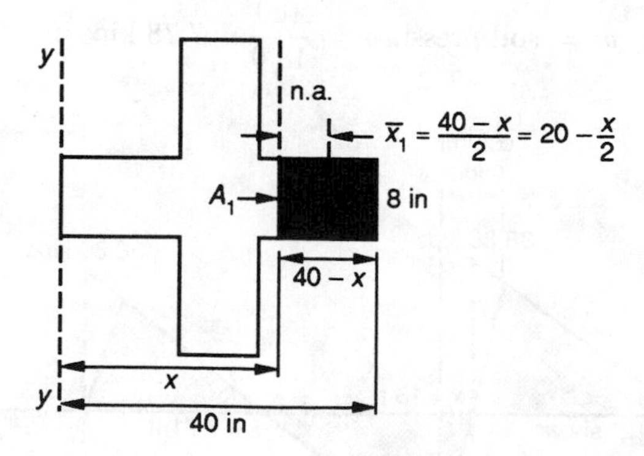

$$A_1 = (40-x)(8) = 320 - 8x$$

$$\overline{x}_1 = x + \frac{40-x}{2} = 20 - \frac{x}{2}$$

Note that x is measured from the y-y axis, but $\overline{x}$ is measured from the neutral axis.

Each steel bar contributes a transformed area of

$$nA_b = (20)(0.31) = 6.2 \text{ in}^2$$

Assume three bars in the east arm are in compression.

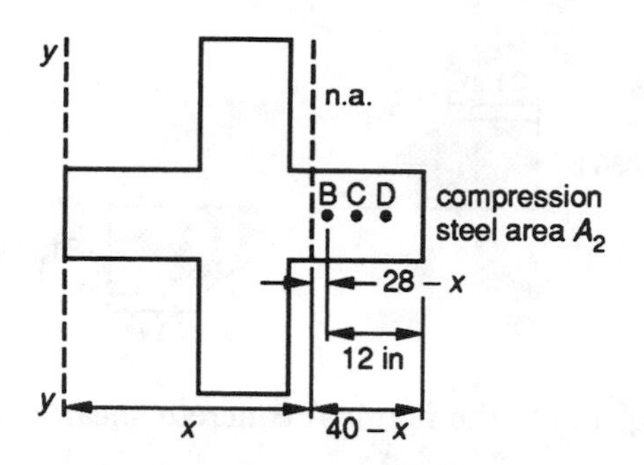

$$\begin{aligned} A_{2B} &= 6.2 \\ \overline{x}_{2B} &= 40 - x - 12 = 28 - x \quad \text{[from neutral axis]} \\ A_{2C} &= 6.2 \\ \overline{x}_{2C} &= 32 - x \quad \text{[from neutral axis]} \\ A_{2D} &= 6.2 \\ \overline{x}_{2D} &= 36 - x \quad \text{[from neutral axis]} \end{aligned}$$

The steel in tension is

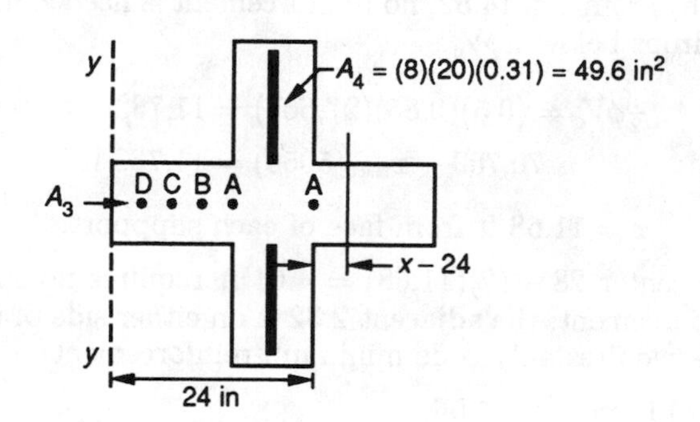

$$\begin{aligned} A_{2A} &= 6.2 \\ \overline{x}_{2A} &= x - 24 \\ A_{3A} &= 6.2 \\ \overline{x}_{3A} &= x - 16 \\ A_{3B} &= 6.2 \\ \overline{x}_{3B} &= x - 12 \\ A_{3C} &= 6.2 \\ \overline{x}_{3C} &= x - 8 \\ A_{3D} &= 6.2 \\ \overline{x}_{3D} &= x - 4 \\ A_4 &= 49.6 \\ \overline{x}_4 &= x - 20 \end{aligned}$$

The neutral axis is located where

$$\sum_{\text{compression}} A_i\bar{x}_i = \sum_{\text{tension}} A_i\bar{x}_i$$

$$(320-8x)\left(20-\frac{x}{2}\right) + (6.2)\left[(28-x)+(32-x)+(36-x)\right]$$
$$= (6.2)\left[(x-24)+(x-16)+(x-12) + (x-8)+(x-4)+(49.6)(x-20)\right]$$
$$= (6.2)(5x-64)+(49.6)(x-20)$$
$$4x^2 - 419.2x = -8384$$
$$x^2 - 104.8x = -2096$$
$$x = 26.91 \text{ in [round to 27.0; location of neutral axis]}$$

(This is not significantly affected by the initial assumption about the loaction of the neutral axis.)

To find the centroidal moment of inertia, make a table.

element	A	$\bar{x}$	$(\bar{x})^2$	I
A_1				$\frac{bh^3}{3} = 5859$
A_{2A}	6.2	3	9	55.8
A_{2B}	6.2	1	1	6.2
A_{2C}	6.2	5	25	155.0
A_{2D}	6.2	9	81	502.2
A_{3A}	6.2	11	121	750.2
A_{3B}	6.2	15	225	1395.0
A_{3C}	6.2	19	361	2238.2
A_{3D}	6.2	23	529	3279.8
A_4	49.6	7	49	2430.2
total				16,671 in^4

The extreme masonry stress is

$$f_m = \frac{M_c}{I} = \frac{(64{,}800)(12)(40-27)}{16{,}671}$$
$$= 606.4 \text{ psi}$$

f_m is limited by $0.18f'_m$ (special inspection required).

$$(0.18)(1500) = 270 \text{ psi}$$

not acceptable

Use bar A_{3D} to find maximum steel stress.

$$f_{st} = \frac{(20)(64{,}800)(12)(23)}{16{,}671} = 21{,}456 \text{ psi}$$

f_{st} is limited to 24,000 psi for grade 60 steel.

acceptable

7. (a) Refer to p. 10-6. From Eq. 10.12,

$$\epsilon = \frac{100}{200} = 0.5 \text{ ft}$$

Assume 150 pcf concrete, 110 pcf soil, and a footing thickness of 20 in.

The net allowable soil pressure is

$$p_{a,\text{net}} = 6000 - \left(\frac{20}{12}\right)(150) = 5750 \text{ psi}$$

For a footing with no moment,

$$A = \frac{200{,}000}{5750} = 34.78 \text{ ft}^2$$
$$B = \sqrt{34.78} = 5.90 \text{ ft} \quad \text{[round to 6 ft]}$$

From Eq. 10.15,

$$p_{\max} = 5750 = \left(\frac{200{,}000}{6B}\right)\left[1+\frac{(6)(0.5)}{B}\right]$$
$$B = 7.98 \text{ ft}$$

Use 6 ft × 8.25 ft footing.

(b) From Eq. 10.15,

$$p_{\max} = \left[\frac{200{,}000}{(6)(8.25)}\right]\left[1+\frac{(6)(0.5)}{8.25}\right] + \left(\frac{20}{12}\right)(150+110)$$
$$= \boxed{5943 \text{ psf}}$$

$$p_{\min} = \left[\frac{200{,}000}{(6)(8.25)}\right]\left[1-\frac{(6)(0.5)}{8.25}\right] + \left(\frac{20}{12}\right)(150+110)$$
$$= \boxed{3004 \text{ psf}}$$

(c) Disregarding dead load and factoring the live load, the variation in ultimate pressure under the footing is

$$\Delta p_u = (1.7)\left[\frac{5943-3004}{(8.25)(12)}\right] = \frac{10{,}103-5107}{99}$$
$$= 50.46 \text{ psf/in}$$

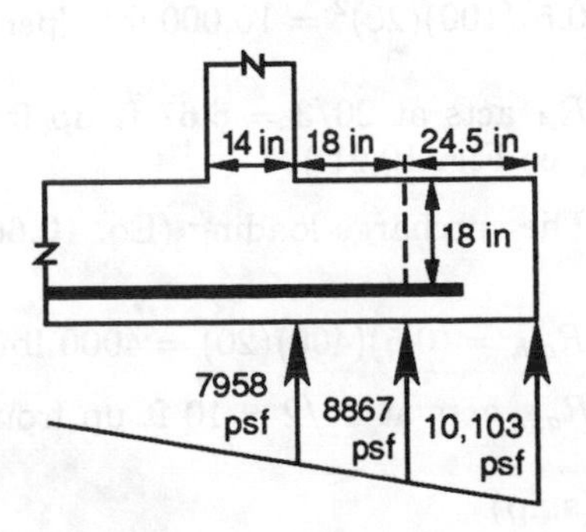

Assume $d = 18$ in. At the critical line (see Fig. 14.20), the soil pressure is

$$10{,}103 - (24.5)(50.46) = 8867 \text{ psf}$$

The required ultimate shear at the critical line is

$$V_u = \left(\frac{24.5}{12}\right)(8867) + \left(\frac{1}{2}\right)\left(\frac{24.5}{12}\right)(10{,}103 - 8867)$$
$$= 19{,}365 \text{ lbf} \quad \text{[per foot of width]}$$

The ultimate shear strength is calculated from Eq. 14.129.

$$\phi V_c = (2)(0.85)\sqrt{3000} = 93.1 \text{ psi}$$

The depth required is

$$d = \frac{19{,}365}{(93.1)(144)(1)}$$
$$= \boxed{1.444 \text{ ft } (17.3 \text{ in}) \quad \text{[round to 18 in]}}$$

(d) The pressure at the critical flexure section (see Fig. 14.21) is

$$10{,}103 - (50.46)(42.5) = 7958 \text{ psf}$$

The moment at the critical line is

$$M_u = \left(\frac{1}{2}\right)(7958)\left(\frac{42.5}{12}\right)^2 + \left(\frac{1}{2}\right)\left(\frac{42.5}{12}\right)[(50.46)(42.5)]\left(\frac{2}{3}\right)\left(\frac{42.5}{12}\right)$$
$$= 49{,}910 + 8967$$
$$= \boxed{58{,}877 \text{ ft-lbf} \quad \text{[per foot of wall]}}$$

8. (a) Refer to the procedure on p. 10-18.

step 1: $H = 18.25 + 1.75 = 20$ ft

The soil pressure resultant (from Eq. 10.60) is

$$R_A = \left(\tfrac{1}{2}\right)(0.5)(100)(20)^2 = 10{,}000 \text{ lbf} \quad \text{[per foot of wall]}$$

R_A acts at $20/3 = 6.67$ ft up from point G (see Fig. 10.21.)

The surcharge loading (Eq. 10.66) is

$$R_{q,h} = (0.5)(400)(20) = 4000 \text{ lbf}$$

$R_{q,h}$ acts at $20/2 = 10$ ft up from point G.

step 2: (skip)

step 3:

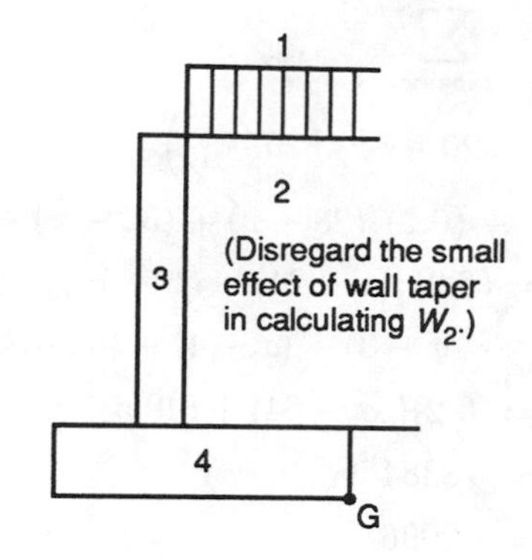

element	length or area	q or γ	F or W	r_i	moment
1	8	400	3200	4	12,800
2	146	100	14,600	4	58,400
3	22.8	150	3420	≈ 8.5	29,070
4	24.5	150	3675	7	25,725
totals			24,895		125,995 ft-lbf

step 4:

$$\sum M_G = (10{,}000)(6.67) + (4000)(10) + 125{,}995$$
$$= 232{,}695 \text{ ft-lbf} \quad \text{[round to 232,700]}$$

(M_G is not used in this problem.)

step 5: From Eqs. 10.81 and 10.82,

$$r_R = \frac{232{,}700}{24{,}895} = 9.35 \text{ ft}$$
$$\epsilon = 9.35 - \left(\tfrac{1}{2}\right)(14) = 2.35 \text{ ft}$$

(ϵ is not used in this problem.)

step 6: Refer to Fig. 10.21.

element	F	$14 - r_i = x_i$	moment
1	3200	10	32,000
2	14,600	10	146,000
3	3420	5.5	18,810
4	3675	7	25,725
total			222,535 ft-lbf

$$\sum R_{A,h,y} = (10{,}000)(6.67) + (4000)(10)$$
$$= 106{,}700 \text{ ft-lbf}$$

$$F_{\text{overturning}} = \frac{222{,}535}{106{,}700}$$
$$= \boxed{2.09 \quad \text{[acceptable]}}$$

(b) Refer to the procedure on p. 14-34.

$$\text{heel weight} = (8)(1.75)(1)(150) = 2100 \text{ lbf} \quad \text{[per foot of width]}$$

$$V_u = (1.4)(2100 + 14{,}600) + (1.7)(3200) = 28{,}820 \text{ lbf}$$

The nominal shear strength of concrete is given by Eq. 14.153.

$$V_n = \phi V_c = (2)(0.85)\sqrt{3000} = 93.1 \text{ psi}$$

The required heel thickness without requiring shear reinforcement is

$$\frac{28{,}820}{(93.1)(12)} = 25.8 \text{ in} \quad \text{[round to 26 in]}$$

To include 3-in cover, use a heel thickness of

$$t_b = \boxed{30 \text{ in}}$$

(c) The heel weight becomes

$$(8)\left(\frac{30}{12}\right)(1)(150) = 3000 \text{ lbf} \quad \text{[per foot of width]}$$

Take moments about the stem face.

element	F	distance	M
1	3200	4	12,800
2	14,600	4	58,400
heel	3000	4	12,000

The ultimate moment on the heel at the stem face is

$$M_u = (1.4)(58{,}400 + 12{,}000) + (1.7)(12{,}800) = 120{,}320 \text{ ft-lbf}$$

From Eq. 14.31 (or 14.32),

$$R_u = \frac{(120{,}320)(12)}{(0.90)(12)(26)^2} = 197.8 \text{ psi}$$

From Eq. 14.29, solving for ρ (or see Eq. 14.142),

$$\rho = \left[\frac{(0.85)(3000)}{60{,}000}\right]\left[1 - \sqrt{1 - \frac{(2)(197.8)}{(0.85)(3000)}}\right] = 0.0034$$

- Check:

$$\rho_{min} = \frac{200}{60{,}000} = 0.0033 \quad \text{[acceptable]}$$

#8 steel has an area of 0.79 in^2.

$$\text{spacing} = \frac{0.79}{(0.0034)(26)} = 8.9 \text{ in}$$

$$\boxed{\text{Use one \#8 bar every 9 in.}}$$

9. (a) Refer to pp. 12-31 and 12-32.

$$M_{DL} = \frac{(1.5)(18)^2}{8} + (8)(6) = 108.75 \text{ ft-kips}$$

$$M_{LL} = \frac{(2)(18)^2}{8} + (13)(6) = 159 \text{ ft-kips}$$

$$M_u = (1.4)(108.75) + (1.7)(159) = 422.6 \text{ ft-kips}$$

(b) Refer to p. 14-11.

step 1: To achieve minimum beam depth as required by the problem, choose the maximum allowable reinforcement.
From Eq. 14.10 with $\beta = 0.85$,

$$\rho_{\text{balanced}} = \left[\frac{(0.85)(0.85)(3000)}{60{,}000}\right]\left(\frac{87{,}000}{87{,}000 + 60{,}000}\right) = 0.0214$$

$$\rho_{max} = (0.75)(0.0214) = 0.016$$

$$\rho_{min} = \frac{200}{60{,}000} = 0.0033 \quad \text{[acceptable]}$$

step 2: $m = \dfrac{60{,}000}{(0.85)(3000)} = 23.53$

$$R_u = (0.016)(60{,}000)\left[1 - (0.5)(0.016)(23.53)\right] = 779.3 \text{ psi}$$

step 3: $bd^2 = \dfrac{(422.6)(12)(1000)}{(0.90)(779.3)} = 7230 \text{ in}^3$

step 4: b was given as 16 in.

$$d = \sqrt{\frac{7230}{16}} = 21.26 \text{ in} \quad \text{[round to 21.25 in]}$$

step 5: (skip)

step 6: $A_{st} = (0.016)(16)(21.25) = 5.44 \text{ in}^2$

step 7: Get as close as possible to 5.44 in^2 without going over. To keep below ρ_{max}, use four #10 bars (5.08 in^2).
(At this point, it may be desirable to recalculate ρ and resize the beam.)

step 8:

$$d_c = \frac{1.270}{2} + 0.375 + 1.5 = 2.51 \text{ in} \quad \text{[round to 2.75 in]}$$

(c) The factored uniform load is

$$(1.4)(1.5) + (1.7)(2) = 5.5 \text{ kips/ft}$$

The factored concentrated load is

$$(1.4)(8) + (1.7)(13) = 33.3 \text{ kips}$$

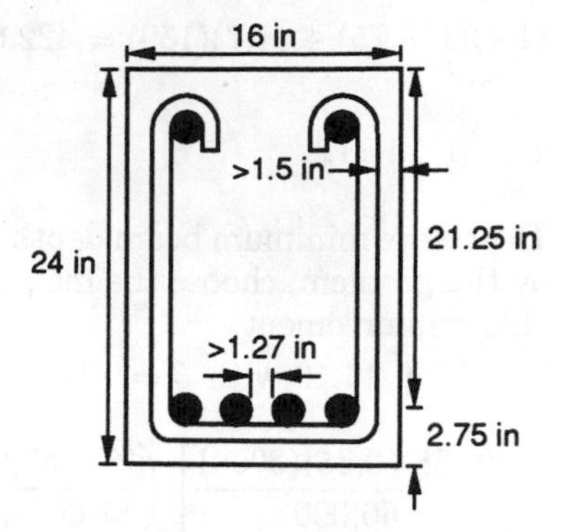

The factored reactions are

$$R = \frac{(5.5)(18) + (2)(33.3)}{2} = 82.8 \text{ psi}$$

The shear diagram is

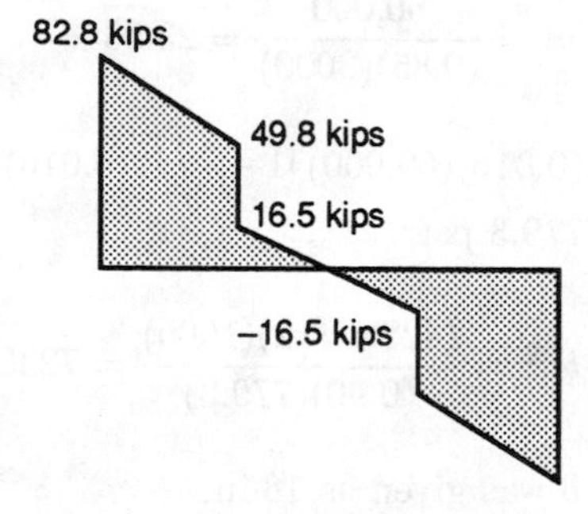

At a distance of 21.25 in from the support, the shear is

$$V_u = 82.8 - \left(\frac{21.25}{12}\right)(5.5) = 73.1 \text{ kips}$$

From Eq. 14.48,

$$V_c = 2\sqrt{3000}(16)(21.25) = 37{,}245 \text{ lbf} \quad [37.2 \text{ kips}]$$

From Eq. 14.50, the required steel strength is

$$V_{st} = \frac{73.1}{0.85} - 37.2 = 48.8 \text{ kips}$$

Check $V_{st,max}$ from Eq. 14.51.

$$\begin{aligned} V_{st,max} &= 8\sqrt{3000}(16)(21.25) \\ &= 148{,}981 \text{ lbf } (149 \text{ kips}) \quad [\text{acceptable}] \end{aligned}$$

From Eq. 14.57, using $A_v = (2)(0.11)$ for #3 bars, the spacing is

$$s = \frac{(2)(0.11)(60{,}000)(21.25)}{48{,}800} = 5.75 \text{ in}$$

Start the first stirrup $d/2 = 10$ in from support face.

(It may be desirable to recalculate spacing required at $x = 3$ ft and continue that new spacing to $x = 6$ rather than using $s = 5.75$ in over the entire 6 ft.)

At $x = 6$, the shear reinforcement required is

$$V_{st} = \frac{16.5}{0.85} - 37.2 = -17.8 \quad [\text{zero}]$$

Between $x = 6$ ft and $x = 12$ ft, only minimum reinforcement is required (ACI 11.5.5.1).

10. $d = 40 - 2 - \dfrac{2.5}{2} = 36.75$ in

Check the deep beam.

$$\frac{l}{d} = \frac{(12)(12)}{36.75} = 3.9 < 5 \quad [\text{a deep beam}]$$

For deep beams, Sec. 11.8.6 of ACI-318 applies.

From Eq. 14.75, V_u and M_u are evaluated at

$$x = (0.5)(4) = 2 \text{ ft } (24 \text{ in})$$

$$24 \text{ in} < d = 36.75 \text{ in} \quad [\text{acceptable}]$$

The service dead load of the beam is

$$\frac{(24)(40)(150)}{144} = 1000 \text{ lbf/ft} \quad [1 \text{ kip/ft}]$$

Dead load will be omitted.

The factored reaction is

$$\begin{aligned} R &= (1.7)(126) = 214.2 \text{ kips} \\ V_u &= 214.2 \text{ kips} \quad [2 \text{ ft from face}] \\ M_u &= (214.2)(2) = 428.4 \text{ ft-kips} \end{aligned}$$

See footnote 42, p. 14-20. From ACI-318, Sec. 11.8.6, the deep beam multiplier is

$$\begin{aligned} &\left(3.5 - 2.5\frac{M_u}{V_u d}\right) \\ &= 3.5 - (2.5)\left[\frac{428.4}{(214.2)\left(\frac{36.75}{12}\right)}\right] \\ &= 1.87 < 2.5 \quad [\text{acceptable}] \end{aligned}$$

$$\rho_w = \frac{(8)(1)}{(24)(36.75)} = 0.00907$$

$$\begin{aligned} v_c = \frac{V_c}{bd} &= (1.87)\Bigg(1.9\sqrt{3000} \\ &\quad + (2500)(0.00907)\left[\frac{(214.2)(36.75)}{(12)(428.4)}\right]\Bigg) \\ &= \boxed{259.5 \text{ psi}} \end{aligned}$$

- Check:

$$\text{v}_{c,\text{max}} = 6\sqrt{3000} = 328.6 \text{ psi}$$

Since 328.6 psi > 259.5 psi, it is acceptable.

(b)

$$\begin{aligned} V_c &= \text{v}_c bd \\ &= (259.5)(24)(36.75) = 228{,}879 \text{ lbf} \quad [228.9 \text{ kips}] \end{aligned}$$

The required shear is

$$V_n = \frac{V_u}{\phi} = \frac{214.2}{0.85} = 252.0 \text{ kips}$$

From Eq. 14.78,

$$\begin{aligned} V_{n,\text{max}} &= \left(\tfrac{2}{3}\right)(10 + 3.9)\sqrt{3000}(24)(36.75) \\ &= 447{,}665 \text{ lbf} > V_n \quad [\text{acceptable}] \end{aligned}$$

Since $V_n > V_c$, shear reinforcement is needed.

$$V_{\text{st}} = 252.0 - 228.9 = 23.1 \text{ kips} \quad [\text{not much}]$$

Using $l_n/d = 3.9, V_{\text{st}} = 23{,}100$ lbf, $d = 36.75$ in, and $f_y = 60{,}000$ psi, from Eq. 14.83,

$$\begin{aligned} V_{\text{st}} = 23{,}100 &= \left[\left(\frac{A_\text{v}}{s}\right)\left(\frac{1+3.9}{12}\right) + \left(\frac{A_{\text{v}h}}{s_2}\right)\left(\frac{11-3.9}{12}\right)\right](60{,}000)(36.75) \\ 0.01047 &= \left[\left(\frac{A_\text{v}}{s}\right)(0.408) + \left(\frac{A_{\text{v}h}}{s_2}\right)(0.592)\right] \end{aligned}$$

There are many configurations that work. From Eqs. 14.79 and 14.80,

$$s_{\text{max}} = \text{minimum}\left\{\begin{matrix} \dfrac{36.75}{5} = 7.35 \text{ in} \\ 18 \text{ in} \end{matrix}\right\} = 7.35 \text{ in}$$

$$s_{2,\text{max}} = \text{minimum}\left\{\begin{matrix} \dfrac{36.75}{3} = 12.25 \text{ in} \\ 18 \text{ in} \end{matrix}\right\} = 12.25 \text{ in}$$

With $40 - 2 - 2.5 - 2 - 0.375 = 33.1$ in, try a spacing of $s_2 = 11$ between horizontal bars.

Use #5 bars. Then, from Eq. 14.81,

$$\begin{aligned} A_{\text{v}h,\text{min}} &= (0.0025)(24)(11) = 0.66 \text{ in}^2 \\ \text{actual } A_{\text{v}h} &= (2)(0.31) = 0.62 \text{ in}^2 \end{aligned}$$

(If this was not acceptable, closer spacing of 8 in could have been used.)

Use #4 stirrups with 6-in spacing.

$$A_\text{v} = (2)(0.20) = \boxed{0.40 \text{ in}^2}$$

$$\left(\frac{0.40}{6}\right)(0.408) + \left(\frac{0.62}{11}\right)(0.592) = 0.0606$$

0.0606 > 0.01047—almost six times as much shear reinforcing as is required.

It is difficult to meet the maximum spacing requirements without using excess steel.

Perhaps the beam should rely on welded wire fabric for this small amount of shear reinforcement.

(c) **s and s_2 spacing should be maintained across the entire length of beam.**

STEEL DESIGN AND ANALYSIS

Untimed

1. Assume the beam will have a dead weight of 60 lbf/ft. The shear and moment diagrams are

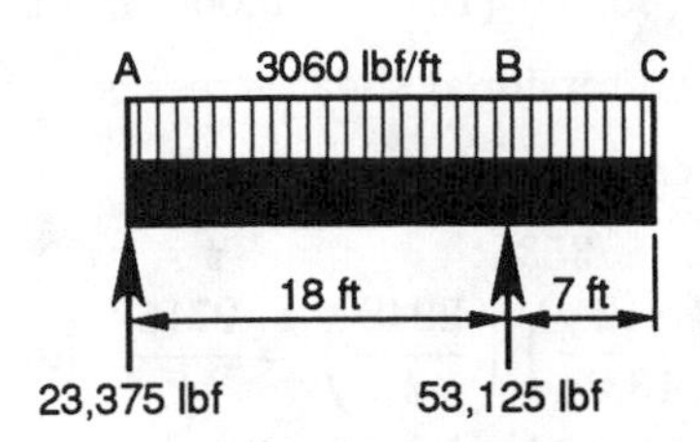

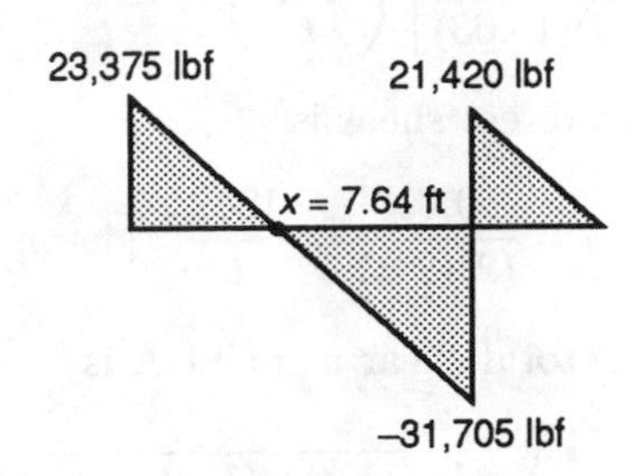

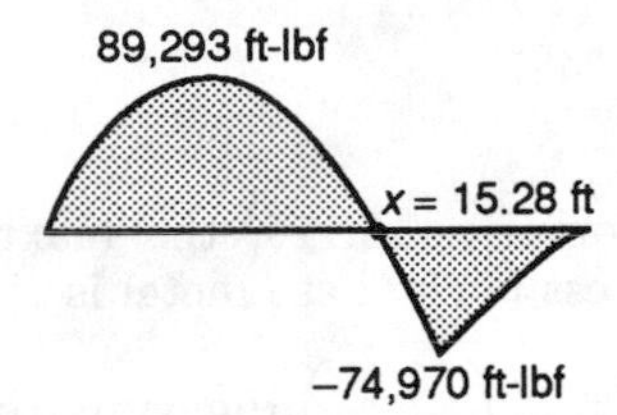

(Since M_{max} does not occur at the end of the cantilever span, C_b does not apply.)

From the beam selection chart, choose **W14 × 43.**

This will meet the requirements of 89.3 ft-kips with an unbraced length of 18 ft. Check the shear stress as a final step.

Notice that $L_b > L_u$. This is acceptable at lower stress levels. (See Fig. 15.10.)

2. Assume the column weight will be about 2000 lbf. The column load is, then, 160,000 + 2000 = 162,000 lbf. The end constraint coefficients are

$$K_x = K_y = 1.0$$

The unbraced lengths are

$$L_x = 30 \text{ ft}$$
$$L_y = 15 \text{ ft}$$
$$L_{e,\text{min}} = (1.0)(15) = 15 \text{ ft}$$

From the *AISC Manual of Steel Construction* column tables, try W8 × 40.

$$P = 701 \text{ kips} \quad [\text{for } L = 15 \text{ ft}]$$
$$\frac{r_x}{r_y} = 1.73$$

Since 1.73 < 30/15 = 2.0, the strong axis controls. The equivalent unbraced length for the strong axis is

$$\frac{30}{1.73} = 17.3 \text{ ft}$$

For W8 × 40, P = 153 kips (for L = 17.3). This is not enough, so try W10 × 39.

$$P = 162 \text{ kips} \quad [\text{for } L = 15 \text{ ft}]$$
$$\frac{r_x}{r_y} = 2.16$$
$$r_y = 1.98 \text{ in}$$
$$A = 11.5 \text{ in}^2$$

Since 2.16 > 2.0, the weak axis controls. Therefore, the strong axis load does not have to be checked.

use W10 × 39

3. Since the bolt group is symmetrical, the centroid is located between bolts 4 and 11.

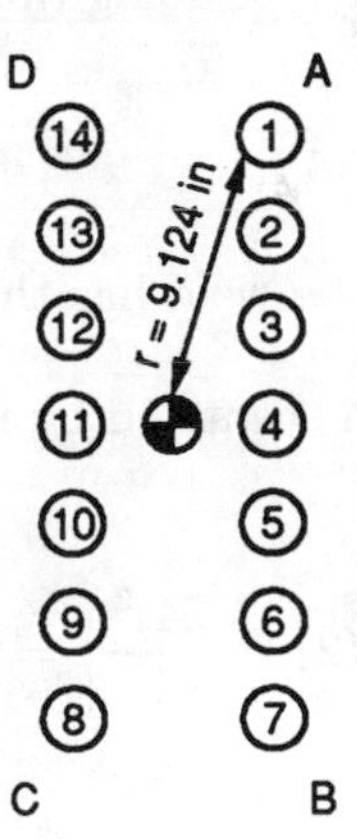

For each bolt,

$$A_i = \left(\frac{\pi}{4}\right)(0.75)^2 = 0.4418 \text{ in}^2$$
$$J_i = \left(\frac{\pi}{32}\right)(0.75)^4 = 0.03106 \text{ in}^4$$

The solution is tabulated.

bolt	r	r^2A
3	3.354	4.97
2	6.185	16.90
1	9.124	36.78
total		58.65
4	1.5	0.99

$$J = (14)(0.03106) + (4)(58.65) + (2)(0.99) = 237.0 \text{ in}^4$$

The torsional shear stress in bolts A, B, C, and D is

$$f_{v,t} = \frac{Tr}{J} = \frac{(30{,}000)(17)(9.124)}{237} = 19{,}634 \text{ psi}$$

$$f_{v,t,y} = \left(\frac{1.5}{9.124}\right)(19{,}634) = 3227.9 \text{ psi}$$

$$f_{v,t,x} = \left(\frac{9}{9.124}\right)(19{,}634) = 19{,}367.2 \text{ psi}$$

There is also a direct vertical shear stress shared equally among all bolts.

$$f_{v,d} = \frac{30{,}000}{(14)(0.4418)} = 4850.3 \text{ psi}$$

The stress in bolts A and B is

$$\sqrt{(19{,}367.2)^2 + (3227.9 + 4850.3)^2} = \boxed{20{,}984 \text{ psi}}$$

The stress in bolts C and D is

$$\sqrt{(19{,}367.2)^2 + (3227.9 - 4850.3)^2} = \boxed{19{,}435 \text{ psi}}$$

(A reduced eccentricity (Eqs. 15.102 and 15.103) could have been used if the bolts were known to be low-strength.)

4. There are two ways of doing this problem.

Method 1:

Proceed as in Ex. 15.17.

step 1: Assume the weld has thickness t.

step 2: Locate the centroid by inspection.

step 3:

$$I_x = (2)\left[\frac{t(24)^3}{12}\right] = 2304t$$

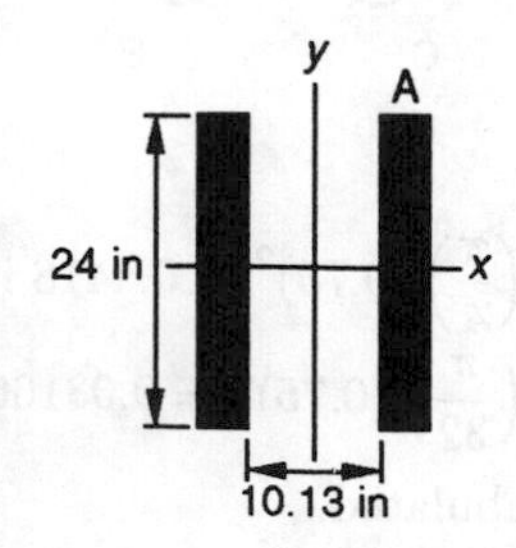

step 4: $b_f = 10.13$ in [for W14 × 82]

$$I_y = (2)\left[\frac{24t^3}{12} + (24t)\left(\frac{10.13 + t}{2}\right)^2\right]$$

$$\approx 1232t \quad [\text{assuming } t^2 = t^3 = 0]$$

step 5: $J = 2304t + 1232t = 3536t$

step 6: The maximum shear stress will occur at point A.

$$r = \sqrt{\left(\frac{10.13}{2}\right)^2 + (12)^2} = 13.03 \text{ in}$$

step 7: The applied moment is

$$(50{,}000)(16) = 800{,}000 \text{ in-lbf}$$

step 8: The torsional shear stress is

$$f_{v,t} = \frac{(800{,}000)(13.03)}{3536t} = \frac{2948}{t}$$

$$f_{v,t,x} = \left(\frac{12}{13.03}\right)\left(\frac{2948}{t}\right) = \frac{2715}{t} \quad \text{[to the right]}$$

$$f_{v,t,y} = \left[\frac{10.13}{(2)(13.03)}\right]\left(\frac{2948}{t}\right) = \frac{1146}{t} \quad \text{[down]}$$

step 9: The direct shear is

$$f_{v,d} = \frac{50{,}000}{(2)(24t)} = \frac{1042}{t} \quad \text{[down]}$$

step 10: The total shear at point A is

$$f = \left(\frac{1}{t}\right)\sqrt{(2715)^2 + (1146 + 1042)^2} = \frac{3487}{t}$$

step 11: From Eq. 15.110, the maximum allowable stress in the base metal is

$$f_a = \frac{R_w}{t_w} = \left(\frac{0.4}{0.707}\right)(36{,}000) = 20{,}368 \text{ psi}$$

The weld size is

$$t = \frac{3487}{20{,}368} = 0.171 \text{ in}$$

step 12:

$$w = \frac{0.171}{0.707} = 0.24 \text{ in}$$

Use the minimum size $\frac{5}{16}$-in weld (see Table 15.8) since the flange thickness is less than $\frac{7}{8}$ in.

Method 2:

Use AISC Table XIX (Eccentric Loads on Weld Groups).

$$l = 2 \text{ ft (24 in)}$$

$$al = 16 \text{ in}$$

$$a = \frac{al}{l} = \frac{16 \text{ in}}{24 \text{ in}} = 0.666$$

$$kl = b_f = 10.13 \text{ in}$$

$$k = \frac{kl}{l} = \frac{10.13 \text{ in}}{24 \text{ in}} = 0.422$$

	k		
	0.4	0.422	0.5
0.6	0.795		0.834
a 0.666	0.737	0.746	0.777
0.7	0.708		0.748

$C \approx 0.746$ [double interpolation]

$C_1 = 1.0$ [E70 electrodes]

$P = CC_1Dl$ [kips]

$$D = \frac{P}{CC_1 l} = \frac{50 \text{ kips}}{(0.746)(1)(24)} = 2.79 \quad \text{[say 3]}$$

$$w = D \times \left(\tfrac{1}{16} \text{ in}\right) = \boxed{\tfrac{3}{16}\text{-in weld}}$$

5. This solution is found through trial and error. Other shapes may be more economical.

Each column carries 275,000 lbf. Neglect self-weight.

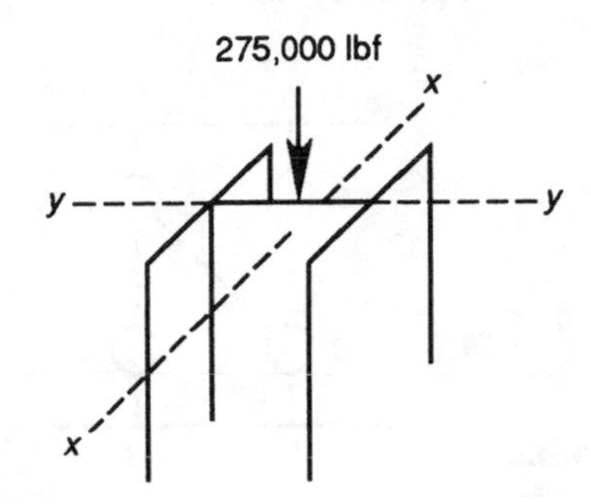

From the *AISC Manual of Steel Construction* column selection table, try a W12 × 65, which has an allowable load for bending about the minor axis of

$$P_{21} = \left(\tfrac{1}{2}\right)(P_{20} + P_{22})$$
$$= \left(\tfrac{1}{2}\right)(294 + 277) = 285.5 \text{ kips}$$
$$I = 533 \text{ in}^4$$
$$A = 19.1 \text{ in}^2$$
$$\frac{r_x}{r_y} = 1.75$$

The column is braced against bending in the major axis by the 40-ft girders, but the column table is for bending in the major axis, so this is just a start.

The effective length factor K in this problem is most easily evaluated by use of Fig. 15.15.

For the W12×96 horizontal beams, $I = 833 \text{ in}^4$. For an interior column,

$$G_{\text{top}} = \frac{\frac{533}{21}}{(2)\left(\frac{833}{40}\right)} = 0.609$$

Since the bottom end is built-in, $G_{\text{bottom}} = 1.0$.

From the alignment chart (sidesway permitted),

$$K_x = 1.25$$

The effective length for major axis bending is

$$KL = (1.25)(21) = 26.25 \text{ ft}$$

The table is for minor axis bending only, however. KL relative to the minor axis is given by Eq. 15.25.

$$L'_x = \frac{K_x L_x}{\frac{r_x}{r_y}} = \frac{26.25}{1.75} = 15 \text{ ft}$$

Since 15 ft < 21 ft, the effective length for minor axis bending is critical. W12 × 65 is acceptable.

Check the actual compressive stress.

$$f_a = \frac{P}{A} = \frac{275{,}000}{19.1} = 14{,}398 \text{ psi}$$

The slenderness ratio for major axis bending is

$$\frac{KL_x}{r_x} = \frac{(1.25)(21)\left(12 \frac{\text{in}}{\text{ft}}\right)}{5.28} = 59.7$$

From Table C-36, Part 3, of the *AISC Manual of Steel Construction*,

$$f_{\max} = 17.40 \text{ ksi} \quad \text{[acceptable]}$$

$$\boxed{\text{use W12} \times 65}$$

6. Eye bolt design is covered in the *AISC Manual of Steel Construction*, Sec. D3.3.

step 1: The allowable stress on the net area of the eye is

$$F_{te} = (0.45)(36{,}000) = 16{,}200 \text{ psi}$$

step 2: The allowable stress on the bar is

$$\text{minimum}\left\{\begin{array}{l} F_{tg} = (0.6)(36{,}000) \\ F_t = (0.5)(58{,}000) \end{array}\right\} = 21{,}600 \text{ psi}$$

step 3: The bar cross-sectional area is

$$A = \frac{300{,}000}{21{,}600} = 13.89 \text{ in}^2$$

Try a $1\tfrac{1}{2}$-in plate. The width is

$$b = \frac{13.89}{1.5} = 9.26 \text{ in}$$

Use $PL\,1\tfrac{1}{2} \times 9.5$ for actual area of 14.25 in^2.

step 4: Check the b/t ratio.

$$\frac{9.5}{1.5} = 6.33 < 8 \quad \text{[acceptable]}$$

step 5: The net area at the eye is

$$\frac{300{,}000}{16{,}200} = 18.52 \text{ in}^2$$

Check the minimum area.

$$1.33bt = (1.33)(14.25) = 18.95 \text{ in}^2$$

18.95 in^2 controls.

step 6: The eye width is

$$C = \frac{18.95}{(2)(1.5)} = 6.32 \text{ in} \quad \text{[round to 6.5 in]}$$

step 7: Check the maximum area.

$$\frac{(2)(6.5)(1\frac{1}{2})}{14.25} = 1.37 \quad \text{[acceptable]}$$

step 8: Calculate the minimum pin diameter.

$$\text{pin diameter} = \left(\tfrac{7}{8}\right)(9.5) = 8.31 \text{ in}$$

step 9: The maximum hole diameter is

$$d \approx 8.31 + \tfrac{1}{32} = 8.31 + 0.03125 = 8.34 \text{ in}$$

Make the hole 8.5-in diameter. Make the pin 8.5 − 0.03 = 8.47 in.

step 10: The minimum external radius is

$$R = 8.5 + (2)(6.5) = 21.5 \text{ in}$$

step 11:

$$F_{p,\text{max}} = (0.90)(36{,}000) = 32{,}400 \text{ psi}$$

$$f_p = \frac{300{,}000}{(8.47)(1.5)} = 23{,}613 \text{ psi} \quad \text{[acceptable]}$$

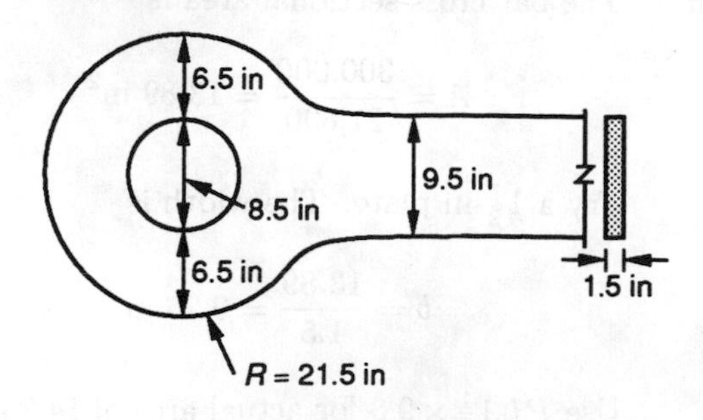

7. Proceed as in Ex. 15.11.

step 1: The allowable stress on the gross section is

$$F_t = (0.6)(50{,}000) = 30{,}000 \text{ psi}$$

step 2: Assume A588 plate. The allowable stress on the net effective section (see p. 12-34) is

$$F_t = (0.5)(70{,}000) = 35{,}000 \text{ psi}$$

step 3: The gross plate area is

$$(6)\left(\tfrac{5}{8}\right) = 3.75 \text{ in}^2$$

step 4: The standard hole diameter is

$$\tfrac{7}{8} + \tfrac{1}{16} = \tfrac{15}{16} \text{ in}$$

The failure paths to be investigated are EFGH and EFCD. Although insufficient information is given to evaluate EFCD, path EFGH obviously controls.

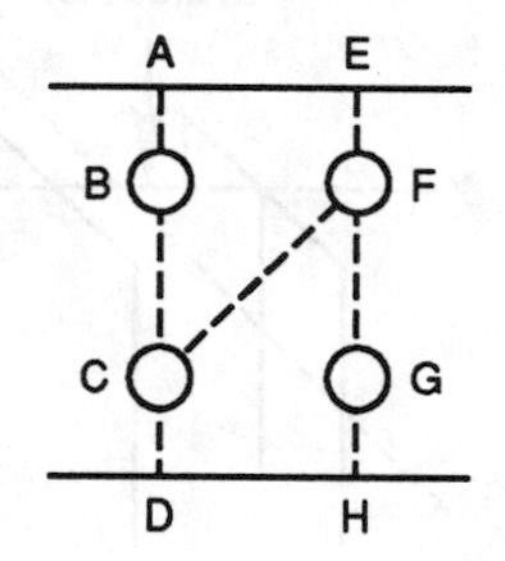

The net effective area is

$$\left(\tfrac{5}{8}\right)\left[6 - (2)\left(\tfrac{15}{16}\right)\right] = 2.58 \text{ in}^2$$

step 5: The capacity is

$$\text{minimum}\begin{Bmatrix} (3.75)(30{,}000) \\ (2.58)(35{,}000) \end{Bmatrix} = 90{,}300 \text{ lbf}$$

step 6: Check the shear capacity of the four bolts. From Table 15.7,

$$F_v = 17{,}500 \text{ psi}$$

$$V = (4)\left(\frac{\pi}{4}\right)\left(\tfrac{7}{8}\right)^2(17{,}500) = 42{,}092 \text{ lbf}$$

Since 42,092 lbf < 90,300 lbf, shear capacity governs.

step 7: Check the stress if the second row carries two-fourths of the load.

$$\text{stress} = \frac{\left(\frac{2}{4}\right)(42{,}092)}{2.58} = 8157 \text{ psi} \quad \text{[acceptable]}$$

step 8: Check bearing stress. The ultimate strength of grade 50 steel is approximately 65 psi. From Eq. 15.30,

$$F_p = (1.50)(65{,}000) = 97{,}500 \text{ psi}$$

The area in bearing is

$$(4 \text{ bolts}) \left(\tfrac{5}{8}\right) \left(\tfrac{7}{8}\right) = 2.1875 \text{ in}^2$$

The bearing stress is

$$f_p = \frac{42{,}092}{2.1875} = 19{,}242 \text{ psi}$$

$$f_p < F_p \quad \text{[acceptable]}$$

$$\text{capacity} = \boxed{90{,}300 \text{ lbf}}$$

8. *step 1:* The reactions are each

$$\frac{(1000)(20)}{2} = 10{,}000 \text{ lbf}$$

step 2: From p. 12.31, the maximum moment is

$$\frac{(1000)(20)^2}{8} = 50{,}000 \text{ ft-lbf}$$

step 3: From the beam selection chart, try W14 × 22 to meet 50 ft-kips and $L_b = 5.5$ ft.

$$t_w = 0.230 \text{ in}$$
$$d = 13.74 \text{ in}$$
$$I = 199 \text{ in}^4$$

step 4: Check for shear stress.

$$F_v = (0.40)(36{,}000) = 14{,}400 \text{ psi}$$

$$f_v = \frac{10{,}000}{(0.230)(13.74)} = 3164 \text{ psi} \quad \text{[acceptable]}$$

step 5: Check deflection.

$$y_{\text{allowable}} = \frac{(20)\left(12 \ \frac{\text{in}}{\text{ft}}\right)}{240} = 1 \text{ in}$$

From p. 12-31,

$$y_{\text{actual}} = \frac{5wL^4}{384EI} = \frac{(5)\left(\frac{1000}{12}\right)[(20)(12)]^4}{(384)(2.9 \times 10^7)(199)}$$
$$= 0.624 \text{ in} \quad \text{[acceptable]}$$

9. *step 1:* Assume the beam weight will be 50 lbf/ft.

step 2: The maximum deflection will be

$$y = \frac{(24)(12)}{300} = 0.96 \text{ in}$$

step 3: The required moment of inertia is

$$I_x = \frac{5wL^4}{384Ey} = \frac{(5)\left(\frac{850}{12}\right)[(24)(12)]^4}{(384)(2.9 \times 10^7)(0.96)}$$
$$= 227.9 \text{ in}^4$$

step 4: From the moment of inertia selection table, choose W14 × 26.

$$I_x = 245 \text{ in}^4$$
$$S = 35.3 \text{ in}^3$$

step 5: The maximum moment will be

$$M_{\text{max}} = \frac{(826)(24)^2(12)}{8} = 713{,}664 \text{ in-lbf}$$

step 6: The allowable stress (since the bracing is complete) is

$$(0.66)(36{,}000) = 23{,}760 \text{ psi}$$

step 7: The actual stress is

$$f_b = \frac{713{,}664}{35.3} = 20{,}217 \text{ psi} \quad \text{[acceptable]}$$

$$\boxed{\text{use W14} \times 26}$$

10. For A441 steel, W12 × 58,

$$F_y = 50{,}000 \text{ psi} \quad [\text{affects } L_c]$$
$$A = 17.0 \text{ in}^2$$
$$S = 78.0 \text{ in}^3$$
$$L_c = 9 \text{ ft}$$
$$r_x = 5.28 \text{ in}$$
$$r_y = 2.51 \text{ in}$$

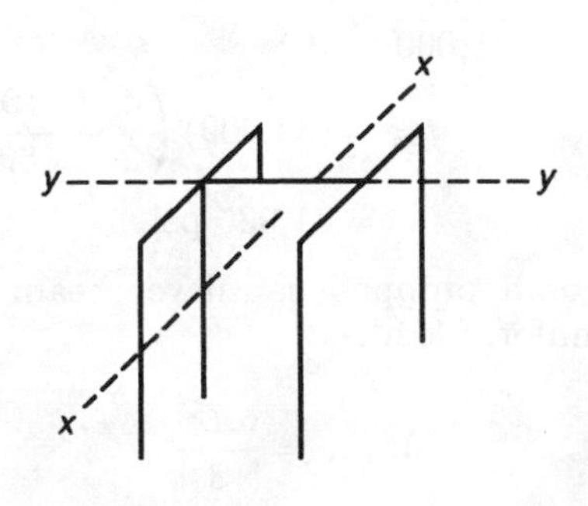

(a) For maximum lateral loading with bending in the x-x plane, $F_b = 0.66F_y$. This is allowed only if $L_b < L_c$.

Bracing is required every 9 ft.

(b)

step 1: $f_a = \dfrac{17{,}000 + (16)(58)}{17.0} = 1055 \text{ psi}$

step 2: From Table 15.5 for bending in the x-x plane, $K_x = 0.8$.

The slenderness ratio for bending in the x-x plane is

$$\frac{KL}{r_x} = \frac{(0.8)(9)(12)}{5.28} = 16.4 \text{ in}$$

step 3: From Table 15-5 for bending in the y-y plane, $K_y = 2.1$.

The slenderness ratio is

$$\frac{KL}{r_y} = \frac{(2.1)(9)(12)}{2.51} = 90.36 \text{ in}$$

step 4: From Table C-50, Part 3, of the *AISC Manual of Steel Construction*,

$$\frac{KL}{r} = \text{maximum}\left\{\begin{matrix}16.4 \text{ in} \\ 90.36 \text{ in}\end{matrix}\right\} = 90.36 \text{ in}$$

$$F_a = 16{,}860 \text{ psi}$$

step 5: If bracing is every 9 ft, then

$$F_{bx} = (0.66)(50{,}000) = 33{,}000 \text{ psi}$$

step 6:

$$\frac{f_a}{F_a} = \frac{1055}{16{,}860} = 0.0626 < 0.15$$

Eq. 15.47 can be used.

step 7: Since AISC Specifications allow a one-third increase in stress for wind and earthquake loads,

$$\frac{f_a}{F_a} + \frac{f_{bx}}{F_{bx}} \leq \frac{4}{3}$$

$$\frac{1055}{16{,}860} + \frac{f_{bx}}{33{,}000} = \frac{4}{3}$$

$$f_{bx} = (33{,}000)\left(\frac{4}{3} - \frac{1055}{16{,}860}\right)$$

$$= 41{,}935 \text{ psi}$$

step 8: For a propped cantilever beam carrying a uniform load,

$$M_{\text{max}} = \frac{wL^2}{8}$$

Since $f = \dfrac{M}{s}$,

$$41{,}935 = \frac{wL^2}{8s} = \frac{w\left[(16)(12)\right]^2}{(8)(78.0)}$$

$$w = \boxed{709.8 \text{ lbf/in} \quad [8518 \text{ lbf/ft}]}$$

Timed

1. (a) Ignoring the self-weight of the beam, find the reactions. Taking clockwise moments as positive,

$$\sum M_{\text{hinge to left}} = 20R_A - \left(\tfrac{1}{2}\right)(1.4)(20)^2 = 0$$

$$R_A = 14 \text{ kips}$$

$$\sum M_D = (45)(14) + 20R_C + \left(\tfrac{1}{2}\right)(1.4)(5)^2 - \left(\tfrac{1}{2}\right)(1.4)(45)^2 = 0$$

$$R_C = 38.5 \text{ kips}$$

$$R_D = (50)(1.4) - 14 - 38.5$$

$$= 17.5 \text{ kips}$$

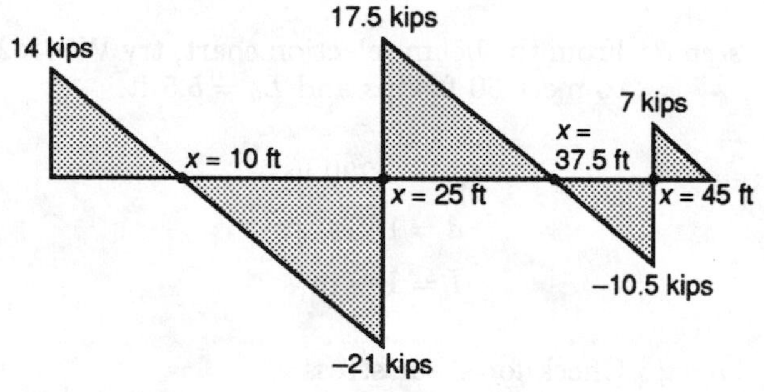

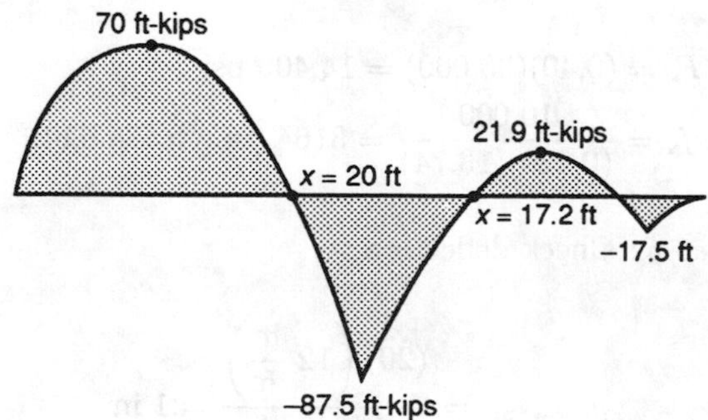

(b) $L_u = 25$ ft

(Some interpretations recognize that the flange is discontinuous at the hinge and use $L_u = 20$.)

$$M_{\text{max}} = 87.5 \text{ ft-kips}$$

From the beam selection chart, use

W18 × 60 [must be W18]

Note that this exceeds the beam's $L_u = 13.3$ ft, but is permitted at the lower bending stress.

At this point, it would be appropriate to include the self-weight and check the shear stress.

(c) For W18 × 60, $\frac{d}{A_f} = 3.47$.

On span CD, $M_1 = 17.5$ and $M_2 = 87.5$.

$$\frac{M_1}{M_2} = \frac{-17.5}{87.5} = -0.2 \quad \text{[reverse curvature]}$$

From App. D,

$$C_b = 1.75 + (1.05)(-0.2) + (0.3)(-0.2)^2 = 1.55$$

(Table 6 in the *AISC Manual of Steel Construction*, in the Numerical Values section, says $C_b = 1.55$.) From Eq. 15.4,

$$L_c = \frac{(20{,}000)(1.55)}{(3.47)(42)} = \boxed{212.7 \text{ in}}$$

2. (a) Proceed as in Ex. 15.12. Evaluate the 17.5-ft lower column.

For a W14 × 82 member,

$$A = 24.1 \text{ in}^2$$
$$I_x = 882 \text{ in}^2$$
$$L_c = 9.1 \text{ ft} \quad \text{[50 ksi]}$$
$$L_u = 20.2 \text{ ft} \quad \text{[50 ksi]}$$
$$r_x = 6.05 \text{ in}$$
$$S_x = 123 \text{ in}^3$$

For the W14 × 53 member,

$$I_x = 541 \text{ in}^4$$

The axial stress is

$$f_a = \frac{525}{24.1} = 21.8 \text{ ksi}$$

Use the alignment chart to obtain K. With no sidesway,

$$G_{\text{A}} = 10 \quad \text{[pinned]}$$

$$G_{\text{B}} = \frac{\left(\frac{882}{10} + \frac{882}{17.5}\right)}{\left(\frac{541}{36} + \frac{541}{36}\right)} = 4.6$$

$$K = 0.94$$

The maximum slenderness ratio (Eq. 15.16) is

$$\text{SR} = \frac{KL}{r_x} = \frac{(0.94)(17.5)(12)}{6.05} = 32.6 \quad \text{[round to 33]}$$

(Note that the problem stated that the weak axis bending controlled.)

From Table C-50, Part 3, of the *AISC Manual of Steel Construction*,

$$F_a = 26.77 \text{ ksi} \quad \text{[50 ksi steel]}$$

Since $F_a > f_a$, further checking as a beam-column is required.

$$\frac{f_a}{F_a} = \frac{21.8}{27.03} = 0.807 > 0.15$$

So, the axial compression is large. Use Eqs. 15.48 and 15.49.

$$f_{bx} = \frac{M_x}{S_x} = \frac{(225)(12)}{123} = 22.0 \text{ ksi}$$

Since $L_c < L_b < L_u$, $F_b = (0.60)(50) = 30$ ksi. From Eq. 15.48,

$$\frac{21.8}{(0.6)(50)} + \frac{22.0}{30} = 1.46 > 1.0$$

The column is inadequate.

(The beams also could be checked.)

(b) Follow the procedure on p. 15-24.

step 1: Since the column is not known, conservatively estimate $K = 1$.

$$KL = (1)(17.5) = 17.5 \text{ ft}$$

step 2: From the $KL = 18$ column, use $m = 1.7$. (Since $M_y = 0$ and W14 shape is to be used, only one iteration will be required.)

step 3: (skip)

step 4: From Eq. 15.52,

$$P_{\text{eff}} = 525 + (225)(1.7) = 907.5 \text{ kips}$$

step 5: Try $\boxed{\text{W14} \times 132.}$ [Read footnote 32.]

capacity = 898 kips at 18 ft [50 ksi]

Prove this is adequate.

$$A = 38.8 \text{ in}^2$$
$$I_x = 1530 \text{ in}^4$$
$$L_c = 13.2 \text{ ft}$$
$$L_u = 34.4 \text{ ft}$$
$$r_x = 6.28 \text{ in}$$
$$S_x = 209 \text{ in}^3$$
$$B_x = 0.186$$

$$f_a = \frac{525}{38.8} = 13.5 \text{ ksi}$$

$$G_A = 10$$

$$G_B = \frac{\left(\frac{1530}{10} + \frac{1530}{17.5}\right)}{\left(\frac{541}{36} + \frac{541}{36}\right)} = 8.0$$

From the alignment chart, $K = 0.96$.

$$\text{SR} = \frac{KL}{r_x} = \frac{(0.96)(17.5)(12)}{6.28} = 32.1$$

$$F_a = 26.9 \text{ ksi}$$

$$\frac{f_a}{F_a} = \frac{13.5}{26.9} = 0.502 > 0.15$$

$$f_{bx} = \frac{(225)(12)}{209} = 12.92 \text{ ksi}$$

Since $L_c < L_b < L_u$,

$$F_b = (0.6)(50) = 30 \text{ kips}$$

From Eq. 15.48,

$$\frac{13.5}{(0.6)(50)} + \frac{12.92}{30} = 0.88 < 1.0 \quad \text{[acceptable]}$$

One more check needs to be made. Although C_m could be calculated (Eq. 15.51), choose $C_m = 1$ (conservative shortcut).

From Table 8 of the Numerical Values section of the *AISC Manual of Steel Construction* (or Eq. 15.50),

$$F'_e = 145.83 \text{ ksi}$$

From Eq. 15.49,

$$\frac{13.5}{26.9} + \frac{(1)(12.92)}{\left(1 - \frac{13.5}{145.83}\right)(30)} = 0.976 < 1.0 \quad \text{[acceptable]}$$

3. Each angle carries $\left(\frac{1}{2}\right)(140) = 70$ kips. Check the tensile stress.

$$A = 3.25 \text{ in}^2$$

$$F_a = (0.6)(36) = 21.6 \text{ ksi}$$

$$f_a = \frac{70}{3.25} = 21.5 \text{ ksi} < 21.6 \text{ ksi} \quad \text{[acceptable]}$$

(a) Since thickness $= \frac{1}{2}$ in, the maximum weld thickness is

$$w = \tfrac{1}{2} - \tfrac{1}{16} = \tfrac{7}{16} \text{ in} \quad \text{[see p. 15-36]}$$

$$t_e = (0.707)\left(\tfrac{7}{16}\right) = 0.309 \text{ in}$$

From Eq. 15.110,

$$R_w = \text{minimum}\begin{Bmatrix} (0.309)(0.3)(70) = 6.49 \text{ kips/in} \\ \left(\frac{7}{16}\right)(0.4)(36) = 6.3 \text{ kips/in} \end{Bmatrix} = 6.3 \text{ kips/in}$$

Refer to Fig. 15.35. (Read footnote 53, but balance the weld group anyway.) For the angle,

$$d = 4 \text{ in}$$

$$y = 1.33 \text{ in}$$

From Eq. 15.113,

$$P_2 = (6.3)(4) = 25.2 \text{ kips}$$

From Eq. 15.112,

$$P_3 = (70)\left(1 - \frac{1.33}{4}\right) - \frac{25.2}{2} = 34.1 \text{ kips}$$

$$L_3 = \frac{34.1}{6.3} = 5.41 \text{ in} \quad \left[\text{round to } 5\tfrac{1}{2} \text{ in}\right]$$

$$P_1 = 70 - 25.2 - 34.1 = 10.7 \text{ kips}$$

$$L_1 = \frac{10.7}{6.3} = 1.70 \text{ in} \quad \left[\text{round to } 1\tfrac{3}{4} \text{ in}\right]$$

- Check the minimum weld length:

$$L_1 \geq \begin{Bmatrix} 4 \times \frac{7}{16} = 1.75 \text{ in} \\ 1\frac{1}{2} \text{ in} \end{Bmatrix} = 1.75 \text{ in} \quad \text{[acceptable]}$$

(b)

$$P_x = (140)(\cos 40°) = 107.2 \text{ kips}$$

$$P_y = (140)(\sin 40°) = 90.0 \text{ kips}$$

For $\frac{7}{8}$-in bolts,

$$A = \left(\frac{\pi}{4}\right)\left(\frac{7}{8}\right)^2 = 0.6013 \text{ in}^2$$

The shear stress in the bolts is

$$f_v = \frac{90}{(6)(0.6013)} = 24.9 \text{ ksi}$$

The tensile stress in the bolts is

$$f_t = \frac{107.2}{(6)(0.6013)} = 29.7 \text{ ksi}$$

- Slip-critical type connection:

For A325 bolts in shear-only configurations,

$$F_v = 17.5 \text{ ksi} \quad \text{[Table 15.7]}$$

From *AISC Manual of Steel Construction*, Sec. J3.6 (for combined shear and tension),

$$F'_v = F_v\left(1 - \frac{f_t A_b}{T_b}\right)$$

Assume bolt tension = 39 kips (minimum preload for $\frac{7}{8}$-in bolts, *AISC Manual of Steel Construction*, Table J3.7).

$$F'_v = (17.5)\left[1 - \frac{(29.7)(0.6013)}{39}\right] = 9.5 \text{ ksi}$$

Since 24.9 ksi > 9.5 ksi, **it is not acceptable.**

- Bearing type connection:

From *AISC Manual of Steel Construction*, Table J3.3,

$$F_t = \sqrt{(44)^2 - 2.15f_v^2} = \sqrt{(44)^2 - (2.15)(24.9)^2} = 24.6 \text{ ksi}$$

Since 29.7 ksi > 24.6 ksi, **it is not acceptable.**

Check the bolt locations.

When prying action is absent, all bolts will be stressed equally. To avoid prying action, the line of action of the force ($y = 1.33$ in from the short side of the angle) should coincide with the bolt group centroid. This does not appear to be the case here.

4. Follow the procedure on p. 15-15 for column analysis.

step 1: This is not a tabulated combination section, so determine the composite section properties.

$$\sum A_i = 35.1 + 14.7 = 49.8 \text{ in}^2$$

$$\bar{y} = \frac{\sum A_i \bar{y}_i}{\sum A_i} = \frac{(35.1)\left(\frac{18.97}{2}\right) + (14.7)[18.97 + (0.716 - 0.798)]}{49.8} = 12.26 \text{ in}$$

From the parallel axis theorem,

$$I_x = 2190 + (35.1)\left(\frac{18.97}{2} - 12.26\right)^2 + 11.0 + (14.7)\,[18.97 + (0.716 - 0.798 - 12.26)]^2 = 3117.07 \text{ in}^4$$

$$r_x = \sqrt{\frac{I}{A}} = \sqrt{\frac{3117.07}{49.8}} = 7.91 \text{ in}$$

$$I_y = 253 + 404 = 657 \text{ in}^4$$

$$r_y = \sqrt{\frac{657}{49.8}} = 3.63 \text{ in}$$

step 2: $K = 1$.

step 3: From Eq. 15.16, the slenderness ratios are

$$\text{SR}_x = \frac{(1)(20)(12)}{7.91} = 30.34$$

$$\text{SR}_y = \frac{(1)(10)(12)}{3.63} = 33.06$$

Since $33.06 > 27.5$, SR_y controls.

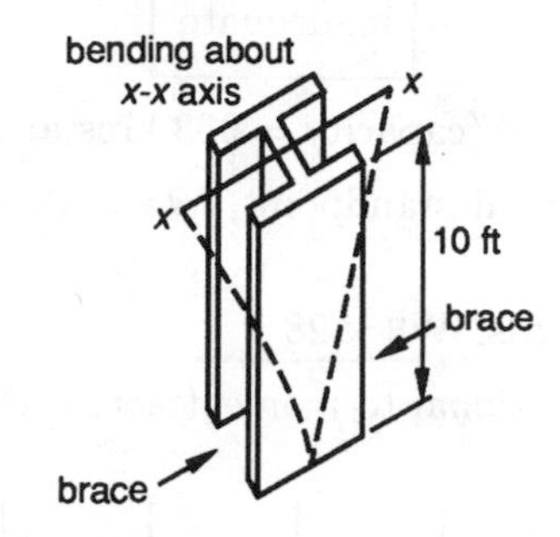

step 4: $C_c = 126.1$.

step 5: The problem asked for buckling load, not allowable load.

From Eq. 15.18,

$$F_a = \frac{\text{buckling stress}}{\text{factor of safety}}$$

$$\text{buckling stress} = \left[1 - \frac{(33.06)^2}{(2)(126.1)^2}\right](36) = 34.8 \text{ ksi}$$

$$\text{buckling load} = (34.8)(49.8) = 1733 \text{ kips}$$

5. Neglect self-weight.

$$R_L = R_R = \frac{(9)(20)}{2} = 90 \text{ kips}$$

- Top chords: W8 × 28

The force in horizontal members is proportional to the moment across that panel. The moment on the truss is maximum at center.

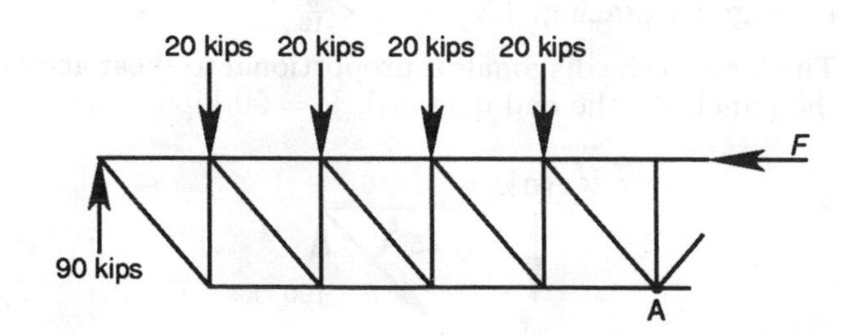

Taking clockwise moments as positive,

$$\sum M_{\mathrm{A}} = (90)(50) - (20)(10 + 20 + 30 + 40) - 10F$$
$$= 0$$
$$F = 250 \text{ kips (compression)}$$

$K_x = K_y = 1$

From the column table, capacity = 132 kips at KL = 10 ft, which is

inadequate

Choose W8 × 58 (capacity = 303 kips at $KL = 10$ ft).

(From a practical standpoint, W8 × 48 might also be chosen.)

- Bottom chords: W8 × 28

Force is proportional to moment across the panel.

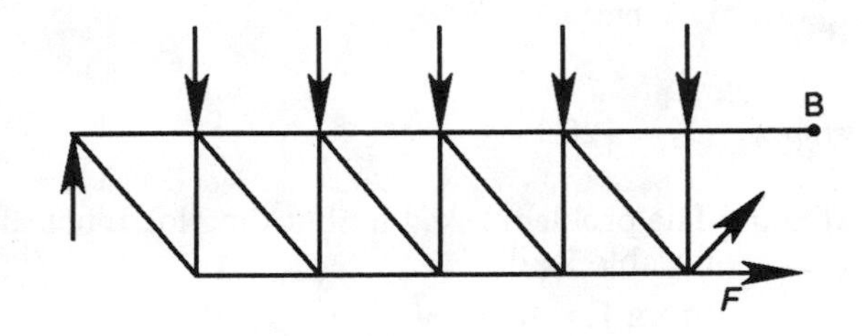

$$\sum M_{\mathrm{B}} = (90)(60) - (20)(10 + 20 + 30 + 40 + 50)$$
$$- 10F = 0$$
$$F = 240 \text{ kips (tension)}$$
$$\text{SR} = 74.1 < 240 \quad \text{[acceptable—see p. 15-20]}$$
$$F_t = (0.6)(36) = 21.6 \text{ ksi}$$
$$f_t = \frac{240}{8.25} = 29.1 \text{ ksi}$$

inadequate

The required area is

$$A = \frac{240}{21.6} = 11.11 \text{ in}^2$$

This disregards net section reductions. Try W8 × 40 (11.7 in²).

- Diagonal bracing: L$3\frac{1}{2} \times 3\frac{1}{2} \times \frac{5}{16}$

The force in the diagonals is proportional to shear across the panel. At the end diagonal, $V = 90$ kips.

V = 90 kips
45°
F
90 kips

$$F = \sqrt{(90)^2 + (90)^2} = 127.3 \text{ (tension)}$$
$$L = (1.41)(10) = 14.1 \text{ ft}$$

From the double-angle tables,

$$r_x = 1.08 \text{ in}$$
$$A = 4.18 \text{ in}^2$$

$$\text{SR} = \frac{(1)(14.1)(12)}{1.08} = 156.7 < 240 \quad \text{[acceptable]}$$
$$F_a = (0.6)(36) = 21.8 \text{ ksi}$$
$$f_a = \frac{127.3}{4.18} = 30.45 \text{ ksi}$$

inadequate

The required area is

$$A = \frac{127.3}{21.8} = 5.84 \text{ in}^2$$

The area of L$3\frac{1}{2} \times 3\frac{1}{2} \times \frac{1}{2}$-in pair is

$$(2)(3.25) = 6.5 \text{ in}^2$$

This is sufficient if the angle is available.

- Vertical members: L$4 \times 4 \times \frac{3}{8}$

The force in the vertical member is proportional to vertical force in the adjacent diagonal.

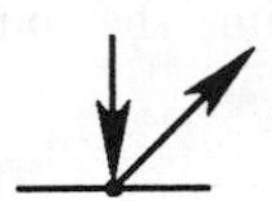

By inspection, at first vertical member,

$$F = 90 \text{ kips (compression)}$$

From the double-angle tables, capacity = 76 kips.

inadequate

Try L$4 \times 4 \times \frac{1}{2}$ (capacity = 99 kips at $KL = 10$ ft).

6. (a) First, locate the neutral axis.

$$\overline{y} = \frac{\sum A_i y_i}{\sum A_i} = \frac{(35.1)\left(\frac{18.97}{2}\right) + (9.96)(18.97 + 0.787)}{35.1 + 9.96}$$
$$= 11.76 \text{ in}$$

$$I_c = 2190 + (35.1)\left(\frac{18.97}{2} - 11.76\right)^2 + 8.13$$
$$+ (9.96)(18.97 + 0.787 - 11.76)^2$$
$$= 3016.8 \text{ in}^4$$

The distance to extreme fiber for compression is

$$c = 18.97 + 3.4 - 11.76 = 10.61 \text{ in}$$

For tension,

$$c = 11.76 \text{ in} \quad \text{[the larger } c\text{]}$$

The area for shear is

$$A = (18.97)(0.655) = 12.43 \text{ in}^2$$

(This assumes the channel does not carry shear.)

It is not clear if Table 15.4 applies. If it does, the load P should be increased 10–25%.

From p. 12-20 (moving loads), the moment is maximum when the loads are positioned as shown.

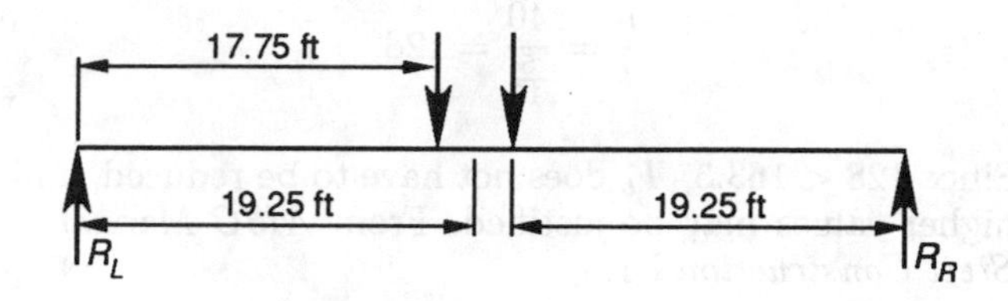

$$R_R = P\left(\frac{20.75}{40} + \frac{17.75}{40}\right) = 0.9625P$$

$$R_L = 2P - R_R = 1.0375P$$

$$M_{max} = (19.25)(0.9625P) = 18.53P \text{ ft-kips}$$

From *AISC Manual of Steel Construction* F1.3,

$$F_b = (0.6)(36) = 21.6 \text{ ksi (tension)}$$

$$F_v = (0.40)(36) = 14.4 \text{ ksi}$$

From $f = \frac{Mc}{I}$,

$$P = \frac{(21.6)(3016.8)}{(18.53)(12)(11.76)} = \boxed{24.92 \text{ kips}}$$

Check the shear. Placing the load group at one beam end,

$$V = 24.92 + \left(\frac{40-3}{40}\right)(24.92) = 47.97 \text{ kips}$$

$$f_{v,max} = \frac{V}{A} = \frac{47.97}{12.43} = 3.86 \text{ ksi} \quad \text{[acceptable]}$$

(b) *AISC Manual of Steel Construction* Section B10 requires the welds to resist the total horizontal shear.

For the W18,

$$t_f = 1.06 \text{ in}$$

For the C15,

$$t_w = 0.40 \text{ in}$$

From Table 15.8, $w_{min} = \frac{5}{16}$ in.

Minimum weld length is

$$\text{maximum}\left\{\begin{array}{l}(4)\left(\frac{5}{16}\right) \text{ in} \\ 1\frac{1}{2} \text{ in}\end{array}\right\} = 1\tfrac{1}{2} \text{ in}$$

From Eq. 12.24,

$$f_v = \frac{QV}{Ib}$$

$$\begin{aligned} Q &= (\text{area of C15})(\text{distance to its centroid}) \\ &= (9.96)(18.97 - 11.76 + 0.787) \\ &= 79.65 \text{ in}^3 \end{aligned}$$

$$b = 2t_e = (2)(0.707)\left(\tfrac{5}{16}\right) = 0.442 \text{ in}$$

$$V = (48.5)(1000) = 48{,}500 \text{ lbf}$$

$$f_v = \frac{(79.65)(48{,}500)}{(3016.8)(0.442)} = 2897 \text{ psi}$$

This shear is carried by the weld.

$$F_v = \text{minimum}\left\{\begin{array}{l}(0.30)(70) = 21 \text{ ksi} \\ \left(\frac{0.40}{0.707}\right)(36) = 20.4 \text{ ksi}\end{array}\right\} = 20.4 \text{ ksi}$$

The fraction of length requiring weld is

$$\frac{2.897}{20.4} = 0.142$$

Use a $1\frac{1}{2}$-in-long weld spaced every 10 in.

(10 in is less than the maximum 24 in permitted for rolled shapes by *AISC Manual of Steel Construction* E4.)

$$\frac{1.5}{10} = 0.15 \quad \text{[acceptable]}$$

(c)

$$t_w = 0.655 \text{ in}$$

$$k = 1.75$$

From Eq. 15.8,

$$\begin{aligned} N_{min} &= \frac{48.5}{(0.66)(36)(0.655)} - 2.5 \\ &= 0.62 \text{ in} \quad \text{[round to 1 in]} \end{aligned}$$

Since $N > 0$, the web crippling requirements will not be met only if the reaction is truly concentrated.

($N_{min} < 0$ from Eq. 15.7, so the interior load is no problem.)

7. Find the moment of inertia of the plate girder with and without the concrete flange.

shape	A	$\overline{y}$	$\overline{y}A$	I_c
$\frac{5}{8} \times 14$	8.75	41.0625	359.297	0.2848
$\frac{5}{16} \times 40$	12.5	20.75	259.375	1666.67
$\frac{3}{4} \times 16$	12	0.375	4.5	0.5625
subtotals	33.25		623.172	1667.52
transformed concrete	48.0	45.375	2178	144
totals	81.25 in^2		2801.2 in^3	1811.52 in^4

Work with the steel acting alone before the concrete hardens.

$$\overline{y} = \frac{623.172}{33.25} = 18.74 \text{ in}$$

$$\begin{aligned} I &= 1667.52 + (12)(18.74 - 0.375)^2 \\ &\quad + (12.5)(20.75 - 18.74)^2 \\ &\quad + (8.75)(41.0625 - 18.74)^2 \\ &= 10{,}125.4 \text{ in}^4 \end{aligned}$$

The distances to the extreme fibers are

$$c = 18.74 \text{ in (tension)}$$
$$c = 22.64 \text{ in (compression)}$$

The section modulus referred to the bottom of steel is

$$S = \frac{I}{c} = \frac{10{,}125.4}{18.74} = 540 \text{ in}^3$$

The maximum moment due to 1 kip/ft^2 dead load is

$$M_D = \frac{wL^2}{8} = \frac{(1)(80)^2}{8} = 800 \text{ ft-kips}$$

The shear due to dead load is

$$V_D = \frac{wL}{2} = \frac{(1)(80)}{2} = 40 \text{ kips}$$

Check the width-thickness ratios.

From Eqs. 15.66 and 15.67, for the tension flange,

$$\frac{95}{\sqrt{36}} = 15.83$$
$$b = \left(\tfrac{1}{2}\right)(16) = 8 \text{ in}$$
$$\frac{b}{t} = \frac{8}{\frac{3}{4}} = 10.7$$
$$10.7 < 15.83 \quad \text{[acceptable]}$$

For the compression flange,

$$b = \left(\tfrac{1}{2}\right)(14) = 7 \text{ in}$$
$$\frac{b}{t} = \frac{7}{\frac{5}{8}} = 11.2$$
$$\frac{95}{\sqrt{36}} = 15.83$$
$$11.2 < 15.83 \quad \text{[acceptable]}$$

For the web, use Eq. 15.68.

$$\frac{14{,}000}{\sqrt{(36)(36 + 16.5)}} = 322.0$$
$$\frac{h}{t} = \frac{40}{\frac{5}{16}} = 128$$
$$128 < 322 \quad \text{[acceptable]}$$

Check the bending stress.

From Eq. 15.63,

$$\frac{760}{\sqrt{F_b}} = \frac{760}{\sqrt{(0.60)(36)}} = 163.5$$

Calculate h/t for the web.

$$\frac{h}{t} = \frac{40}{\frac{5}{16}} = 128$$

Since $128 < 163.5$, F_b does not have to be reduced, and higher values may be justified. From *AISC Manual of Steel Construction* F1.3,

$$r_T = \sqrt{\frac{I_{y,\text{flange}} + I_{\frac{1}{3}\text{ of web in compression}}}{A_{\text{flange}} + A_{\frac{1}{3}\text{ of web in compression}}}}$$

Disregard the contribution of one-third of the web to the moment of inertia. Approximately half of the web is in compression.

$$\begin{aligned} r_T &= \sqrt{\frac{\left(\frac{1}{12}\right)\left(\frac{5}{8}\right)(14)^3 + 0}{8.75 + \left(\frac{1}{2}\right)\left(\frac{12.5}{3}\right)}} \\ &= 3.63 \text{ in} \end{aligned}$$

$$\frac{l}{r_T} = \frac{(20)(12)}{3.63} = 66.1$$

The moment diagram due to the dead load is

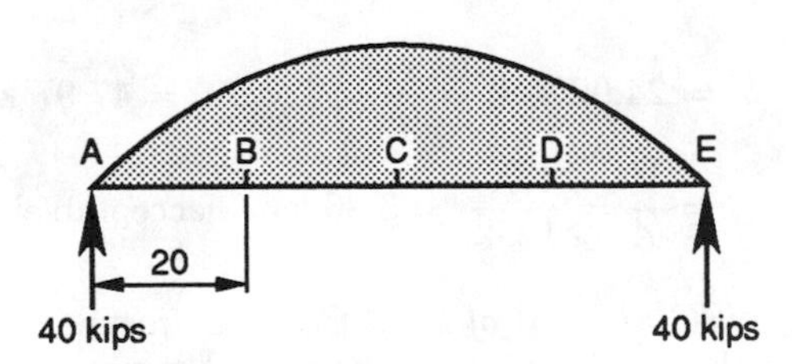

$C_b > 1$ in all 4 panels.

$$M_B = (40)(20) - \left(\tfrac{1}{2}\right)(1)(20)^2 = 600 \text{ ft-kips}$$
$$M_C = 800 \text{ ft-kips}$$

From App. D,

$$C_{b,\text{span A-B}} = 1.75 + (1.05)\left(\frac{-0}{600}\right) + (0.3)\left(\frac{-0}{600}\right)^2 = 1.75$$

$$C_{b,\text{span B-C}} = 1.75 + (1.05)\left(\frac{-600}{800}\right) + (0.3)\left(\frac{-600}{800}\right)^2 = 1.13$$

Use $C_b = 1.13$ as the more conservative value.

From *AISC Manual of Steel Construction* F1-6,

$$\sqrt{\frac{102{,}000C_b}{F_y}} = \sqrt{\frac{(102{,}000)(1.13)}{36}} = 56.6$$

$$\sqrt{\frac{510{,}000C_b}{F_y}} = \sqrt{\frac{(510{,}000)(1.13)}{36}} = 126.5$$

Since $56.6 < 66.1 < 126.5$,

$$F_b = \left[\frac{2}{3} - \frac{F_y\left(\frac{l}{r_T}\right)^2}{(1530 \times 10^3)\,C_b}\right]F_y = \left[\frac{2}{3} - \frac{(36)(66.1)^2}{(1530 \times 10^3)(1.13)}\right](36) = 20.7 \text{ ksi}$$

Check the bending stress.

$$f_b = \frac{Mc}{I} = \frac{(800)(12)(22.64)}{10{,}125.4} = 21.47 \text{ ksi}$$

$$21.47 \text{ ksi} > 20.7 \text{ ksi} \quad \text{[not acceptable]}$$

Check the shear stress.

No data on stiffeners were given, so assume the worst case. From Eqs. 15.60 and 15.62,

$$k = 5.34$$

$$\frac{h}{t} = 128 > \frac{548}{\sqrt{36}} = 91.3$$

$$F_v = \frac{83{,}150}{(128)^2} = 5.08 \text{ ksi}$$

$$f_v = \frac{V}{A_{\text{web}}} = \frac{40}{(40)\left(\frac{5}{16}\right)} = \frac{40}{12.5} = 3.2 \text{ ksi} < 5.08 \text{ ksi} \quad \text{[acceptable]}$$

- After the concrete hardens:

$$\bar{y} = \frac{2801.2}{81.25} = 34.48 \text{ in}$$

All of the concrete is in compression. The distances to the extreme fibers are

$$c = \left\{\begin{array}{ll} 6.895 \text{ in} & \text{[steel in compression]} \\ 13.895 \text{ in} & \text{[concrete in compression]} \end{array}\right\} = 34.48 \text{ in (tension)}$$

$$\begin{aligned} I &= 1811.52 + (8.75)(41.0625 - 34.48)^2 \\ &\quad + (12.5)(34.48 - 20.75)^2 \\ &\quad + (12)(34.48 - 0.375)^2 \\ &\quad + (48)(45.375 - 34.48)^2 \\ &= 24{,}202.5 \text{ in}^4 \end{aligned}$$

The transformed section modulus referred to the tension flange is

$$S = \frac{I}{c} = \frac{24{,}202.5}{34.48} = 701.9 \text{ in}^3$$

From the live load, the maximum shears and moments are

$$V_t = 40 + \frac{(1)(80)}{2} = 80 \text{ kips}$$

$$M_L = (20)(40) = 800 \text{ ft-kips}$$

$$M_t = 800 + 800 = 1600 \text{ ft-kips}$$

Maximum S_{tr} allowed without shoring, per *AISC Manual of Steel Construction*, is

$$S_{\text{tr,max}} = \left[1.35 + (0.35)\left(\frac{M_L}{M_D}\right)\right]S = \left[1.35 + (0.35)\left(\frac{800}{800}\right)\right](540) = 918 \text{ in}^3$$

$$701.9 \text{ in}^3 < 918 \text{ in}^3 \quad \text{[acceptable]}$$

Check the shear stress.

$$f_v = \frac{V}{A_{\text{web}}} = \frac{80 \text{ kips}}{12.5 \text{ in}^2} = 6.4 \text{ ksi}$$

$$F_v = 5.08 \text{ ksi}$$

$$6.4 \text{ ksi} > 5.08 \text{ ksi} \quad \text{[not acceptable]}$$

Check the compressive stress in steel.

$$f_b = \frac{Mc}{I} = \frac{(1600)(12)(6.895)}{24{,}202.5} = 5.47 \text{ ksi}$$

$$5.47 \text{ ksi} < 20.7 \text{ ksi} \quad \text{[acceptable]}$$

Check the tensile stress in steel.

$$F_{b,\text{tension}} = (0.6)(60) = 36 \text{ ksi}$$

$$f_b = \frac{(1600)(12)(34.48)}{24{,}202.5} = 27.35 \text{ ksi}$$

$$27.35 \text{ ksi} < 36 \text{ ksi} \quad \text{[acceptable]}$$

Check the concrete compressive stress.

Assume 3000 psi concrete. From *AISC Manual of Steel Construction* I2.2,

$$F_c = (0.45)(3000) = 1.35 \text{ ksi}$$

Only the live load is applied after curing.

$$f_c = \left(\frac{1}{10}\right)\left[\frac{(800)(12)(13.895)}{24{,}202.5}\right] = 0.55 \text{ ksi}$$

$$0.55 \text{ ksi} < 1.35 \text{ ksi} \quad \text{[acceptable]}$$

Several aspects of this design are unacceptable.

8. Chapter N of the *AISC Manual of Steel Construction* implies plastic analysis. Loads given are already ultimate.

First, determine the ultimate moment.

- Failure by beam collapse:

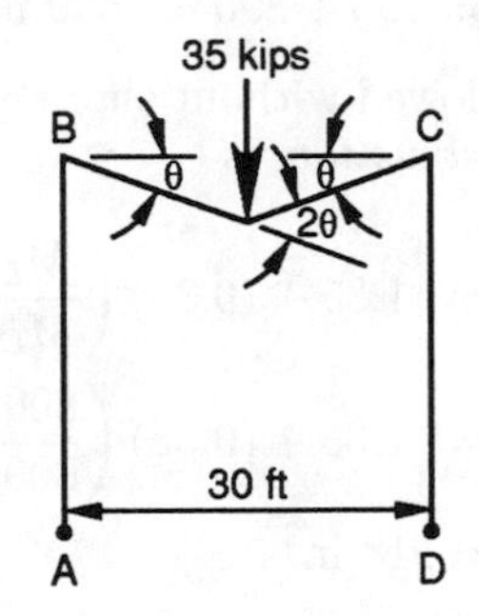

By virtual work,

$$M_p(4\theta) = (35)\left(\frac{30}{2}\right)\theta$$

$$M_p = 131.25 \text{ ft-kips}$$

- Failure by sway collapse:

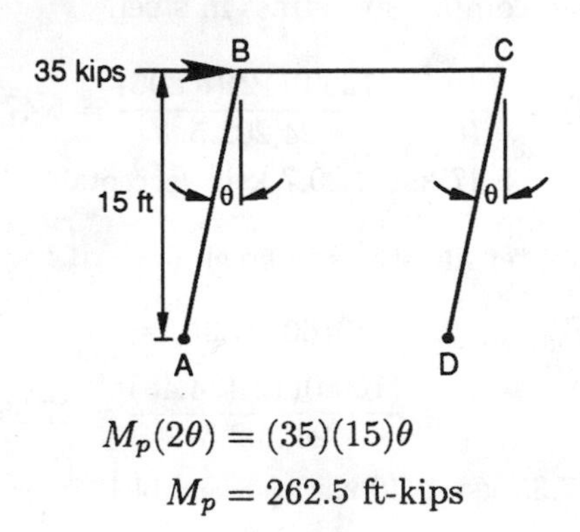

$$M_p(2\theta) = (35)(15)\theta$$

$$M_p = 262.5 \text{ ft-kips}$$

- Failure by beam and sway collapse:

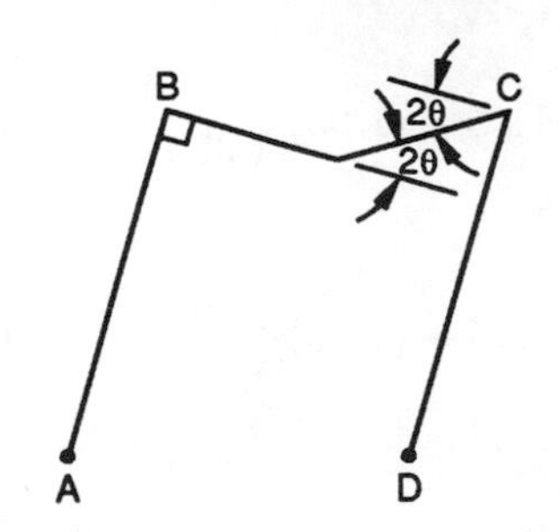

$$4M_p\theta = (35)\left(\frac{30}{2}\right)\theta + (35)(15)\theta$$

$$M_p = 262.5 \text{ ft-kips}$$

Next, get the reactions at the supports. Taking clockwise moments as positive,

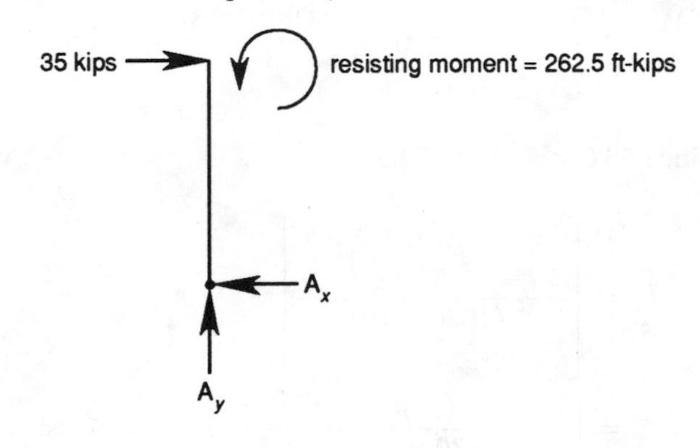

$$\sum M_B = 15A_x - 262.5 = 0$$

$$A_x = 17.5 \text{ kips}$$

From $\sum F_x = 0$,

$$D_x = 35 - 17.5 = 17.5 \text{ kips}$$

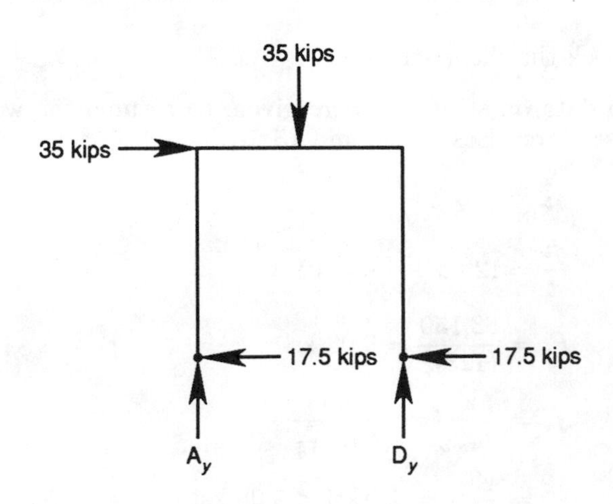

$$\sum M_A = (35)(15) + (35)(15) - 30D_y = 0$$

$$D_y = 35 \text{ kips}$$

$$A_y = 0 \text{ kips}$$

Try W21 × 44. From the plastic design selection table,

$$M_p = 286 \text{ ft-kips}$$
$$P_y = 468 \text{ kips}$$
$$r_x = 8.06 \text{ in}$$

From *AISC Manual of Steel Construction* Eq. N4-3,

$$\frac{P}{P_y} + \frac{M}{1.18M_P} \leq 1.0$$
$$\frac{35}{468} + \frac{262.5}{(1.18)(286)} = 0.853$$

Assume $K = 2$ (Table 15.5, pinned—rotation fixed, translation free).

$$\frac{Kl}{r} = \frac{(2)(15)(12)}{8.06} = 44.67 \quad \text{[round to 45]}$$

From *AISC Manual of Steel Construction* Table C-36,

$$F_a = 18.78 \text{ ksi}$$
$$P = 35 \text{ kips}$$
$$A = 13 \text{ in}^2$$
$$C_m = 0.85 \quad [\textit{AISC Manual of Steel Construction}, \text{Sec. H1}]$$
$$P_e = \frac{23}{12} A F'_e$$
$$F'_e = 73.74 \text{ ksi} \quad [\textit{AISC Manual of Steel Construction}, \text{Table 8}]$$
$$P_e = \left(\frac{23}{12}\right)(13)(73.74) = 1837.4 \text{ kips}$$
$$M_m = 286 \text{ ft-kips}$$

From *AISC Manual of Steel Construction* Eq. N4-2,

$$\frac{P}{1.7AF_a} + \frac{C_m M}{\left(1 - \frac{P}{P_e}\right) M_m} \leq 1.0$$
$$\frac{35}{(1.7)(13)(18.78)} + \frac{(0.85)(262.5)}{\left(1 - \frac{35}{1837.4}\right)(286)} = 0.084 + 0.795 = 0.88 < 1.0$$

Final check: W21 × 44 is starred in the plastic design table, so a check against *AISC Manual of Steel Construction* Sec. N7 is required.

$$\frac{b_f}{2t_f} = 7.2 < 8.5 \quad \text{[for 36 ksi steel—acceptable]}$$

The ratio of axial to yield loads in member CD is

$$\frac{P}{P_y} = \frac{35}{468} = 0.0748 < 0.27$$

From *AISC Manual of Steel Construction* N7-1,

$$\frac{d}{t} \leq \left(\frac{412}{\sqrt{F_y}}\right)\left[1 - (1.4)\left(\frac{P}{P_y}\right)\right]$$
$$59.0 \leq \left(\frac{412}{\sqrt{36}}\right)\left[1 - (1.4)\left(\frac{35}{468}\right)\right]$$
$$59.0 \leq 61.5 \quad \text{[acceptable]}$$

use W21 × 44

9. (a) All vertical loads act simultaneously with the wind load. The total axial load is

$$17 + 21 + 31 = 69 \text{ kips}$$

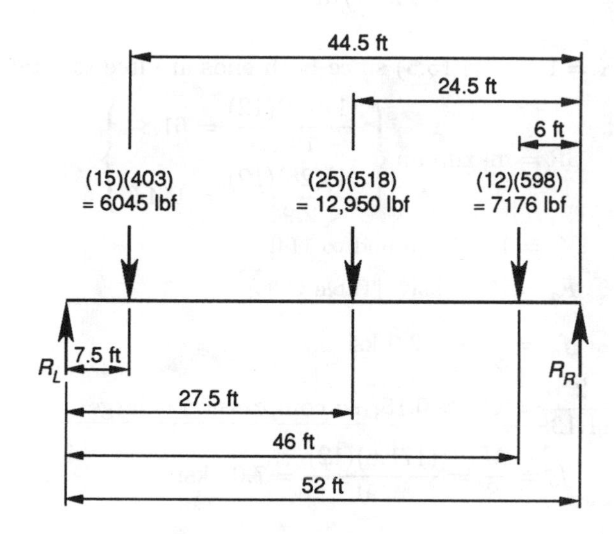

Find the reactions. Taking clockwise moments as positive,

$$\sum M_{R_L} = (6.045)(7.5) + (12.950)(27.5) + (7.176)(46) - 52R_R = 0$$
$$R_R = 14.07 \text{ kips}$$
$$\sum M_{R_R} = 52R_L - (7.176)(6) - (12.950)(24.5) - (6.045)(44.5) = 0$$
$$R_L = 12.10 \text{ kips}$$

The shear diagram is

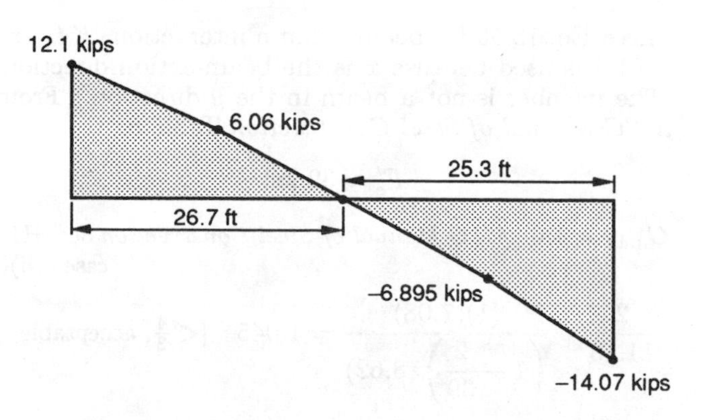

$$M_{\text{max},x=26.7 \text{ ft,to the left}} = (12.1)(26.7) - (6.045)(26.7 - 7.5) - \left(\tfrac{1}{2}\right)(0.518)(26.7 - 15)^2 = 171.6 \text{ ft-kips}$$

W24 is specified. Column tables cannot be used. Try W24 × 117.

$$A = 34.4 \text{ in}^2 \qquad d = 24.26 \text{ in}$$
$$r_x = 10.1 \text{ in} \qquad b_f = 12.800 \text{ in}$$
$$r_y = 2.94 \text{ in} \qquad I_x = 3540 \text{ in}^4$$
$$S = 291 \text{ in}^3$$
$$\frac{d}{A_f} = 2.23 \text{ 1/in}$$

$K = 1$ (Table 15.5) since both ends are free to rotate.

$$\text{SR} = \text{maximum}\begin{Bmatrix} \dfrac{(1)(52)(12)}{10.1} = 61.8 \\ \dfrac{(1)(28)(12)}{2.94} = 114.3 \end{Bmatrix} = 114.3 \quad \text{[round to 114]}$$

$$F_a = 11.13 \text{ ksi} \quad \text{[Table C-36]}$$

$$f_a = \frac{69}{34.4} = 2.0 \text{ ksi}$$

$$\frac{2}{11.13} = 0.18 > 0.15, \text{ so compression is large.}$$

$$f_b = \frac{M}{S} = \frac{(171.6)(12)}{291} = 7.08 \text{ ksi}$$

Use *AISC Manual of Steel Construction* Eq. F1-8 as a shortcut to finding F_b. Assume no support of compression flange. (If the column is inside the building, the wall sheathing may be connected to the tension flange.)

Use $C_b = 1$ since there is no lateral support.

$$F_b = \frac{(12 \times 10^3)(1)}{(52)(12)(2.23)} = 8.62 \text{ ksi}$$

From Eq. 15.48, using the one-third wind load increase,

$$\frac{2}{(0.6)(36)} + \frac{7.08}{8.62} = 0.91 \quad \left[\text{acceptable since} < \tfrac{4}{3}\right]$$

Check Eq. 15.50 for beam-column interaction. $KL_x/r_x = 61.8$ is used because x is the beam-action direction. (The member is not a beam in the y-direction.) From *AISC Manual of Steel Construction* Table 8,

$$F'_e \approx 39 \text{ ksi}$$

$$C_{\text{max}} = 1.0 \; [\textit{AISC Manual of Steel Construction} \text{ Sec. H1, case c(ii)}]$$

$$\frac{2}{11.13} + \frac{(1)(7.08)}{\left(1 - \dfrac{2}{39}\right)(8.62)} = 1.045 \quad \left[< \tfrac{4}{3}, \text{ acceptable}\right]$$

Use W24 × 117.

An exact deflection calculation is difficult since the load is nonuniform. Assume 598 lbf/ft over the entire column (limiting case). From p. 12-31, case 5,

$$\frac{\text{deflection}}{L} = \frac{5wL^3}{384EI} = \frac{(5)\left(\dfrac{598}{12}\right)(52 \times 12)^3}{(384)(2.9 \times 10^7)(3540)} = 0.001536 = \frac{1}{651}$$

This is less than the general guidelines given on p. 15-5.

(b) Since the concrete floor (support) is apparently not limited in size, the maximum value of F_p can be used (Eq. 15.36).

$$F_p = 0.70 f'_c$$
$$A_1 = \frac{17{,}000 + 21{,}000}{(0.70)(3000)} = 18.1 \text{ in}^2$$

This is much smaller than the column footprint ($b_f \times d$). Any reasonable design that allowed room for bolts would be acceptable. It is not possible to optimize the design. Use 28 in × 16 in.

10. (a) Assume adequate stiffeners as required to achieve $F_v = 0.40F_y$.

$$F_v = (0.40)(36) = 14.4 \text{ ksi}$$

The shear is

$$V = \frac{40w}{2} = 20w$$

From Eq. 15.55,

$$f_v = \frac{20w}{(48)(0.375)} = 14.4 \text{ ksi}$$
$$w = 12.96 \text{ kips/ft}$$

The girder weight per foot is

$$(1 \text{ ft})\left[\left(\frac{\frac{3}{8}}{12}\right)\left(\frac{16 + 16 + 48}{12}\right)\right]\left(490 \; \frac{\text{lbf}}{\text{ft}^3}\right) = 102 \text{ lbf/ft}$$

$$w_{\text{load}} = 12.96 - 0.1 = \boxed{12.86 \text{ kips/ft}}$$

(b) Assume stiffener spacing $a = 4$ ft $= 48$ in. (There is an alternate interpretation—that of no stiffeners in the central 32 ft. However, this type of construction would be unlikely.)

$$\frac{a}{h} = \frac{48}{48} = 1$$

From Eqs. 15.59 or 15.60,

$$k = 4.00 + \frac{5.34}{(1)^2} = 9.34$$

Assume $C_v < 0.8$. From Eq. 15.57,

$$C_v = \frac{(45{,}000)(9.34)}{(36)\left(\frac{48}{0.375}\right)^2} = 0.71$$

Assume A36 stiffeners. Use stiffeners in pairs. From Eq. 15.75, the stiffener steel area is

$$A_{st} = \left(\frac{1-0.71}{2}\right)\left[1 - \frac{(1)^2}{\sqrt{1+(1)^2}}\right]\left(\frac{36}{36}\right)(1)(48)\left(\tfrac{3}{8}\right)$$
$$= 0.76 \text{ in}^2$$

Total width of $\frac{3}{8}$-in-thick plates is

$$\frac{0.76}{\frac{3}{8}} = 2.0 \text{ in}$$

Check the minimum moment of inertia. From Eq. 15.78,

$$I_{st} \geq \left(\frac{48}{50}\right)^4 = 0.85 \text{ in}^4$$

From Eq. 15.77,

$$b_{st} = \sqrt[3]{\frac{(12)(0.85)}{\frac{3}{8}}} = 3.0 \text{ in}$$

Use $1\frac{1}{2}$-in-wide stiffeners on each side.

TRAFFIC ANALYSIS, TRANSPORTATION, AND HIGHWAY DESIGN

Untimed

1. Use App. B to convert the CBR values to approximate R values.

	CBR	R
base	90	86
subbase	40	70
subgrade	5	28

The actual thickness of asphalt is known (3 in or 0.25 ft). Check the adequacy of this value. From Eq. 16.37,

$$GE_{\text{asphalt required}} = (0.0032)(7.5)(100 - 86) = 0.34$$

From Table 16.18, $G_f = 2.01$. From Eq. 16.38,

$$t_{\text{asphalt}} = \frac{0.34}{2.01} = 0.17 \text{ ft} < 0.25 \text{ ft} \quad \text{[acceptable]}$$

$$\text{actual } GE_{\text{asphalt}} = (2.01)(0.25) = 0.50$$

- For the base:

$$GE_{\text{base+asphalt}} = (0.0032)(7.5)(100 - 70) = 0.72$$

$$GE_{\text{base}} = 0.72 - 0.50 = 0.22$$

From Table 16.18, $G_f = 1.10$.

$$t_{\text{base}} = \frac{0.22}{1.10} = 0.2 \text{ ft}$$

(2 in = 0.17 ft minimum practical thickness, so 0.20 ft is acceptable.)

- For the subbase:

$$GE_{\text{total}} = (0.0032)(7.5)(100 - 28) = 1.73$$

$$GE_{\text{subbase}} = 1.73 - 0.50 - 0.22 = 1.01$$

Since $G_{f,\text{subbase}} = 1.0$ (Table 16.18),

$$t_{\text{subbase}} = 1.01 \text{ ft}$$

layer	thickness
asphalt	0.25 ft
base	0.20 ft
subbase	1.01 ft

subgrade

There are practical considerations that might change the design.

2. *step 1:* $p_t = 2.5$

step 2: Average axles per vehicle:

$$\frac{22{,}000}{8800} = 2.5$$

Total number of axle loads per day:

$$(2200)(2.5) = 5500$$

For the outside lane,

$$(5500)(0.75) = 4125 \text{ axles/day}$$

Use App. E, assuming SN = 3 (see p. 16-16, Method A).

axle load	base	fraction	equivalence factor	EAL per day
single axles				
<8k	4125	0.363	≈ 0.004	6
8k–16k	4125	0.284	0.23	269
16k–20k	4125	0.129	1.00	532
tandem axles				
8k–16k	4125	0.045	≈ 0.02	4
16k–20k	4125	0.105	0.11	48
20k–24k	4125	0.040	0.23	38
24k–30k	4125	0.031	0.49	63
30k–34k	4125	0.003	0.89	11
total				971 per day

Since Fig. 16.11 has daily EAL, it is not necessary to convert to 20-year EAL.

step 3: Assume $R = 0.8$ (within 0.2–1.0 range).

step 4: From App. B, for CBR = 5, $S \approx 4.8$.

step 5: From Fig. 16.11, $\overline{SN} = 4.3$.

Since 4.3 > 3.0, repeat from step 2 with SN = 4. (Actually, this step is not necessary since equivalency factors are not very sensitive to SN.)

axle load	base	fraction	equivalence factor	EAL per day
single axles				
<8k	4125	0.363	≈ 0.003	4
8k–16k	4125	0.284	0.21	246
16k–20k	4125	0.129	1.00	532
tandem axles				
8k–16k	4125	0.045	≈ 0.02	4
16k–20k	4125	0.105	0.09	39
20k–24k	4125	0.040	0.21	35
24k–30k	4125	0.031	≈ 0.46	59
30k–34k	4125	0.003	0.89	11
total				930 per day

From Fig. 16.11, $\overline{\text{SN}} = 4.3$ (no change) and $\text{SN} = 4.2$.

step 6: From Table 16.17,

$$\begin{array}{ll} \text{subbase:} & a_3 = 0.14 \\ \text{base:} & a_2 = 0.34 \\ \text{surface:} & a_1 = 0.44 \end{array}$$

step 7: From Eq. 16.35,

$$0.44t_1 + 0.34t_2 + 0.14t_3 = 4.2$$

Try

$$\boxed{\begin{array}{ll} \text{surface:} & t_1 = 4 \text{ in} \\ \text{base:} & t_2 = 5 \text{ in} \\ \text{subbase:} & t_3 = 7 \text{ in} \end{array}}$$

- Check:

$$(0.44)(4)+(0.34)(5)+(0.14)(7) = 4.44 > 4.2 \text{ [acceptable]}$$

3. (a) $\left(60 \ \frac{\text{mi}}{\text{hr}}\right)\left(\frac{5280 \ \frac{\text{ft}}{\text{mi}}}{3600 \ \frac{\text{sec}}{\text{hr}}}\right) = 88 \text{ ft/sec}$

initial separation $= (20)(6) = 120$ ft

separation after reaction time $= 120 - (0.5)(88) = 76$ ft

From Newton's second law,

$$\begin{aligned} F_f &= \text{frictional retarding force} = ma \\ &= W\mu = mg\mu \end{aligned}$$

$$ma = mg\mu$$

$$a = g\mu = (32.2)(0.6) = 19.32 \text{ ft/sec}^2$$

$$\text{v}_o = 88 \text{ ft/sec}$$

$$s = 76 \text{ ft}$$

$$a = -19.32 \text{ ft/sec}^2$$

From Table 16.1,

$$\text{v} = \sqrt{(88)^2 + (2)(-19.32)(76)} = \boxed{69.3 \text{ ft/sec}}$$

(b) The rule is one car length (20 ft) per 88/60 = 14.67 ft/sec. Working backward, knowing that terminal velocity must be zero,

$$0 = \sqrt{\text{v}^2 - (2)(19.32)\left[(20)\left(\frac{\text{v}}{14.67}\right) - 0.5\text{v}\right]}$$

$$\text{v}^2 - 33.36\text{v} = 0$$

$$\text{v} = \boxed{33.36 \text{ ft/sec (22.7 mph)}}$$

(c) If a similar simple car-length rule is to apply at 60 mph,

$$0 = (88)^2 - (2)(19.32)\left[(x)\left(\frac{88}{14.67}\right) - (0.5)(88)\right]$$

$$x = \boxed{40.7 \text{ ft} \quad \text{[approximately two car lengths]}}$$

4. From Eq. 16.15,

(a) $185 = \dfrac{\text{v}^2_{\text{mph}}}{(30)(f - 0.03)}$

By trial and error, using the test values of f,

$$\text{v} \approx \boxed{49 \text{ mph}}$$

(b) Use Table 16.2 for a wet road surface. Assume new tires. By trial and error,

$$\text{v} \approx \boxed{41 \text{ mph}}$$

5. Spot speed is an instantaneous value at a particular point and is not a function of the number of vehicles on the highway. Use space mean speed.

$$30 \text{ mph} = (30)\left(\frac{5280}{3600}\right) = 44 \text{ ft/sec}$$

$$29 \text{ mph} = (29)\left(\frac{5280}{3600}\right) = 42.53 \text{ ft/sec}$$

(a) From Eq. 16.27,

$$\text{headway} = \frac{80}{42.53} = \boxed{1.88 \text{ sec/vehicle}}$$

(b) $\text{density} = \dfrac{5280}{80} = \boxed{66 \text{ vehicles/mi}}$

(c) From Eq. 16.28,

$$\text{volume} = \frac{(3600)(42.53)}{80} = \boxed{1914 \text{ vehicles/hr (vph)}}$$

(d) From 1985 *Highway Capacity Manual* (p. 8-4), maximum stable capacity is 2800 pcph (two directions).

(e)

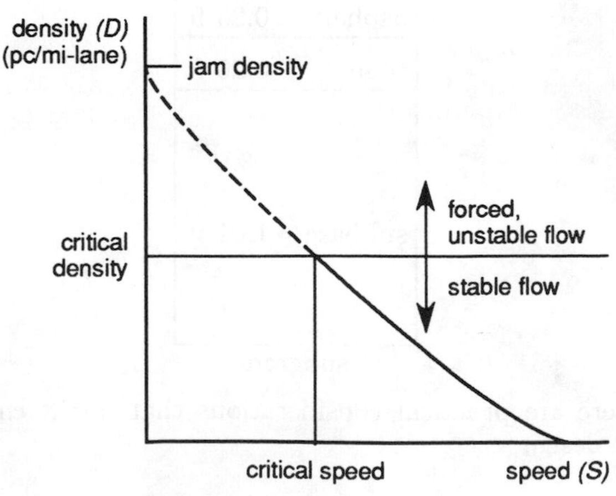

(f)

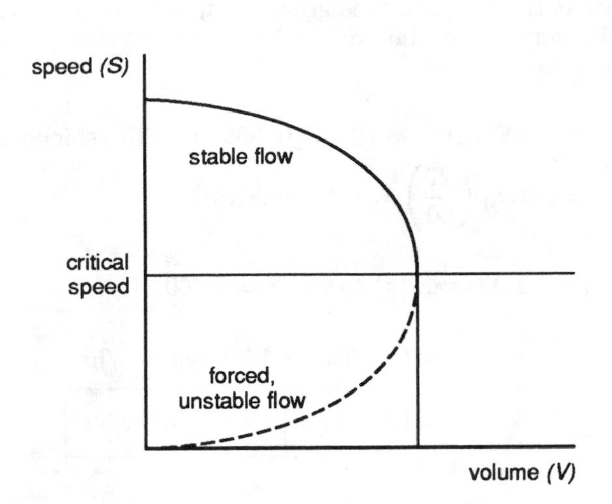

(g)

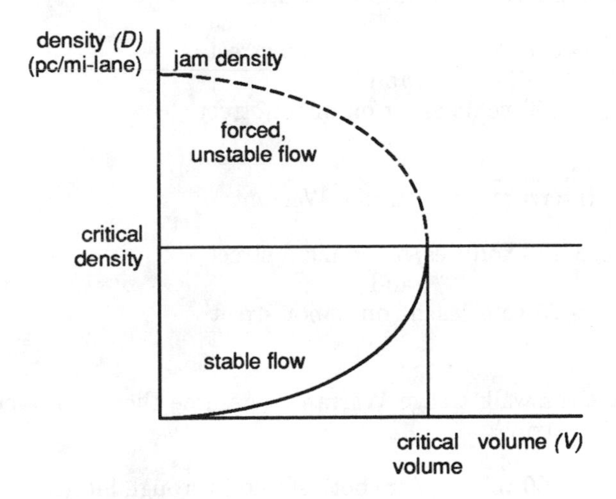

(h) Only volume has units of time. Density is a function of the number of cars in a given distance, but has nothing to do with speed. The cars could be stopped and still have a high density.

Volume throughput is more accurate.

(i) | 2000 vehicles/hr (vph) | [See Table 16.9.]

6.

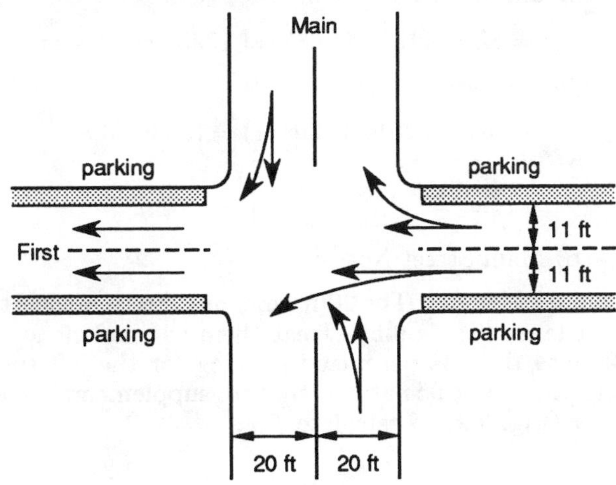

The notation used comes from Chap. 9 of the 1985 *Highway Capacity Manual.* Assume that lost time includes the yellow time so that effective green time is the actual green time.

(a) First Street:

Treat as one lane group. Find the saturation flow rate. From Table 9-3, the default number of parking maneuvers is 20 for each side of the street, so $N_m = 40$. Adjustment factors are

$$f_W = 0.97 \quad \text{[Table 9-5]}$$
$$f_{\text{HV}} = 0.965 \quad \text{[Table 9-6]}$$
$$f_p = 0.85 \quad \text{[Table 9-8, using } N_m = 40 \text{ (Table 9-3)]}$$
$$f_{\text{RT}} = 0.985 \quad \text{[Table 9-11]}$$
$$f_{\text{LT}} = 0.995 \quad \text{[Table 9-12, Case 4]}$$

Therefore,

$$\begin{aligned} s &= (1800)(2)(0.97)(0.965)(0.85)(0.985)(0.995) \\ &= 2807 \text{ vehicles/hr} \end{aligned}$$

Lane group capacity (Eq. 9-1) is

$$\begin{aligned} c &= (s)(\text{effective green ratio}) = (2807)\left(\frac{27}{60}\right) \\ &= 1263 \text{ vehicles/hr} \end{aligned}$$

Maximum delay for level of service E is 60 sec (Table 9-1). The delay equation is Eq. 9-18.

$$\begin{aligned} d &= 0.38C\left[\frac{\left(1-\frac{g}{C}\right)^2}{1-\left(\frac{g}{C}\right)(X)}\right] \\ &\quad + 173X^2\left[(X-1)+\sqrt{(X-1)^2+\left(\frac{16X}{c}\right)}\right] \end{aligned}$$

Solving for X (the volume/capacity ratio) by trial and error, with $d = 60$, $C = 60$, $g = 27$, and $c = 1263$, yields $X = 1.057$.

Maximum flow rate is

$$v = cX = (1316)(1.057) = 1335 \text{ vehicles/hr}$$

Maximum service flow is

$$V = v(\text{peak hour factor}) = (1335)(0.85)$$

$$= \boxed{1135 \text{ vehicles/hr}}$$

(b) Main Street North:

Proceed as in (a). The 20-ft lane must be treated as two 10-ft lanes (Table 9-5). Treat them as one lane group. Because there is permitted phasing for the left turns, perform the calculations in the supplemental worksheet (Fig. 9-9) to calculate f_{LT}.

$$S_{op} = \frac{1800N_o}{1 + P_{LT_o}}\left(\frac{400 + v_M}{1400 - v_M}\right) = 3600$$

$$Y_o = \frac{v_o}{S_{op}} = 0.111$$

$$g_u = \frac{g - CY_o}{1 - Y_o} = 23$$

$$f_s = \frac{875 - 0.625v_o}{1000} = 0.625$$

$$P_L = P_{LT}\left[1 + \frac{(N-1)g}{f_s g_u + 4.5}\right] = 0.243$$

$$g_q = g - g_u = 4$$

$$P_T = 1 - P_L = 0.757$$

$$g_f = 2\frac{P_T}{P_L}\left(1 - P_T^{0.5g_q}\right) = 2.7$$

$$E_L = \frac{1800}{1400 - v_o} = 1.8$$

$$f_m = \frac{g_f}{g} + \frac{g_u}{g}\left[\frac{1}{1 + P_L(E_L - 1)}\right] + \left(\frac{2}{g}\right)(1 + P_L)$$

$$= 0.905$$

$$f_{LT} = \frac{f_m + N - 1}{N} = 0.95$$

(Note that $P_L < 1.0$, confirming that the lane group will be in equilibrium and the two lanes should be treated as one lane group.)

The other adjustment factors are $f_W = 0.93$ and $f_{HV} = 0.975$ (Table 9-6).

$$s = (1800)(2)(0.93)(0.975)(0.95) = 3101 \text{ vehicles/hr}$$

$$c = (3101)\left(\frac{27}{60}\right) = 1395 \text{ vehicles/hr}$$

Maximum delay for level of service B is 15 sec (Table 9-1). Solving for volume/capacity ratio, with $d = 15$, $C = 60$, $g = 27$, and $c = 3101$, yields $X = 0.35$.

$$v = (1395)(0.35) = 488 \text{ vehicles/hr}$$

$$V = (488)(0.85) = \boxed{415 \text{ vehicles/hr}}$$

(c) Main Street South:

The 20-ft lane must again be treated as two 10-ft lanes. Treat them as one lane group. Adjustment factors are the same as for Main Street North, except that $f_{LT} = 1$ and $f_{RT} = 0.985$.

$$s = (1800)(2)(0.93)(0.975)(0.985) = 3215 \text{ vehicles/hr}$$

$$c = (3215)\left(\frac{27}{60}\right) = 1447 \text{ vehicles/hr}$$

Solve for X as in (a) to get $X = 1.060$.

$$v = (1447)(1.060) = 1534 \text{ vehicles/hr}$$

$$V = (1534)(0.85) = \boxed{1304 \text{ vehicles/hr}}$$

7. At least one of the following warrants is required to justify signaling for this intersection (see p. 16-30). Only brief outlines are given. For more detail, see *Manual on Uniform Traffic Control Devices.*

Assume one lane in each of four directions.

- High Traffic Volume Warrant:

  > 500 vehicles/hr on main street
  and
  > 150 vehicles/hr on minor street

- Interruption of Traffic Warrant:

  > 750 vehicles/hr on main street
  and
  > 75 vehicles/hr on minor street

- Crosswalk Usage Warrant: (Assume there is no median strip.)

  > 600 vehicles/hr (both streets) through intersection
  and
  > 150 pedestrians/hr (highest crosswalk usage)

- School Crossing Warrant:

  Depends on pedestrian and vehicle volumes. Requires a custom study.

- Platooning Warrant:

  If signals are currently widely spaced, do a speed study to see if vehicle speeds are excessive.

  Note: Signals should not be closer than 1000 ft.

- Accidents Warrant:

  five or more accidents per year, each with more than \$100 in damage
  and
  80% of volumes in any of the first three warrants
  and
  less restrictive measures have been tried

- Systems Warrant:

  on a major route
  and
  need exists to concentrate or regulate
  and
  > 800 vehicles/hr on a typical weekend

- Combination Warrant:

  Two or more of the first three warrants are 80% satisfied.

Traffic-actuated controllers should be considered when

- there are low, fluctuating, or unbalanced traffic volumes
- there are high side street flows and delays during peak hours only
- only the accident or crosswalk warrant is satisfied
- only one direction of the two-way traffic is to be controlled

Timed

1. (a) Headway is 5 min and the station wait is 1 min. The train has 4 min to accelerate, travel, and decelerate.

$$4 \text{ min} = 240 \text{ sec}$$

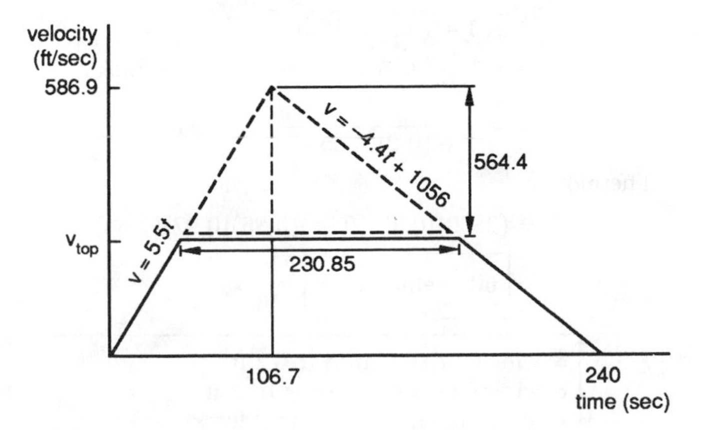

Find v_{top}. (This is not the maximum possible speed of the train.)

- Acceleration line:

$$v = 5.5t$$

- Deceleration line:

$$v = -4.4t + b$$
$$0 = (-4.4)(240) + b$$
$$b = 1056 \text{ ft/sec}$$

The intersection is at

$$5.5t = -4.4t + 1056$$
$$t = 106.7 \text{ sec}$$

Velocity at $t = 106.7$ is

$$v = (5.5)(106.7) = 586.9 \text{ ft/sec}$$

The triangle's aspect ratio is

$$\frac{586.9}{240} = 2.445$$

The area under the curve is the distance traveled. The area under the entire curve is

$$\left(\tfrac{1}{2}\right)(240)(586.9) = 70{,}428 \text{ ft}$$

The area under the dotted triangle = 1 mi or 5280 ft. The area under the bottom section is

$$70{,}428 - 5280 = 65{,}148 \text{ ft}$$

By similar triangles,

$$\frac{\text{area}}{\text{height}} = \frac{70{,}428}{586.9} = \frac{65{,}148}{x}$$
$$x = 542.9 \text{ ft}$$
$$65{,}148 = \tfrac{1}{2}bh = \left(\tfrac{1}{2}\right)(b)(2.445b)$$
$$b = \sqrt{\frac{(2)(65{,}148)}{2.445}} = 230.85 \text{ sec}$$
$$h = (2.445)(230.85) = 564.4 \text{ ft/sec}$$
$$v_{top} = 586.9 - 564.4$$
$$= \boxed{22.5 \text{ ft/sec (15.3 mph)}}$$

(b)
$$\left(\frac{60 \text{ trains}}{5 \text{ hr}}\right)(5 \text{ cars})\left(220 \ \frac{\text{passengers}}{\text{car}}\right)$$
$$= \boxed{13{,}200 \text{ passengers/hr}}$$

(c)
$$\frac{(1 \text{ mi})\left(60 \ \frac{\text{min}}{\text{hr}}\right)}{4 \text{ min}} = \boxed{15 \text{ mph}} \quad \text{[time mean speed]}$$

(d) $\dfrac{(1 \text{ mi})\left(60 \dfrac{\text{min}}{\text{hr}}\right)}{5 \text{ min}} = \boxed{12 \text{ mph}}$ [space mean speed]

(e) The basis is time spent by a car in the system. Nevertheless, the term *space mean speed* is used.

(f) Nothing would be saved since the maximum speed is never reached. Only an increase in acceleration or deceleration would benefit the system, and even then only if the headway were changed. If the headway is not changed, no mechanical improvements will benefit the system.

2. (a)

$$\text{A: } \frac{(4)(1{,}000{,}000)}{(820)(365)} = 13.4 \text{ accidents/mil vehicles}$$

$$\text{B: } \frac{(5)(1{,}000{,}000)}{(1200)(365)} = 11.4 \text{ accidents/mil vehicles}$$

$$\text{C: } \frac{(7)(1{,}000{,}000)}{(1070)(365)} = 17.9 \text{ accidents/mil vehicles}$$

$$\text{D: } \frac{(6)(1{,}000{,}000)}{(1400)(365)} = 11.7 \text{ accidents/mil vehicles}$$

- Ranking from worst to best:

C, A, D, B

(b)

$$1: \frac{(1)(1{,}000{,}000)}{(1900)(365)(1.5)} = 0.96 \text{ accidents/mi-yr}$$

$$2: \frac{(14)(1{,}000{,}000)}{(2000)(365)(1.35)} = 14.2 \text{ accidents/mi-yr}$$

$$3: \frac{(18)(1{,}000{,}000)}{(5500)(365)(4.5)} = 1.99 \text{ accidents/mi-yr}$$

$$4: \frac{(11)(1{,}000{,}000)}{(3000)(365)(0.53)} = 18.95 \text{ accidents/mi-yr}$$

$$5: \frac{(30)(1{,}000{,}000)}{(4000)(365)(2.48)} = 8.29 \text{ accidents/mi-yr}$$

- Ranking from worst to best:

4, 2, 5, 3, 1

(c) Calculations are performed in order to rank highway improvement priorities. However, this assumes that all locations produce accidents of equal costs. Actually, some locations may produce a higher fraction of fatalities than others.

(d) Weight by cost of accident.

3. The notation used comes from Chap. 8 of the 1985 *Highway Capacity Manual.* Operational analysis for specific grades is followed.

(a) At level of service D, speed = 40 mph (Table 8-2).

$$\frac{\text{v}}{c} = 1.00 \quad \text{[Table 8-7, with 40\% no-passing zones]}$$

$$f_d = 0.78 \quad \text{[Table 8-8]}$$

$$f_w = 0.85 \quad \text{[Table 8-5, using average shoulder width of 4 ft]}$$

$$P_p = 0.8$$

$$P_{T/\text{HV}} = \frac{0.05}{0.05 + 0.15} = 0.25$$

$$E = 3.7 \text{ and } E_o = 1.3 \quad \text{[Table 8-9]}$$

$$I_p = (0.02)(E - E_o) = 0.048$$

$$f_g = \frac{1}{1 + P_p I_p} = \frac{1}{1 + (0.8)(0.048)} = 0.96$$

$$E_{\text{HV}} = 1 + (0.25 + P_{T/\text{HV}})(E - 1) = 1 + (0.25 + 0.25)(2.7) = 2.35$$

$$f_{\text{HV}} = \frac{1}{1 + (1 - P_p)(E_{\text{HV}} - 1)} = \frac{1}{1 + (0.2)(1.35)} = 0.79$$

The maximum hourly service flow (volume) is

$$\text{SF} = (2800)\left(\frac{\text{v}}{c}\right) f_d f_w f_g f_{\text{HV}} = (2800)(1.00)(0.78)(0.85)(0.96)(0.79) = \boxed{1408 \text{ vehicles/hr}}$$

(b) At level of service B, speed = 50 mph (Table 8-2). Factors that change are

$$\frac{\text{v}}{c} = 0.55 \quad \text{[Table 8-7]}$$

$$E = 6.7 \text{ and } E_o = 1.6 \quad \text{[Table 8-9]}$$

$$I_p = (0.02)(E - E_o) = 0.102$$

$$f_g = \frac{1}{1 + (0.8)(0.102)} = 0.92$$

$$E_{\text{HV}} = 1 + (0.25 + 0.25)(6.7 - 1) = 3.85$$

$$f_{\text{HV}} = \frac{1}{1 + (0.2)(3.85 - 1)} = 0.64$$

Therefore,

$$\text{SF} = (2800)(0.55)(0.78)(0.85)(0.92)(0.64) = \boxed{601 \text{ vehicles/hr}}$$

(c)
- widening the lanes to 12 ft
- widening the shoulder to 6 ft
- allowing use of the shoulder
- realigning roadway to improve sight distance
- adding a passing lane or a truck climbing lane

(d) On the worksheet for specific grades (p. 8-33), plot the results from (a) and (b): (speed = 50, SF = 601) and (speed = 40, SF = 1408). A line through those points intersects the capacity line where

$$\text{SF} = \boxed{1755}$$

4. 25 mph = terminal speed at time of accident (not original speed).

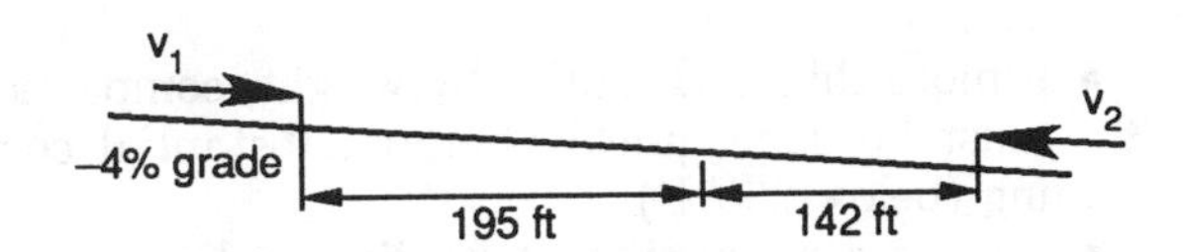

(a) Use Eq. 16.15.

- Impala:

If there had been no collision, the skidding distance would have been

$$d_{v_1} = \frac{v_1^2}{(30)(0.48 - 0.04)} = \frac{v_1^2}{13.2}$$

At 25 mph, the skidding distance would be

$$d_{25,1} = \frac{(25)^2}{13.2} = 47.3 \text{ ft}$$

The actual skid was 195 ft, so

$$195 = d_{v_1} - 47.35$$
$$d_{v_1} = 242.35 \text{ ft}$$

Then,

$$242.35 = \frac{v_1^2}{13.2}$$
$$v_1 = \boxed{56.6 \text{ mph}}$$

- Honda:

$$d_{v_2} = \frac{v_2^2}{(30)(0.48 + 0.04)} = \frac{v_2^2}{15.6}$$
$$d_{25,2} = \frac{(25)^2}{15.6} = 40.06 \text{ ft}$$
$$d_{v_2} = 142 + 40.06 = 182.06$$
$$182.06 = \frac{v_2^2}{15.6}$$
$$v_2 = \boxed{53.3 \text{ mph}}$$

(b) The speed of impact and coefficient of friction are the only two variables. Vary each by 10% and see how velocities change.

If the speed at impact is $(1 + 0.10)(25) = 27.5$ mph,

$$v_1 = 57.7 \quad \text{[1.6\% change]}$$

If the coefficient of friction is $(1 + 0.10)(0.48) = 0.528$,

$$v_1 = 59.0 \quad \text{[4.2\% change]}$$

$$\boxed{\text{The coefficient of friction is more sensitive to error.}}$$

5. At 35 mph, stopping sight distance is given by Eq. 16.13. Assume the worst case (wet pavement, worn tires). From Table 16.2,

$$f = 0.34 \quad \text{[Table 16.2]}$$

(a) Case 1: Eastbound car sees southbound car and skids to a stop.

If sight is into the intersection only (that is, drivers cannot see each other until the other is into the intersection), add 2.3 sec (the time to accelerate 10 ft from a stop into the intersection).

$$S = (1.47)(1 + 2.3)(35) + \frac{(35)^2}{(30)(0.34)}$$
$$= 289.9 \text{ ft}$$

This value corresponds to having a driver at the limit point who sees no cars coming and then, 2.3 sec later, sees a car.

Case 2: Southbound car sees eastbound car (whose driver is not watching) and accelerates across the intersection.

To travel the 60 + 20 ft will take 8.5 sec (20 ft is the car length). The car will have traveled 10 ft by the time it reaches the intersection. Jeopardy time is

$$t = 8.5 - 2.3 = 6.2 \text{ sec}$$

The distance traveled by the eastbound car at 35 mph is

$$(1.47)(6.2)(35) \approx 319$$

Since 319 ft > 289.9 ft,

$$L = \boxed{319 \text{ ft}}$$

(This is actually low. The Department of Defense recommends 470 ft at 35 mph.)

(b) Allow 20 ft for the first space and 24 ft for subsequent spots.

$$1 + \frac{319 - 20}{24} = 13.5$$

$$\boxed{\text{14 places}}$$

6. Accident data can be obtained from:
   - police records (property damage only)
   - municipal records
   - commercial insurance records
   - state disability funds
   - expected remaining earnings estimates
   - accepted federal standards (See Table 16.29)

7. Reasons for using:
   - is a cheaper material in some cases
   - has a higher strength than asphalt for same thickness
   - uses a by-product of desulfurized oil
   - reduces dependence on imported oil
   - is highly durable

Disadvantages:

- sulfur is shipped molten and must be heated to be removed from tanks
- particulate fumes may be irritating (but are not poisonous)
- has a distinct odor during production
- requires mixing apparatus (colloidal mill)
- is combustible (same as asphalt)

Special installation steps:

- eye goggles may be required to prevent irritation
- if more than 25% sulfur by weight, compaction must be done quickly before substantial cooling (below 250°F)
- there are no significant hauling and placement differences

Special repair step:

- cannot use a cold patch of sulfur, so a patch of different composition is required

SURVEYING

Untimed

1.

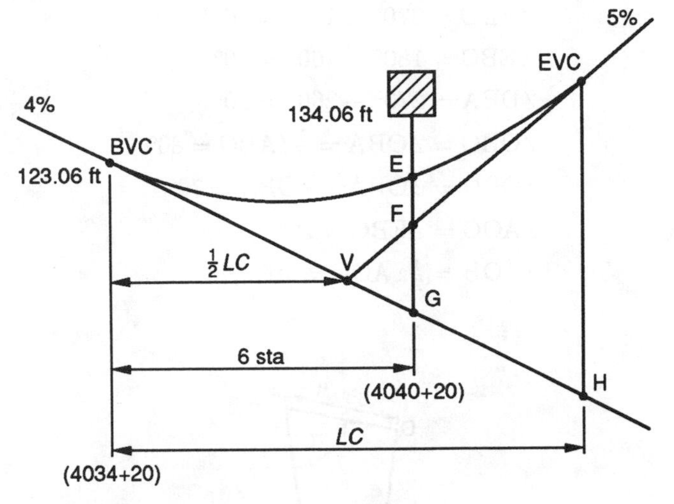

Use a modification of the procedure on p. 17-16.

step 1: The elevation at point E is

$$134.06 - 15 = 119.06 \text{ ft}$$

step 2: The elevation at point G is

$$123.06 - (4)(4040.2 - 4034.2) = 99.06 \text{ ft}$$

$$\text{distance EG} = 119.06 - 99.06 = 20 \text{ ft}$$

step 3: ∠FVG is a 9% divergence.

$$\begin{aligned}\text{distance EVC-H} &= (9)\left(\tfrac{1}{2}\right)(LC)\\ &= 4.5(LC)\end{aligned}$$

step 4:

$$\frac{\text{EG}}{(6)^2} = \frac{\text{EVC-H}}{(LC)^2}$$

$$\frac{20}{(6)^2} = \frac{4.5(LC)}{(LC)^2}$$

$$LC = \boxed{8.1 \text{ sta}}$$

To check, use the procedure on p. 17-16 exactly as written.

V will be at

$$(\text{sta } 4034{+}20) + (\text{sta } 4{+}05) = \text{sta } 4038{+}25$$

The elevation at V is

$$123.06 - (4)(4.05) = 106.86 \text{ ft}$$

From Fig. 17.19,

$$\begin{aligned}d &= (\text{sta } 4040{+}20) - (\text{sta } 4038{+}25)\\ &= \text{sta } 1{+}95\end{aligned}$$

step 3: $\text{EF} = 119.06 - 106.86 - (1.95)(5) = 2.45 \text{ ft}$

step 4:

$$\frac{20}{\left(\frac{8.1}{2} + 1.95\right)^2} = \frac{2.45}{\left(\frac{8.1}{2} - 1.95\right)^2}$$

$$0.555 = 0.555 \quad \text{[check]}$$

(Equation 17.46 could also be used to solve this directly.)

2. From Eq. 17.45,

$$\text{(a) } r_1 = \frac{-2.5 - 5}{3} = -2.5\%/\text{sta}$$

From Eq. 17.47,

$$x = \frac{-5}{-2.5} = 2 \text{ sta}$$

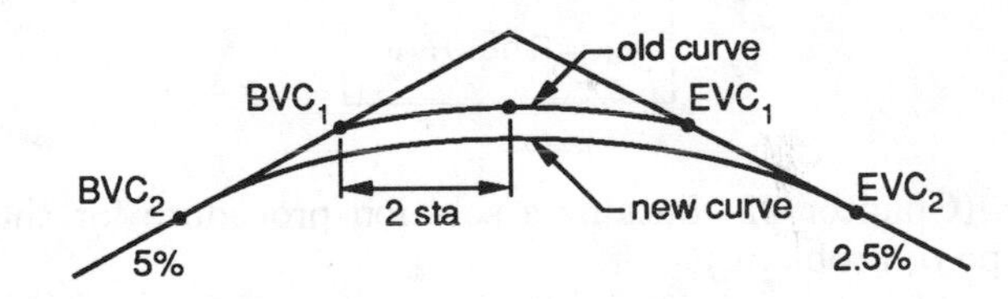

From Eq. 17.46,

$$173.07 = \left(\frac{-2.5}{2}\right)(2)^2 + (5)(2) + \text{elevation BVC}_1$$

$$\text{elevation BVC}_1 = 168.07 \text{ ft}$$

The equation of curve 1 is

$$y_1 = (-1.25)x^2 + 5x + 168.07 \text{ ft}$$

The elevation at the midpoint of curve 1 is

$$\begin{aligned}y &= (-1.25)(1.5)^2 + (5)(1.5) + 168.07\\ &= 172.76 \text{ ft}\end{aligned}$$

BVC_1 and BVC_2 are separated by 150 ft (1.5 sta). The elevation at BVC_2 is

$$168.07 - (5)(1.5) = 160.57 \text{ ft}$$

From Eq. 17.45,

$$r_2 = \frac{-2.5 - 5}{6} = -1.25\%/\text{sta}$$

From Eq. 17.46,

$$y_2 = (-0.625)x^2 + 5x + 160.57$$

The elevation at the midpoint of curve 2 is

$$\begin{aligned} y_2 &= (-0.625)(3)^2 + (5)(3) + 160.57 \\ &= 169.95 \text{ ft} \end{aligned}$$

The midpoint cut is

$$172.76 - 169.95 = \boxed{2.81 \text{ ft}}$$

(b) Measuring x from BVC_2,

$$\boxed{\begin{aligned} y_0 &= 160.57 \text{ ft} \\ y_1 &= 164.95 \text{ ft} \\ y_2 &= 168.07 \text{ ft} \\ y_3 &= 169.95 \text{ ft} \\ y_4 &= 170.57 \text{ ft} \\ y_5 &= 169.95 \text{ ft} \\ y_6 &= 168.07 \text{ ft} \end{aligned}}$$

3. (Chapter 22 contains a solution procedure for this type of problem.)

Refer to the following figure. Due to the long radius, the drawing is not to scale. Bearings are clockwise from the north.

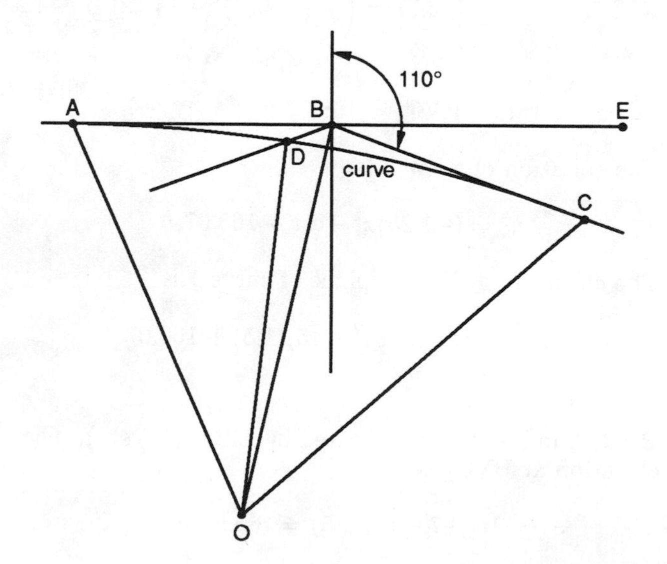

Lines OA, OD, and OC are all radii. Lines AB and OA are perpendicular. (They do not look perpendicular, however, because the location of point O is not drawn to scale.) Similarly, lines CB and OC are also perpendicular. The following angles are needed.

$$\begin{aligned} \angle\text{ABC} &= 270° - 110° = 160° \\ \angle\text{EBC} &= 180° - 160° = 20° \\ \angle\text{DBA} &= 270° - 260° = 10° \\ \angle\text{OBC} &= \angle\text{OBA} = \tfrac{1}{2}\angle\text{ABC} = 80° \\ \angle\text{OBD} &= \angle\text{OBA} - \angle\text{DBA} = 70° \\ \angle\text{AOC} &= \angle\text{EBC} = 20° \\ \angle\text{AOB} &= \tfrac{1}{2}\angle\text{AOC} = 10° \end{aligned}$$

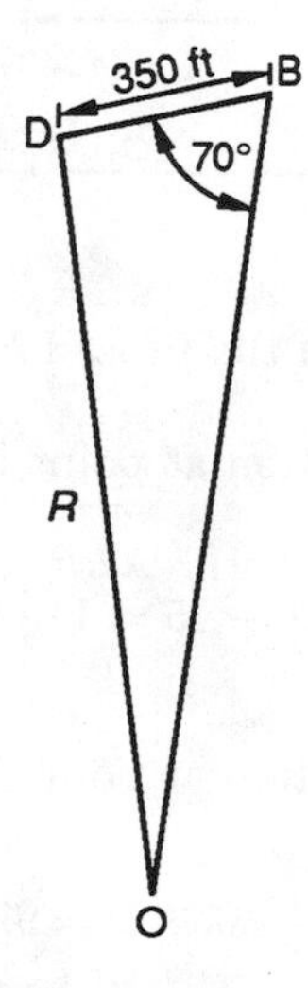

For the triangle ODB from the sine rule,

$$\begin{aligned} \frac{\sin 70°}{R} &= \frac{\sin \angle\text{BDO}}{\text{OB}} \\ \text{OB} &= R\sec\angle\text{AOB} \\ \frac{\sin 70°}{R} &= \frac{\sin \angle\text{BDO}}{R\sec 10°} = \frac{\sin \angle\text{BDO}}{R}(\cos 10°) \\ \sin \angle\text{BDO} &= 0.954 \\ \angle\text{BDO} &= 72.59° \text{ or } 107.41° \end{aligned}$$

Either $\angle\text{DOB} = 180° - 70° - 72.59° = 37.41°$ or $\angle\text{DOB} = 180° - 70° - 107.41° = 2.59°$.

A scale drawing shows that $\angle\text{DOB}$ is very small, so $\angle\text{DOB} = 2.59°$.

Using the sine rule,

$$\begin{aligned} \frac{\sin 70°}{R} &= \frac{\sin 2.59°}{350} \\ R &= \boxed{7278.2 \text{ ft} \quad [\text{use 7278 ft}]} \end{aligned}$$

From Eq. 17.31,

$$T = (7278)\left(\tan\frac{20°}{2}\right) = \boxed{1283.3 \text{ ft}}$$

From Eq. 17.35,

$$LC = (7278)(20)\left(\frac{2\pi}{360}\right) = \boxed{2540.5 \text{ ft}}$$

From Eq. 17.44,

$$50 = (2)(7278)\left(\sin\frac{\theta}{2}\right)$$

$$\theta = 0.3936°$$

$$\text{deflection angle} = \tfrac{1}{2}\theta = \boxed{0.1968°}$$

4. This is essentially a horizontal curve.

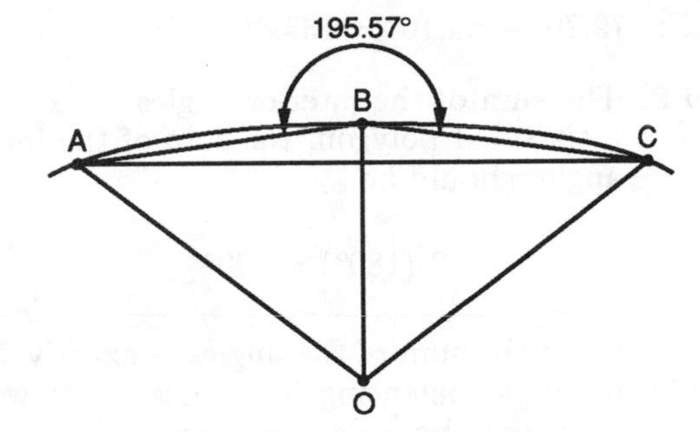

(a) The 195.57° angle is shown above. The interior angle is 360° − 195.57° = 164.43°. The distances AB and BC can be found from the stadia readings (see p. 17-4). From Eq. 17.7,

$$AB = (100)(4.851 - 2.403) + 0 = 244.8 \text{ ft}$$

$$BC = (100)(7.236 - 4.320) = 291.6 \text{ ft}$$

Use the law of cosines to calculate distance AC.

$$(AC)^2 = (244.8)^2 + (291.6)^2 - (2)(244.8)(291.6)(\cos 164.43°)$$

$$AC = 531.5 \text{ ft}$$

Angles BAC and BCA can be found from the law of sines.

$$\frac{\sin(\angle BAC)}{291.6} = \frac{\sin(\angle BCA)}{244.8} = \frac{\sin 164.43°}{531.5}$$

$$\angle BAC = 8.47°$$

$$\angle BCA = 7.10°$$

$$DC = \left(\tfrac{1}{2}\right)(291.6°) = 145.8°$$

$$\angle BOC = 2\angle BAC$$

From the second theorem on p. 17-14,

$$\angle DOC = \tfrac{1}{2}\angle BOC = \angle BAC$$

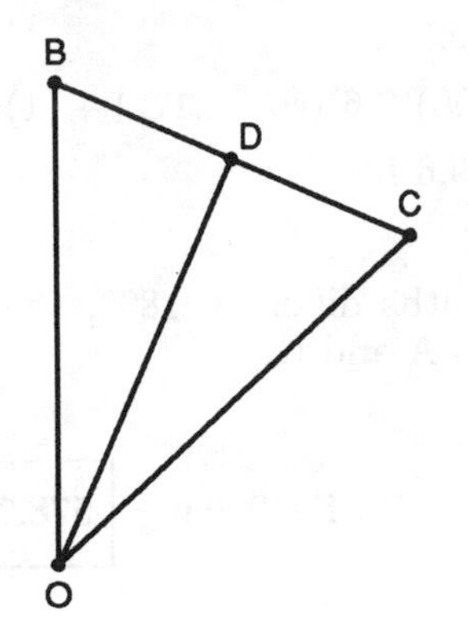

$$\angle DOC = \angle BAC = 8.47°$$

$$R = OC = \frac{145.8}{\sin 8.47°} = \boxed{989.9 \text{ ft}}$$

(b)
$$\text{grade AB} = \frac{3.627 - 4.7}{244.8} = \boxed{-0.00438 \text{ ft/ft}}$$

$$\text{grade BC} = \frac{4.7 - 5.778}{291.6} = \boxed{-0.00370 \text{ ft/ft}}$$

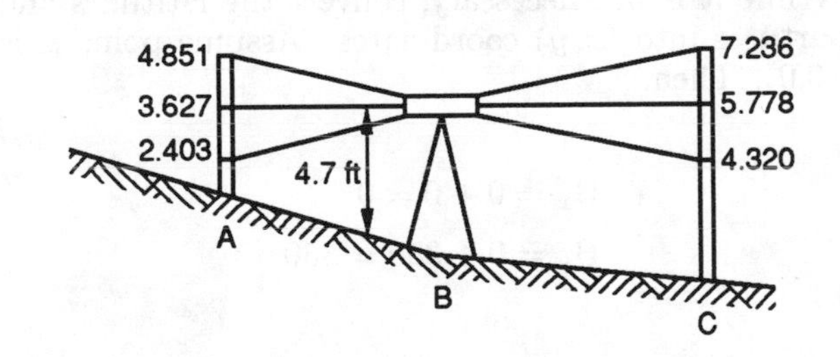

5. Proceed as in Ex. 17.5.

$$(0.03)^2 + (0.01)^2 + (0.09)^2 + (0.05)^2 = 0.0116$$

The weights to be applied are

$$\frac{0.0116}{(0.03)^2} = 12.89$$

$$\frac{0.0116}{(0.01)^2} = 116.0$$

$$\frac{0.0116}{(0.09)^2} = 1.43$$

$$\frac{0.0116}{(0.05)^2} = 4.64$$

The most probable elevation is

$$745 + \frac{(12.89)(0.08) + (116)(0.22) + (1.43)(0.45) + (4.64)(0.17)}{12.89 + 116 + 1.43 + 4.64} = \boxed{745.21 \text{ ft}}$$

6. (a) From Eq. 17.8,

$$H_{\text{scope-A}} = (100)(3.22)[\cos^2(-6.3^\circ)] + (1)[\cos(-6.3^\circ)] = 319.1 \text{ ft}$$

$$H_{\text{scope-B}} = (100)(2.6)(\cos^2 4.17^\circ) + (1)(\cos 4.17^\circ) = 259.6 \text{ ft}$$

Since the azimuths differ by 180°, the telescope is in line with points A and B.

$$\text{AB} = 319.1 + 259.6 = \boxed{578.7 \text{ ft}}$$

(b) From Eq. 17.9,

$$V_{\text{scope-B}} = \left(\tfrac{1}{2}\right)(100)(2.60)[\sin(2)(4.17^\circ)] + (1)(\sin 4.17^\circ) = 18.9 \text{ ft}$$

$$\text{elevation B} = 297.8 + 18.9 + 4.8 - 10.9 = \boxed{310.6 \text{ ft}}$$

7. While it is not necessary, convert the latitudes and departures into (x,y) coordinates. Assume point A is at (0,0). Then,

$$B_x = 0 + 0 = 0$$

$$B_y = 0 + 350 = 350$$

point	(x,y)
A	(0,0)
B	(0,350)
C	(600,900)
D	(1800,650)
E	(2000,−100)
F	(900,−650)

Refer to p. 17-11.

$$\frac{0}{0} \times \frac{0}{350} \times \frac{600}{900} \times \frac{1800}{650} \times \frac{2000}{-100} \times \frac{900}{-650} \times \frac{0}{0}$$

$$\text{full line products} = -1{,}090{,}000$$

$$\text{dotted line products} = 3.04 \times 10^6$$

$$\text{area} = \tfrac{1}{2}|-1{,}090{,}000 - 3{,}040{,}000| = \boxed{2{,}065{,}000 \text{ ft}^2}$$

8. Proceed as on p. 17-8.

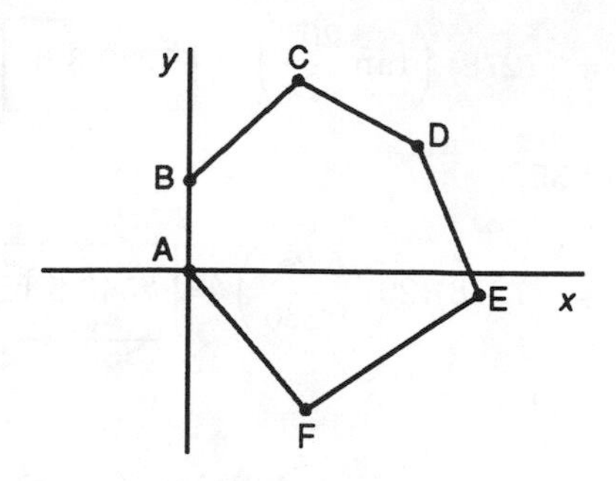

step 1: The interior angles are

A: $90^\circ - 54.10^\circ + 90^\circ = 125.90^\circ$

B: $180^\circ - 45^\circ = 135^\circ$

C: $45^\circ + 69.45^\circ = 114.45^\circ$

D: $90^\circ - 69.45^\circ + 90^\circ + 11.32^\circ = 121.87^\circ$

E: $90^\circ - 11.32^\circ + 90^\circ - 79.70^\circ = 88.98^\circ$

F: $79.70^\circ + 54.10^\circ = 133.8^\circ$

step 2: The sum of the interior angles is 720°. For a six-sided polygon, the sum of the interior angles should be

$$(6-2)(180^\circ) = 720^\circ$$

Since the sum of the angles is exactly 720°, no angle balancing is necessary. However, there may be a closure error.

For leg AB,

$$\text{latitude} = 500 \text{ ft}$$

$$\text{departure} = 0$$

For leg BC,

$$\text{latitude} = (848.6)(\sin 45^\circ) = 600 \text{ ft}$$

$$\text{departure} = (848.6)(\cos 45^\circ) = 600 \text{ ft}$$

Similarly, the following table is prepared (rounding to the nearest 0.1 ft).

leg	latitude	departure
AB	+500.0	0
BC	+600.0	+600.0
CD	−299.9	+800.0
DE	−1000.0	+200.2
EF	−199.9	−1100.0
FA	+385.1	−532.0
totals	−14.7 ft	−31.8 ft

The total perimeter length is 4997.6 ft (round to 5000 ft).

The latitude correction to line AB is

$$\left(\frac{500}{5000}\right)(14.7) = 1.47 \text{ ft}$$

[round to nearest 0.1 to get 1.5 ft]

The departure correction to line AB is

$$\left(\frac{500}{5000}\right)(31.8) = 3.18 \text{ ft} \quad \text{[round to 3.2 ft]}$$

The corrected latitude and departure of leg AB is

latitude: $500 + 1.5 = 501.5$ ft

departure: $0 + 3.2 = 3.2$ ft

Similarly, the following table is prepared.

leg	latitude	departure
AB	501.5	3.2
BC	602.5	605.4
CD	−297.4	805.4
DE	−997.0	206.7
EF	−196.6	−1092.9
FA	387.0	−527.8
totals	0.0	0.0

Unfortunately, the angles and lengths calculated from these latitudes and departures will not be the same as the given angles and lengths. Therefore, corrected angles and lengths must be calculated.

For leg AB, the corrected angle is

$$\arctan\left(\frac{3.2}{501.5}\right) = 0.37°$$

The corrected bearing is N0.37°E.

The corrected length is

$$\sqrt{(3.2)^2 + (501.5)^2} = 501.5 \text{ ft}$$

The following table is similarly prepared.

	corrected	
leg	bearing	length
AB	N0.37°E	501.5
BC	N45.14°E	854.1
CD	S69.73°E	858.6
DE	S11.71°E	1018.2
EF	S79.80°W	1110.4
FA	N53.75°W	654.5

9.
$$\begin{aligned}\text{elevation BM11} &= 179.65 + \sum(BS) - \sum(FS)\\ &= 179.65 + 4.64 + 5.80 + 2.25\\ &\quad - (5.06 + 5.02 + 5.85)\\ &= \boxed{176.41 \text{ ft}}\end{aligned}$$

$$\begin{aligned}\text{elevation BM12} &= 179.65 + 57.5 - 31.2\\ &= \boxed{205.95 \text{ ft}}\end{aligned}$$

10. The plot of the available data shows two possible locations for point E. However, only one is essentially easterly. Point E′ is ruled out since E′-A is not easterly.

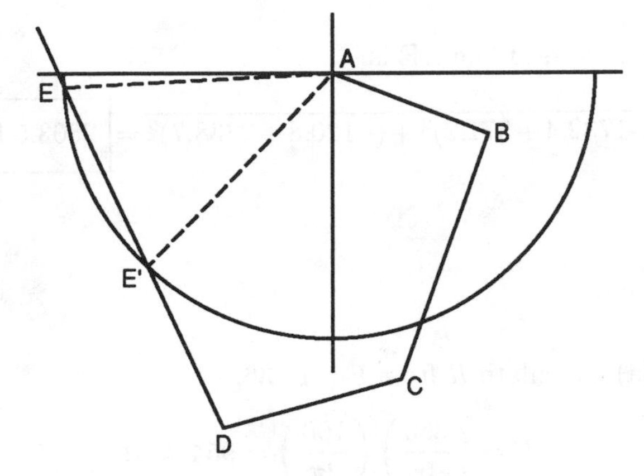

The latitudes and departures of each point are

point	latitude	departure
A	0	0
B	−302.0	1038.3
C	−1533.7	−417.6
D	−503.0	−1191.9
totals	−2338.7	−571.2

The slope of line DE is

$$-\tan(332.37° - 270°) = -1.91$$

The coordinates of point D are (−571.2, −2338.7). The equation of line DE is

$$\begin{aligned}y &= mx + b\\ -2338.7 &= (-1.91)(-571.2) + b\\ b &= -3429.7 \text{ ft}\\ y &= -1.91x - -3429.7\end{aligned}$$

The equation of a circle centered at point A with a radius of 1737.9 ft is

$$x^2 + y^2 = (1737.9)^2$$

Since point E is on both the line and the circle,

$$\begin{aligned}x^2 + (-1.91x - 3429.7)^2 &= (1737.9)^2\\ x^2 + 3.65x^2 + 13{,}101.5x + (3429.7)^2 &= (1737.9)^2\\ x^2 + 2817.5x &= -1.88 \times 10^6\end{aligned}$$

This has solutions $x = -1085.2$ and -1732.4. The larger (negative) x is -1732.4.

From the line equation,

$$y = -(1.91)(-1732.4) - 3429.7 = -120.8 \text{ ft}$$

The angle of line EA is

$$\arctan\left(\frac{120.8}{1732.4}\right) \approx 4.0^\circ$$

The north azimuth of line EA is $90^\circ - 4.0^\circ = 86^\circ$.

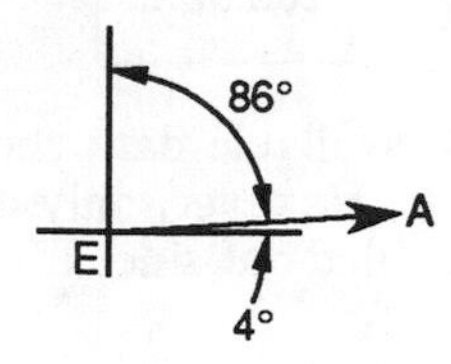

The length of line DE is

$$\sqrt{(-1732.4+571.2)^2 + (-120.8+2338.7)^2} = \boxed{2503.5 \text{ ft}}$$

Timed

1. (a) Calculate R from Eq. 17.36.

$$R = \left(\frac{360}{6}\right)\left(\frac{100}{2\pi}\right) = 954.93 \text{ ft}$$

Calculate the tangent length from Eq. 17.31.

$$T = (954.93)\left(\tan\frac{65^\circ}{2}\right) = 608.36 \text{ ft}$$

$$\begin{aligned} \text{location of PC} &= \text{PVI-T} \\ &= (88+00) - (6+8.36) \\ &= \boxed{\text{sta } 81+91.64} \end{aligned}$$

(b) The tangent length decreases.

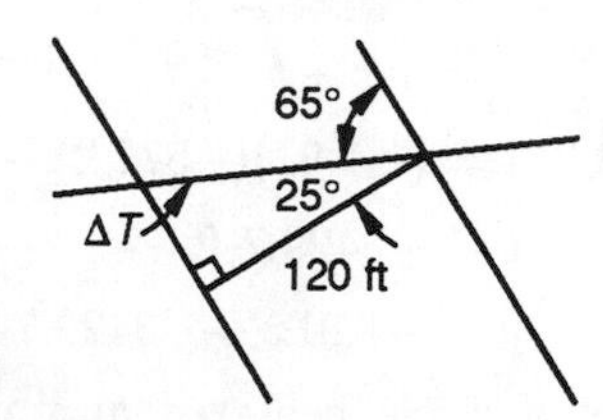

$$\Delta T = \frac{120}{\cos 25^\circ} = 132.41 \text{ ft}$$

The new tangent is $608.36 - 132.41 = 475.95$ ft. From Eq. 17.31,

$$R_{\text{new}} = \frac{475.95}{\tan\left(\frac{65^\circ}{2}\right)} = \boxed{747.09 \text{ ft}}$$

(c) The length of the curve from Eq. 17.35 is

$$LC = (747.09)(65)\left(\frac{2\pi}{360}\right) = 847.55 \text{ ft}$$

PT is located at

$$\begin{aligned} \text{sta PC} + LC &= (81+91.64) + (8+47.55) \\ &= \boxed{\text{sta } 90+39.19} \end{aligned}$$

2. Define the curve mathematically.

$$\begin{aligned} LC &= 94 - 82 = 12 \text{ sta} \\ \text{elevation}_{\text{BVC}} &= 729 + (88 - 82)(4) \\ &= 729 + 24 = 753 \text{ ft} \end{aligned}$$

From Eq. 17.45, the rate of grade per station is

$$r = \frac{g_2 - g_1}{LC} = \frac{3 - (-4)}{12} = \frac{7}{12}\%/\text{sta}$$

From Eq. 17.46,

$$y = \left(\frac{7}{24}\right)x^2 - 4x + 753$$

The points in question have an elevation of $770 - 25 = 745$ ft. Solve for x.

$$745 = \left(\frac{7}{24}\right)x^2 - 4x + 753$$

$$x = 11.29 \text{ sta}, 2.43 \text{ sta}$$

(a) $82+2.43 = \boxed{\text{sta } 84+43}$

(b) $82+11.29 = \boxed{\text{sta } 93+29}$

(c) From Eq. 17.47, the turning point is at

$$x = \frac{-(-4)}{\left(\frac{7}{12}\right)} = 6.86 \text{ ft}$$

$$\begin{aligned} \text{location} &= (82+00) + (6+86) \\ &= \boxed{\text{sta } 88+86} \end{aligned}$$

(d) Putting x back into the equation for y,

$$\begin{aligned} y &= \left(\frac{7}{24}\right)(6.86)^2 - (4)(6.86) + 753 \\ &= \boxed{739.29 \text{ ft}} \end{aligned}$$

3.

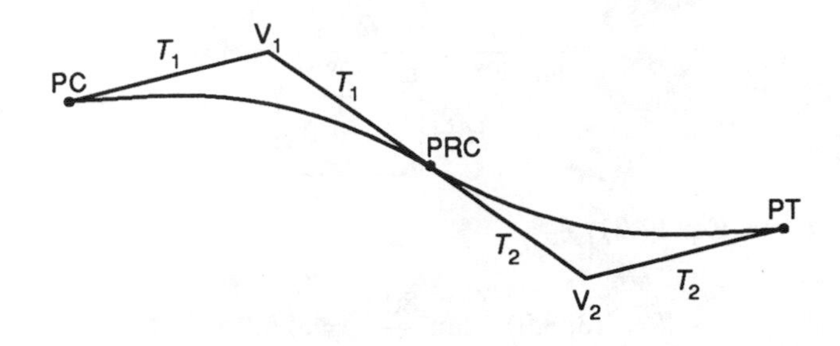

(a) Since the two intersection angles are not the same, the tangents are not the same.

$$T_1 + T_2 = 2100 \text{ ft}$$

From Eq. 17.31,

$$T_1 = R \tan\left(\frac{42^\circ}{2}\right) = 0.384R$$

$$T_2 = R \tan\left(\frac{30^\circ}{2}\right) = 0.268R$$

$$(0.384 + 0.268)R = 2100 \text{ ft}$$

$$R = 3220.9 \text{ ft}$$

$$T_1 = (0.384)(3220.9) = 1236.8 \text{ ft}$$

$$T_2 = (0.268)(3220.9) = 863.2 \text{ ft}$$

From Eq. 17.35,

$$LC_1 = (3220.9)(42)\left(\frac{2\pi}{360}\right) = 2361.04 \text{ ft}$$

$$LC_2 = (3220.9)(30)\left(\frac{2\pi}{360}\right) = 1686.46 \text{ ft}$$

Station of V_1 is 21+25 = 46+00.

$$\text{sta PC} = (46{+}00) - (12{+}36.8) = \boxed{\text{sta } 33{+}63.20}$$

$$\text{sta PRC} = \text{sta PC} + LC_1$$

$$= (33{+}63.20) + (23{+}61.04)$$

$$= \boxed{\text{sta } 57{+}24.24}$$

$$\text{sta PT} = \text{sta PRC} + LC_2$$

$$= (57{+}24.24) + (16{+}86.46)$$

$$= \boxed{\text{sta } 74{+}10.70}$$

(b) The hazards are

- driving off the road due to distraction or inattention
- tipping over (whipping) at high speeds
- difficulty in getting proper superelevation

(c) deflection angle $= \left(\frac{1}{2}\right)\left[\left(\frac{1}{2}\right)(42)\right] = \boxed{10.5^\circ}$

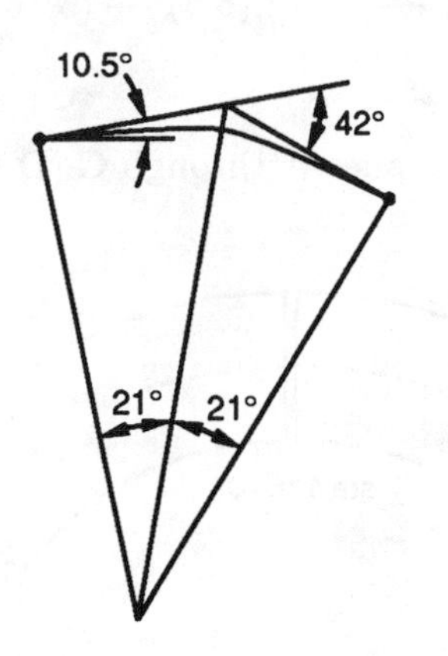

4.

(a) $\text{elevation}_{\text{BVC}} = 1262 - (3)(116 - 110) = 1244 \text{ ft}$

$$LC = 122 - 110 = 12 \text{ sta}$$

From Eq. 17.45,

$$r = \frac{-2 - 3}{12} = -0.417\%/\text{sta}$$

From Eq. 17.46, for the first half-station $x = 0.5$ ft,

$$y = \left(\frac{-0.417}{2}\right)(0.5)^2 + (3)(0.5) + 1244$$

$$= \boxed{1245.5 \text{ ft}}$$

This is repeated 23 times, once for each half-station. For example,

$x = 5.5$,

$$y_{115.5} = \left(\frac{-0.417}{2}\right)(5.5)^2 + (3)(5.5) + 1244$$

$$= 1254.2 \text{ ft}$$

$x = 6.0$,

$$y_{116} = \left(\frac{-0.417}{2}\right)(6)^2 + (3)(6) + 1244$$

$$= 1254.5 \text{ ft}$$

$x = 6.5$,

$$y_{116.5} = \left(\frac{-0.417}{2}\right)(6.5)^2 + (3)(6.5) + 1244$$
$$= 1254.7 \text{ ft}$$

(b) Label the piles A through G. D is the centerline pile.

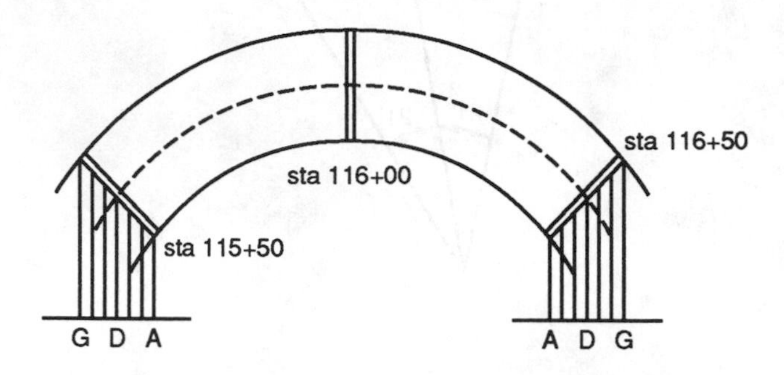

The elevation of the centerline pile (pile D) is found from (a), less 3.5 ft.

This is an 8% superelevation, so for each 10-ft run, the rise is 0.8 ft.

For the first bent at sta 115+50, the top of pile D is at

$$\text{D:} \quad 1254.2 - 3.5 = \boxed{1250.7 \text{ ft}}$$

$$\text{E:} \quad 1250.7 + 0.8 = \boxed{1251.5 \text{ ft}}$$

$$\text{F:} \quad 1250.7 + (2)(0.8) = \boxed{1252.3 \text{ ft}}$$

$$\text{G:} \quad 1250.7 + (3)(0.8) = \boxed{1253.1 \text{ ft}}$$

$$\text{C:} \quad 1250.7 - 0.8 = \boxed{1249.9 \text{ ft}}$$

$$\text{B:} \quad 1250.7 - (2)(0.8) = \boxed{1249.1 \text{ ft}}$$

$$\text{A:} \quad 1250.7 - (3)(0.8) = \boxed{1248.3 \text{ ft}}$$

bent at station	A	B	C	D	E	F	G
115+50	1248.3	1249.1	1249.9	1250.7	1251.5	1252.3	1253.1
116	1248.6	1249.4	1250.2	1251.0	1251.8	1252.6	1253.4
116+50	1248.8	1249.6	1250.4	1251.2	1252.0	1252.8	1253.6

5.

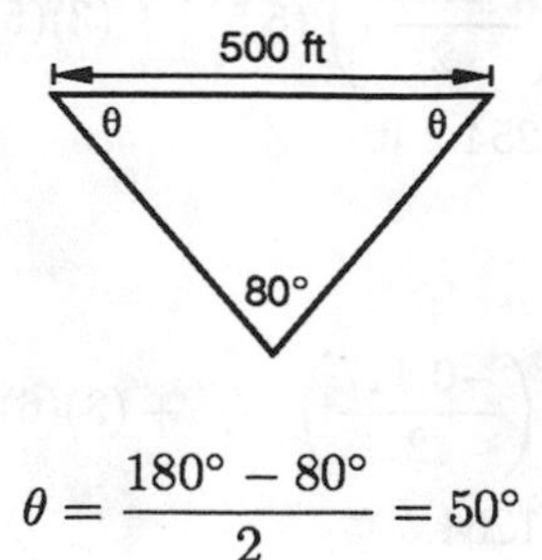

$$\theta = \frac{180° - 80°}{2} = 50°$$

From the sine rule,

$$\frac{500}{\sin 80°} = \frac{R}{\sin 50°}$$
$$R = \boxed{388.9 \text{ ft}}$$

From Eq. 17.31,

$$T = (388.9)\left(\tan\frac{80}{2}\right) = 326.3 \text{ ft}$$
$$T' = T - 10 = 326.3 - 10 = 316.3 \text{ ft}$$

From Eq. 17.31,

$$R' = \frac{316.3}{\tan 40°} = \boxed{377.0 \text{ ft}}$$

6. (b) Use the procedure on p. 17-16.

- Underpass—20 ft clearance:

step 1: $\text{elevation}_E = 510 - 20 = 490 \text{ ft}$

step 2:

$$\text{elevation}_G = \text{elevation}_V - (\text{slope})(x)$$
$$= 482 - (3)(3) = 473 \text{ ft}$$

$$EG = \text{elevation}_E - \text{elevation}_G = 490 - 473 = 17 \text{ ft}$$

step 3: $EF = 490 - 482 - (3)(2) = 2 \text{ ft}$

step 4:

$$\frac{17}{\left(\frac{LC}{2} + 3\right)^2} = \frac{2}{\left(\frac{LC}{2} - 3\right)^2}$$
$$LC = \boxed{12.2 \text{ sta}}$$

(a)

- Overpass—22 ft clearance:

step 1: $\text{elevation}_E = 510 + 22 = 532$

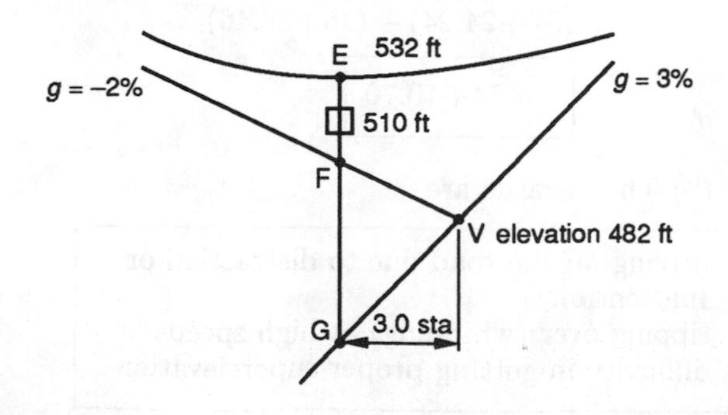

step 2: $EG = 532 - 482 + (3)(3) = 59$

step 3: $EF = 532 - 482 - (3)(2) = 44$

step 4:

$$\frac{59}{\left(\frac{LC}{2}+3\right)^2} = \frac{44}{\left(\frac{LC}{2}-3\right)^2}$$

$$LC = \boxed{82.0 \text{ sta}}$$

7. (b) From Eq. 17.31,

$$T = (1200)\left(\tan\frac{55.5^\circ}{2}\right) = 631.35 \text{ ft}$$

$$LC = (1200)(55.5)\left(\frac{2\pi}{360}\right) = 1162.39 \text{ ft}$$

$$\text{location of PC} = (\text{sta } 182+27.52) - (\text{sta } 6+31.35)$$
$$= \boxed{\text{sta } 175+96.17}$$

$$\text{location of PT} = \text{sta PC} + LC$$
$$= (\text{sta } 175+96.17) + (\text{sta } 11+62.39)$$
$$= \boxed{\text{sta } 187+58.56}$$

(c) From Eq. 17.33,

$$M = (1200)\left(1 - \cos\frac{55.5^\circ}{2}\right)$$
$$= \boxed{138.0 \text{ ft}}$$

(d) From Eq. 17.32,

$$E = (1200)\left(\tan\frac{55.5^\circ}{2}\right)\left(\tan\frac{55.5^\circ}{4}\right)$$
$$= \boxed{155.95 \text{ ft}}$$

(a)

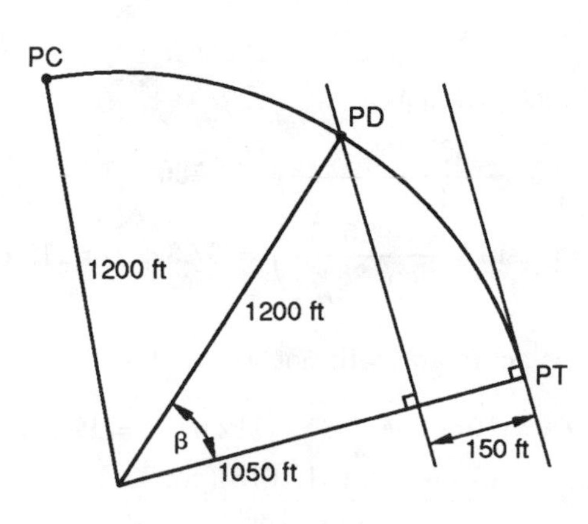

$$\beta = \arccos\left(\frac{1050}{1200}\right) = 28.96^\circ$$

From Eq. 17.35, the curve length from PD to PT is

$$(1200\beta)\left(\frac{2\pi}{360^\circ}\right) = (1200)(28.96^\circ)\left(\frac{2\pi}{360^\circ}\right)$$
$$= 606.54 \text{ ft}$$

$$\text{location}_{\text{PD}} = \text{station}_{\text{PT}} - \text{arc length}$$
$$= 18{,}758.56 - 606.54$$
$$= 18{,}152.02 \text{ sta}$$
$$= \boxed{\text{sta } 181+52.02}$$

8. (d) Use Eqs. 17.45–17.47.

$$r = \frac{2-(-4)}{19} = 0.316\%/\text{sta}$$

$$\text{distance from BVC to V} = \frac{19}{2} = 9.5 \text{ sta}$$

$$\text{elevation}_{\text{BVC}} = 350 + (4)(9.5) = 388 \text{ ft}$$

From Eq. 17.46,

$$y = \left(\frac{0.316}{2}\right)x^2 - 4x + 388$$

Number of stations from BVC to station 106.40 is $106.40 - (105 - 9.5) = 10.9$ stations.

$$y = \left(\frac{0.316}{2}\right)(10.9)^2 - (4)(10.9) + 388 = \boxed{363.17 \text{ ft}}$$

(c) From Eq. 17.47, minimum point occurs at

$$x = \frac{-(-4)}{0.316} = 12.66 \text{ sta}$$

$$y = \left(\frac{0.316}{2}\right)(12.66)^2 - (4)(12.66) + 388$$
$$= \boxed{362.68}$$

(b) The stations of BVC and EVC are

$$105 \pm \frac{19}{2} = \boxed{\text{sta } 95+50, \text{ sta } 114+50}$$

The road must go through the grating, so the low point is at the vertex.

(a) With a symmetrical (equal tangent) parabolic curve, the lowest point will not be at the vertex unless $g_1 = g_2$. Therefore, the curve cannot be equal tangent.

Each tangent of an unequal curve can be thought of as an equal tangent curve with one of the grades equal to zero (i.e., $g = 0\%$).

• BVC to V:

$$g_1 = -4$$
$$g_2 = 0$$

$$(LC)_{\text{BVC-V}} = \text{station}_{\text{V}} - \text{station}_{\text{BVC}} = 105 - \text{station}_{\text{BVC}}$$

From Eq. 17.45,

$$r = \frac{0-(-4)}{(LC)_{\text{BVC-V}}} = \frac{4}{(LC)_{\text{BVC-V}}}$$

The minimum elevation is 385.30 ft at station 105. From Eq. 17.47,

$$x = \frac{-(-4)}{\frac{4}{(LC)_{\text{BVC-V}}}} = (LC)_{\text{BVC-V}}$$

The elevation of BVC is

$$\text{elevation}_{\text{V}} + |g_1|x = 350 + 4(LC)_{\text{BVC-V}}$$

From Eq. 17.46,

$$358.30 = \frac{4(LC)^2}{2(LC)} - 4(LC) + 350 + 4(LC)$$
$$8.30 = 2(LC)$$
$$(LC)_{\text{BVC-V}} = 4.15 \text{ sta}$$
$$\text{location}_{\text{BVC}} = 105 - 4.15 = 100.85 \text{ sta (sta 100+85)}$$
$$\text{elevation}_{\text{BVC}} = 350 + (4)(4.15) = 366.60 \text{ ft}$$

• V to EVC:

(Work with a mirror image to avoid $x = 0$ problems.)

$$g_1 = -2$$
$$g_2 = 0$$
$$(LC)_{\text{V-EVC}} = \text{station}_{\text{V}} - \text{station}_{\text{EVC}} = 105 - \text{station}_{\text{EVC}}$$
$$r = \frac{0-(-2)}{(LC)_{\text{V-EVC}}} = \frac{2}{(LC)_{\text{V-EVC}}}$$
$$x = \frac{-(-2)}{\frac{2}{(LC)_{\text{V-EVC}}}} = (LC)_{\text{V-EVC}}$$
$$\text{elevation}_{\text{EVC}} = 350 + 2(LC)_{\text{V-EVC}}$$

From Eq. 17.46,

$$358.30 = \left[\frac{2}{2(LC)}\right](LC)^2 - 2(LC) + 350 + 2(LC)$$
$$LC = 8.30$$

Working with the actual curve,

$$\text{location}_{\text{EVC}} = 105 + 8.30 = 113.30 \text{ sta (sta 113+30)}$$
$$\text{elevation}_{\text{EVC}} = 350 + (2)(8.30) = 366.6 \text{ ft}$$
$$(LC)_{\text{BVC-EVC}} = 4.15 + 8.30 = \boxed{12.45 \text{ sta}}$$

Note that this is the same answer that you would get using Eqs. 17.45 and 17.47 without recognizing that the curve was not symmetrical. However, the premise would be wrong, and the curve equation $y(x)$ would be incorrect.

9. Follow the procedure given in Chap. 22.

• The road is between the monument and tangent:

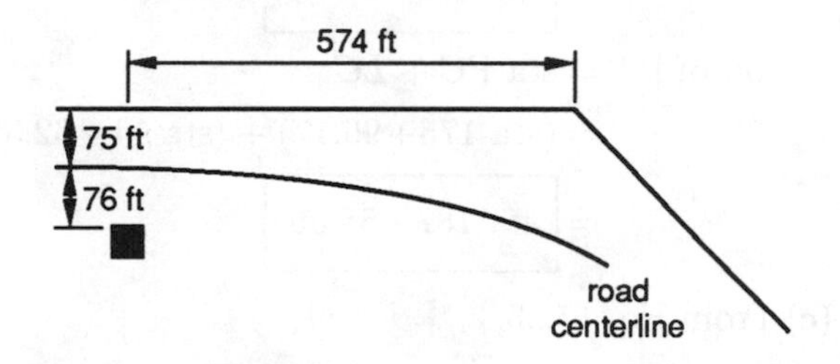

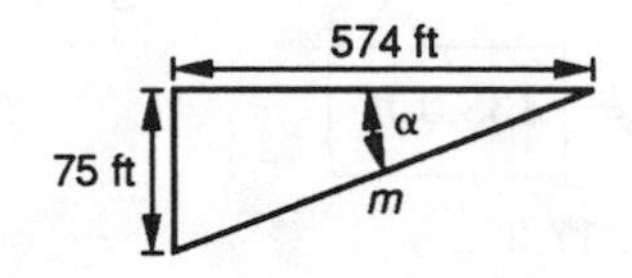

$$\alpha = \arctan\left(\frac{75}{574}\right) = 7.444°$$
$$m = \sqrt{(75)^2 + (574)^2} = 578.88 \text{ ft}$$
$$\Delta = 56° \quad \text{[given]}$$
$$\delta = 90° - \frac{56°}{2} - 7.444° = 54.556°$$
$$\phi = \arcsin\left(\frac{\sin 54.556°}{\cos 28°}\right) = 67.322° \text{ or } 112.678°$$

67.322° is acute and will not work.

$$\theta = 180° - 54.556° - 112.678° = 12.766°$$
$$R = \frac{(578.88)(\sin 112.678)(\cos 28°)}{\sin 12.766°} = \boxed{2134.25 \text{ ft}}$$

- The road is below the monument:

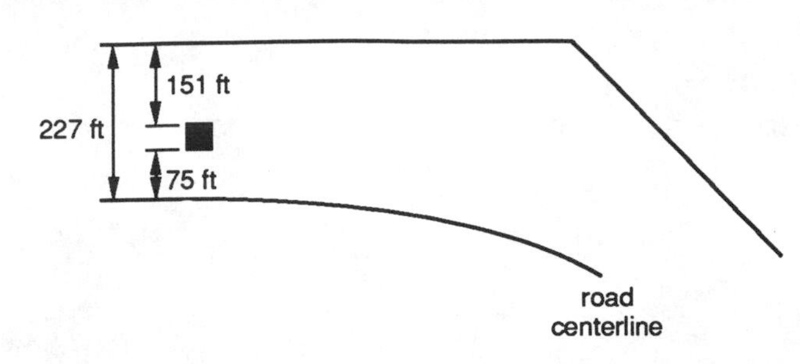

$$\alpha = \arctan\left(\frac{227}{574}\right) = 21.577°$$

$$m = \sqrt{(227)^2 + (574)^2} = 617.256 \text{ ft}$$

$$\Delta = 56°$$

$$\delta = 90° - 28° - 21.577° = 40.423°$$

$$\phi = \arcsin\left(\frac{\sin 40.423°}{\cos 28°}\right) = 47.255° \text{ or } 132.745°$$

47.255° is acute and will not work.

$$\theta = 180° - 40.423° - 132.745° = 6.832°$$

$$R = \frac{(617.256)(\sin 132.745)(\cos 28°)}{\sin 6.832°}$$

$$= \boxed{3364.55 \text{ ft}}$$

This approach introduces a small error because the lines' center-to-monument and center-to-road centerline do not exactly coincide. Therefore, the clearance will not be exactly 1 ft, but the error will be insignificant.